€ 40.00
£10 —

Hay-on-Wye, 30 Mar 08   £10
(The Hay Bk Co. = [neé Mark Westwood Bks])

03

HYPERBOLIC PARTIAL DIFFERENTIAL EQUATIONS III

International Series in
MODERN APPLIED MATHEMATICS AND COMPUTER SCIENCE
Volume 12
Series Editor: Ervin Y. Rodin, Washington University
(Volumes in the Series have also been published as Special Issues of the Journal *Computers and Mathematics with Applications*)

RELATED TITLES

OTHER VOLUMES IN THE SERIES
Bellman, et al. MATHEMATICAL ASPECTS OF SCHEDULING AND APPLICATIONS
Cercone COMPUTATIONAL LINGUISTICS
Choi STATISTICAL METHODS OF DISCRIMINATION AND CLASSIFICATION
Cooper & Cooper INTRODUCTION TO DYNAMIC PROGRAMMING
Hargittai SYMMETRY
Saaty & Alexander THINKING WITH MODELS
Saaty & Kearns ANALYTICAL PLANNING
Suri RESOURCES MANAGEMENT CONCEPTS FOR LARGE SYSTEMS
Witten HYPERBOLIC PARTIAL DIFFERENTIAL EQUATIONS Vols. I and II

OTHER BOOKS OF INTEREST
Anand INTRODUCTION TO CONTROL SYSTEMS
Avula, et al. MATHEMATICAL MODELLING IN SCIENCE AND TECHNOLOGY
Hewitt & Whalley GAS-LIQUID FLOW AND HEAT TRANSFER
Landau, et al. ELECTRODYNAMICS OF CONTINUOUS MEDIA
Marchuk OCEAN TIDES (Mathematical Models and Numerical Experiments)
Mikusinski OPERATIONAL CALCULUS Vol. 1, 2nd Edition
Nowacki THEORY OF ASYMMETRIC ELASTICITY
Washizu VARIATIONAL METHODS IN ELASTICITY AND PLASTICITY, 3rd Edition

RELATED JOURNALS*
Computers and Mathematics with Applications
International Journal of Engineering Science
International Journal of Mechanical Sciences
International Journal of Nonlinear Mechanics
International Journal of Plasticity
Journal of Applied Mathematics and Mechanics
Mathematical Modelling
Nonlinear Analysis

***Sample copies available on request.**

HYPERBOLIC PARTIAL DIFFERENTIAL EQUATIONS III

Edited by
MATTHEW WITTEN

Department of Engineering Mathematics and Computer Science,
Speed Scientific School, University of Louisville, Louisville, KY 40292, USA

*International Series in Modern Applied Mathematics
and Computer Science,* Volume 12

Series Editor
Ervin Y. Rodin
Washington University

PERGAMON PRESS

NEW YORK • OXFORD • TORONTO • SYDNEY • FRANKFURT

Pergamon Press Offices:

U.S.A. Pergamon Press Inc., Maxwell House, Fairview Park,
Elmsford, New York 10523, U.S.A.

U.K. Pergamon Press Ltd., Headington Hill Hall,
Oxford OX3 0BW, England

CANADA Pergamon Press Canada Ltd., Suite 104, 150 Consumers Road,
Willowdale, Ontario M2J 1P9, Canada

AUSTRALIA Pergamon Press (Aust.) Pty. Ltd., P.O. Box 544,
Potts Point, NSW 2011, Australia

FEDERAL REPUBLIC OF GERMANY Pergamon Press GmbH, Hammerweg 6,
D-6242 Kronberg-Taunus, Federal Republic of Germany

BRAZIL Pergamon Editora Ltda., Rua Eça de Queiros, 346,
CEP 04011, São Paulo, Brazil

JAPAN Pergamon Press Ltd., 8th Floor, Matsuoka Central Building,
1-7-1 Nishishinjuku, Shinjuku, Tokyo 160, Japan

PEOPLE'S REPUBLIC OF CHINA Pergamon Press, Qianmen Hotel, Beijing,
People's Republic of China

Copyright © 1986 Pergamon Press Ltd.

ISBN 0-08-034313-9

Published as a special issue of the journal *Computers and
Mathematics with Applications*, Volume 12A, Numbers 4/5
and supplied to subscribers as part of their normal subscription.
Also available to non-subscribers.

Printed in Great Britain by A. Wheaton & Co. Ltd, Exeter

CONTENTS

ABOUT THIS ISSUE

The special issue on hyperbolic partial differential equations is a refereed journal issue whose thrust is current applications, theory and/or applied methods related to hyperbolic partial differential equations, or problems arising out of hyperbolic partial differential equations, in any area of research.

We are interested in all aspects of research in this area: numerical analysis, stability analysis, existence and uniqueness of solutions, periodic solutions, interesting realworld applications or other aspects of research related to hyperbolic partial differential equations.

This journal issue is interested in all types of articles in terms of review, mini-monograph, standard study, or short communication.

(1) *Review papers*. Review papers should be accessible to a more general technical readership. That is, they should be accessible, not only to professionals who work in the given field, but also to senior graduate students and/or professionals who work in related areas. The review should not only provide a concise, as well as comprehensive discussion of the literature, but it should be presented in a manner which leads to new perspectives and/or which suggests new lines of investigation. A bibliographic summary of the literature in the field is not acceptable as a review paper.

(2) *Mini-monographs*. A mini-monograph is a comprehensive and detailed study of a particular problem. This would involve not only a detailed discussion of the realworld relevance of the problem, but also a detailed discussion of the mathematical analysis, as well as a detailed discussion of the relevance and application of the conclusions which are based upon the previous mathematical analysis.

(3) *Standard studies*. A standard study is a concise treatment of a particular problem. It need not be as exhaustive or detailed as a Mini-monograph. This is what most people would normally submit to a journal.

(4) *Short communications*. Because of the annual nature of this special issue, short communications are discouraged. This is due to the fact that we are unable to offer the rapid turnaround time provided by journals specializing in such papers. If, however, you wish to present a short result, which is concisely summarized and not very long, we are willing to consider a short communication.

(5) *Book reviews*. Book reviews on books related to hyperbolic PDEs and problems arising out of hyperbolic PDEs, in any area of research, are solicited from qualified reviewers.

The intent of this journal in general is to provide an interdisciplinary forum for the presentation of results which might not necessarily lie within the confines of other particular journals. As was pointed out before, we are interested in all types of papers, theoretical or applied.

AUTHOR'S GUIDELINES

The annual submission **deadline** for the Advances in Hyperbolic Partial Differential Equaitons issue is the 31st of December of each year. Please submit **four** (4) copies of your paper to Matthew Witten. Each paper should have a list of index or keywords (10–20) with the page of their first appearance in the manuscript. This is for the purposes of providing an index with each volume. For a detailed sample see *Computers and Mathematics with Applications*, Vol. 9, No. 3 (1983) which is the first special issue.

For uniformity of format, please cite **references** in the text as, Jackobson[3], for example. Or, Williams[5–9], in the case where an author is cited with more than one consecutive article. The references should then appear, in the bibliography, in the order that they were cited in the text.

The citation of references should be as follows:

1. J. K. Knowles and E. Reissner, Note on the stress-strain relations for thin elastic shells. *J. Math. Phys.* **37**, 269–282 (1958).
2. H. S. Carslaw and J. C. Jager, *Operational Methods in Applied Mathematics*, 2nd Edn. Oxford University Press, London (1953).

Illustrations should accompany the manuscript and will, in the printed journal, interrupt the test. The author should supply the illustrations on separate pages, but indicate the desired location in the printed text. Line drawings should include all relevant details and should be drawn in black ink on plain white drawing paper or tracing cloth. Good photoprints are acceptable, but blueprints or dye-line prints cannot be used. Drawings, etc. should be about twice the final size required and lettering must be clear and sufficiently large to permit the necessary reduction of size. Please, whenever possible, use the following standard symbols on line drawings as they are most readily available to the printers: ○ ● + × □ ■ △ ▲ ▼ ▽

Photographs should be sent as glossy prints. If words or numbers are to appear on a photograph, two prints should be sent, the lettering being clearly indicated on one print only. Figure legends should be typed on a separate sheet and placed at the end of the manuscript. Authors are requested to supply good quality diagrams and clearly typed tables in a form suitable for direct photographic reproduction.

Computer output should be given on an original print-out and will be reproduced photographically to avoid errors. Glossy prints of the original print-outs are also acceptable.

Because of the international character of the special issue, no rules concerning notation or abbreviation need be observed by the contributors. But, each paper should be self-consistent as to symbols and units which should all be properly defined. The Editor urges all authors to try to adhere to, and be guided by *A Manual for Authors* published by the American Mathematical Society, P.O. Box 6248, Providence, Rhode Island 02904 USA.

All mathematical symbols may be either handwritten or typewritten but no ambiguities should arise. Greek letters and unusual symbols should be identified in the margin. Distinction should be made between capital and lower case letters; between the letter O and the number 0; between the letter l, the number 1, and prime; between the letter k and κ. A vector will be printed boldface and to indicate this, the letter should be underscored with a single wavy line. The numbers identifying mathematical expressions should be placed in parentheses after the equations.

For any further information on this special issue please contact: Matthew Witten, Special Issue Editor—Hyperbolic PDE Issue, Computers and Mathematics With Applications, Department of Engineering Mathematics and Computer Science, Speed Scientific School, University of Louisville, Louisville, KY 40292 USA or phone (502) 588-6304.

Comp. & Maths. with Appls. Vol. 12A, Nos. 4/5, pp. 377–388, 1986
Printed in Great Britain.

A CONSERVATIVE, PIECEWISE-STEADY DIFFERENCE SCHEME FOR TRANSONIC NOZZLE FLOW

LUPING HUANG[†] and TAI-PING LIU[‡]
Department of Mathematics, University of Maryland, College Park, MD 20742, U.S.A.

Abstract—We construct a conservative scheme which approximates gas flow through a duct by discontinuity waves, rarefaction waves and steady waves. Analytical studies on the interaction and stability of these nonlinear elementary waves are used to determine the evolution of the state variables. The scheme is consistent, admissible, and reduces to the Godunov scheme when the duct is uniform. Numerical results show that the scheme is stable and tends to a stable steady flow; it also compares favorably with a fractional Godunov scheme.

1. INTRODUCTION

Consider quasi-one-dimensional gas-dynamics equations for flows through a duct of varying cross section $A(x)$[1]:

$$(\rho)_t + (\rho u)_x = -\frac{A'(x)}{A(x)} \rho u,$$

$$(\rho u)_t + (\rho u^2 + p)_x = \frac{-A'(x)}{A(x)} \rho u^2, \tag{1.1}$$

$$(\rho E)_t + (\rho E u + p u)_x = -\frac{A'(x)}{A(x)} (\rho u E + p u),$$

where ρ, p, u and e are, respectively, density, pressure, velocity and internal energy of the gas and $E = e + u^2/2$, the total energy. For uniform duct, (1.1) is reduced to

$$(\rho)_t + (\rho u)_x = 0,$$

$$(\rho u)_t + (\rho u^2 + p)_x = 0, \tag{1.2}$$

$$(\rho E)_t + (\rho E u + p u)_x = 0.$$

Flows for (1.2) and also supersonic and subsonic flows for (1.1) have been shown to be stable. However, transonic flows for (1.1) may be unstable[2]. Moreover, analysis of asymptotic flows[3] reveals that an expansion wave may reflect as a compression wave at sonic point.

In view of these rich physical phenomena, we construct a conservative scheme using the elementary waves for (1.2) and steady waves, i.e. solutions of

$$(\rho u)_x = -\frac{A'(x)}{A(x)} \rho u,$$

$$(\rho u^2 + p)_x = -\frac{A'(x)}{A(x)} \rho u^2, \tag{1.3}$$

$$(\rho E u + p u)_x = \frac{-A'(x)}{A(x)} (\rho E u + p u)$$

as building blocks. Elementary waves for (1.2) are situated between the meshes, as in the

---

†An exchange visiting scholar from Peking University P.R.O.C.
‡John Simon Guggenheim Fellow.

Godunov scheme[4]. One of the novel features of the scheme is that the cross section $A(x)$ of the duct is approximated by a step function so that steady waves are discretized and concentrated at the center of each mesh. This yields the relatively simple formula (3.1) for the computation of the average value of the state at a new time level. The scheme is of the same degree of consistency and simplicity as the Godunov scheme for conservation laws.

Numerical results show that our scheme is stable, produces sharp shock waves and yields stable steady flow in the time-asymptotic limit. It also compares favorably with a fractional Godunov scheme.

In Sec. 2, we recall some basic formulas for calculating the elementary waves. The scheme is defined in Sec. 3. The key formula, (3.1), is derived in Sec. 4 based on the approximation procedure on the wave propagation (cf. [2]). A much simpler derivation is presented in Sec. 5 by approximating the cross section of the duct by a step function. This has the effect that steady waves are discretized to concentrate at the center of each mesh. Since the entropy is constant along a steady wave, we are able to prove in Sec. 6 that the scheme is admissible, i.e. it satisfies the entropy condition.

Numerical results are presented in Sec. 7. For a diverging duct numerical results show that a linear initial profile converges time-asymptotically to the unique steady flow with the given boundary data. The steady flow contains a standing shock wave and is the same one as in [5]. Our scheme compares favorably with the fractional Godunov scheme in that our scheme yields a stable wave profile sooner, produces a sharper shock wave, and yields a more accurate asymptotic state. For a converging–diverging nozzle we identify two initial profiles, one of them linear, which tend to each of the two stable steady flows with the given boundary data. One of the steady flows possesses a boundary layer at the inflow; the other contains a standing shock wave situated along the diverging portion of the duct.

Recently, there has been intense interest in the calculation of gas flows using higher-order Godunov-type schemes[6,7] and other methods[8,9]; it would be interesting to incorporate our analysis into these schemes. Flow through a duct has also been calculated recently using random-choice methods[10–12].

2. ELEMENTARY WAVES

For notational simplicity, we write (1.1), (1.2) and (1.3), respectively as

$$U_t + F(U)_x = G(x, U), \tag{1.1}$$

$$U_t + F(U)_x = 0, \tag{1.2}$$

$$F(U)_x = G(x, U). \tag{1.3}$$

We will assume that the gas is polytropic:

$$p(\rho, e) = (\gamma - 1)\rho e, \tag{2.1}$$

where $\gamma \geq 1$ is the ratio of specific heats. The characteristic speed λ_i and characteristic vectors $\varkappa_i, i = 1, 2, 3$, are

$$\lambda_1 = u - c, \quad \lambda_2 = u, \quad \lambda_3 = u + c,$$

$$\varkappa_1 = (1, \lambda_1, E + p\rho^{-1} - uc),$$

$$\varkappa_3 = (1, \lambda_3, E + p\rho^{-1} + uc),$$

and $\varkappa_2$ is such that u and p are both invariant along $\varkappa_2$. Here $c = \sqrt{\gamma p \rho^{-1}}$ is the sound speed. There are three kinds of elementary waves for (1.2). An i-rarefaction wave (U_0, U_1), $i = 1, 3$, takes values along the integral curve R_i of $\varkappa_i$:

$$U_1 \in R_i^+(U_0) \equiv \{U : U \text{ is on } R_i \text{ through } U_0 \text{ and } \lambda_i(U_1) \geq \lambda_i(U_0)\}, \quad i = 1, 3.$$

A contact discontinuity (U_0, U_1) takes values along R_2:

$$U_1 \in R_2(U_0) \equiv \text{the integral curve of } \varkappa_2 \text{ through } U_0.$$

An i-shock wave (U_0, U_1), $i = 1, 3$, takes values along the Rankine–Hugoniot curve:

$$U_1 \in S_i^-(U_0) \equiv \{U : \sigma(U - U_0) = F(U) - F(U_0)$$
$$\text{for some scalar } \sigma, \ \lambda_i(U) \le \sigma \le \lambda_i(U_0)\}, \quad i = 1, 3.$$

The Riemann problem (1.2) with

$$U(x, 0) = \begin{cases} U_l & \text{for } x < 0, \\ U_\varkappa & \text{for } x > 0, \end{cases}$$

is solved by finding $U_m \in R_1^+(U_l)US_1^-(U_l)$, $U_n \in R_2(U_m)$ with the property that $U_\varkappa \in R_3^+(U_n)US_3^-(U_n)$ so that the solution is made of the elementary waves just mentioned. Since both u and P are unchanged along R_2 curves, it is convenient to locate the curves R_i and S_i, $i = 1, 3$, on the (u, p)-plane, using (2.1) to eliminate the variables ρ and e. Tedious calculations yield the value of $p_* = p_n = p_m$ as the solution of the following equations:

$$\phi(p_*, p_l, \rho_l) + \phi(p_*, p_\varkappa, \rho_\varkappa) = u_l - u_\varkappa,$$

$$\phi(p_*, p_i, \rho_i) \equiv \begin{cases} \dfrac{p_* - p_i}{p_i c_i \sqrt{\dfrac{(\gamma + 1)p_*}{2\gamma p_i} + \dfrac{\gamma - 1}{2\gamma}}} & \text{when } p_* \ge p_i, \\[3em] \dfrac{2}{\gamma - 1} c_i \left(\left(\dfrac{p_*}{p_i} \right)^{(\gamma - 1)/2\gamma} - 1 \right) & \text{when } p_* < p_i, \end{cases} \quad i = l, \varkappa.$$

The function ϕ is an increasing concave function of p_*. Thus the Newton method may be employed efficiently to find p_*. Having found p_*, the types of i-waves, $i = 1, 3$, are determined and the states U_m and U_n are then easily identified. This procedure follows that of Godunov[4] which is more efficient than some of the other methods for solving the Riemann problem (cf. [13]).

We next turn to the construction of steady waves, i.e. solutions of (1.3). It turns out that the ratio of the cross section of the duct is the only determining factor in locating a state along a steady flow. Suppose that U_1 is a given state in a steady flow situated at the position of the duct with cross section A_1. Then a state U_2 in the same steady flow corresponding to the cross section A_2 can be found from U_1 and A_1/A_2 as follows. The Mach number $M_2 = u_2/c_2$ at U_2 can be found by solving[14]

$$\left(\frac{A_1}{A_2} \right)^2 = \left(\frac{M_2}{M_1} \right)^2 \left[\frac{1 + \dfrac{\gamma - 1}{2}(M_1)^2}{1 + \dfrac{\gamma - 1}{2}(M_2)^2} \right]^{(\gamma + 1)/(\gamma - 1)} . \tag{2.2}$$

The right-hand side of (2.2), as a function of $(M_2)^2$, is convex and has a minimum at $(M_2)^2 = 1$. Thus we may use the Newton method effectively to solve (2.2) for $(M_2)^2$. (As a function of M_2, the right-hand side of (2.2) is not convex and the Newton method may fail.) Since the right-hand side has a minimum at $M_2 = 1$, (2.2) does not have a solution when

$$\left(\frac{A_1}{A_2} \right)^2 > \frac{1}{(M_1)^2} \left[\frac{2}{\gamma + 1} \left(1 + \frac{\gamma - 1}{2}(M_1)^2 \right) \right]^{(\gamma + 1)/(\gamma - 1)} \quad \text{and} \quad A_1 > A_2. \tag{2.3}$$

Since along a steady flow entropy and enthalpy are constant, we have

$$a \equiv p\rho^{-\gamma} \quad \text{and} \quad b \equiv \frac{1}{2}u^2 + \frac{1}{\gamma - 1}c^2 \tag{2.4}$$

are invariant. We write $a_i = a(U_i)$, etc. Writing u^2 as M^2c^2 in the identity $b_1 = b_2$, it follows that

$$c_2 = \left(\frac{b_1}{\frac{1}{2}(M_2)^2 + 1/(\gamma - 1)}\right)^{1/2}. \tag{2.5}$$

Since $(c_2)^2 = \gamma p_2(\rho_2)^{-1}$, it follows from $a_1 = a_2$ that

$$\rho_2 = \frac{(c_2)^2}{\gamma a_1}. \tag{2.6}$$

Finally, having found the values of M_2, c_2 and ρ_2 we have

$$p_2 = a_1\rho_2^\gamma, \quad u_2 = c_2M_2. \tag{2.7}$$

This completes the construction of a steady flow. Even though (2.2) has two solutions for M_2, we set $M_2 > 1$ (or $M_2 < 1$) when $M_1 > 1$ (or $M_1 < 1$). This poses no numerical problem as we use $(M_1)^2$ as the initial guess for $(M_2)^2$ when we use the Newton method to solve (2.2).

3. THE SCHEME

Choose mesh lengths Δx and Δt satisfying the usual Courant–Friedrichs–Lewy condition. At each time level $t = k\Delta t$, $k = 0, 1, 2, \ldots$, the approximate solution $U(x, t) \equiv U_{\Delta x}(x, t)$ is piecewise steady:

$$U(x, k\Delta t) \text{ satisfies (1.3) for } (h - \tfrac{1}{2})\Delta x < x < (h + \tfrac{1}{2})\Delta x,$$
$$h \text{ integers.}$$

Suppose that $U(x, k\Delta t)$ is given. Then $U(x, (k + 1)\Delta t)$ is constructed as follows: Denote by

$$U_{h,k} \equiv U(h\Delta x, k\Delta t), \quad U_{h,k}^{\pm} \equiv U((h \pm \tfrac{1}{2})\Delta x, k\Delta t).$$

Thus $(U_{h,k}^-, U_{h,k}^+)$ is a steady wave over the interval $(h - \tfrac{1}{2})\Delta x < x < (h + \tfrac{1}{2})\Delta x$. As a first step, we solve the Riemann problem $(U_{h-1,k}^+, U_{h,k}^-)$ and $(U_{h,k}^+, U_{h+1,k}^-)$, and denote their solutions by $V_{h,k}(x/t)$ and $V_{h+1,k}(x/t)$, respectively, and set

$$U_{h,k}^l \equiv V_{h,k}(0), \quad U_{h,k}^r \equiv V_{h+1,k}(0)$$

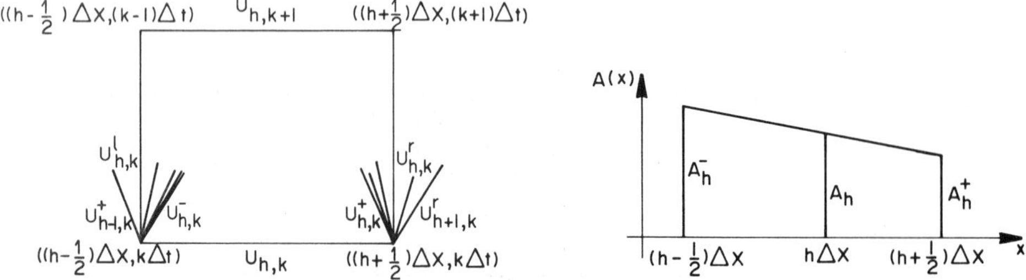

Fig. 1. (a) A typical numerical block in the scheme. (b) Cross section of the duct.

(Fig. 1). We set

$$\bar{U}_{h,k+1} = U_{h,k} + \frac{\Delta t}{\Delta x}\,[F(U_{h,k}^{l}) - F(U_{h,k}^{r}) + F(U_{h,k}^{+}) - F(U_{h,k}^{-})]. \tag{3.1}$$

For convenience, we will choose the mesh points in such a way that the duct is either converging or diverging for each mesh interval $((h - \tfrac{1}{2})\Delta x, (h + \tfrac{1}{2})\Delta x)$. We now describe the procedure for calculating $U_{h,k+1}$ and $U_{h,k+1}^{\pm}$. First assume that the duct is *converging* over the interval $((h - \tfrac{1}{2})\Delta x, (h + \tfrac{1}{2})\Delta x)$ [Fig. 1(b)]:

$$A_h^- > A_h > A_h^+,$$

$$A_h \equiv A(h\Delta x), \quad A_h^{\pm} \equiv A((h \pm \tfrac{1}{2})\Delta x).$$

There are two cases.

Case 1. $A_1 \equiv A_h$, $A_2 \equiv A_h^+$ and $U_1 \equiv \bar{U}_{h,k+1}$ do not satisfy condition (2.3).

Thus the steady wave equations (1.3) can be solved for the interval $x \in ((h - \tfrac{1}{2})\Delta x, (h + \tfrac{1}{2})\Delta x)$ with the given state $\bar{U}_{h,k+1}$ at $x = h\Delta x$. We set

$$U_{h,k+1} \equiv \bar{U}_{h,k+1} \tag{3.2}$$

and calculate $U_{h,k+1}^{\pm}$ from $U_{h,k+1}$ by solving (1.3).

Case 2. $A_1 \equiv A_h$, $A_2 \equiv A_h^+$ and $U_1 = \bar{U}_{h,k+1}$ satisfy condition (2.3).

In this case the above procedure fails because the system (1.3) becomes singular. We set $U_{h,k+1}^{+}$ to be a sonic state which is related to $\bar{U}_{h,k+1}$ by a steady wave. In other words, set $M_2 = 1$ and use (2.4)–(2.7) to calculate U_2 and set

$$U_{h,k+1}^{+} \equiv U_2. \tag{3.3a}$$

With A_2, A_2 and U_2 as given, we may use an analogous procedure to (2.2), (2.4)–(2.7) to find U_1. Note, however, that there are two solutions M_1 of (2.2); we choose the *subsonic* branch:

$$U_{h,k+1} = U_1, \quad M_1 < 1. \tag{3.3b}$$

The state $U_{h,k+1}^{-}$ is then calculated from $U_{h,k+1}$ by solving (1.3). Note that since $A_h^- > A_h > A_h^+$, $U_{h,k+1}$ and $U_{h,k+1}^{-}$ can always be calculated from the given state $U_{h,k+1}^{+}$.

Next we consider the *diverging* duct: $A_1^- < A_1 < A_1^+$. Again we have two cases. In the first case where $A_1 \equiv A_h$, $A_2 \equiv A_h^-$ and $U_1 \equiv \bar{U}_{h,k+1}$ do not satisfy (2.3), we again use (3.2) to find $U_{h,k+1}$ and $U_{h,k+1}^{\pm}$. In the second case, instead of (3.3a) and (3.3b) we set $M_2 = 1$ and use (2.4)–(2.7) to find U_2 and set

$$U_{h,k+1}^{-} \equiv U_2. \tag{3.4a}$$

With A_1, A_2 and U_2 known, find a *supersonic* U_1 using the analogous procedure to (2.2), (2.4)–(2.7) and set

$$U_{h,k+1} \equiv U_1, \quad M_1 > 1 \tag{3.4b}$$

and calculate the state $U_{h,k+1}^{+}$ with the given A_h^+, A_h and $U_{h,k+1}$ by solving (1.3).

The key formula (3.1) will be analyzed and explained in the next two sections from two different perspectives. The rather puzzling procedures (3.3) and (3.4) are the reflection of the following wave phenomena. The second case as described above holds when a supersonic (subsonic) rarefaction wave propagates through a converging (diverging) duct; it reflects as a subsonic (supersonic) compression wave as it reaches the sonic state. That this should be so is a consequence of the study of noninteracting wave pattern discovered by Liu[3]. The use of the analytical studies on the asymptotic states to construct a random-choice method was first proposed for a scalar model by the second author and generalized to gas dynamics in [10]. The rather simple procedure, (3.3) and (3.4), is equivalent to a slightly more complicated one used in [10] up to a second-order numerical error.

The above procedures are for the evaluation of the interior states. In the calculation of gas flows, there has always been controversy as to how to treat the boundary condition. We have performed several numerical experiments and found that the following numerical treatment of the boundary condition yields desirable results. Suppose that the interior points are $x = h\Delta x$, $h = 0, \ldots, N$ and that the inflow boundary is $x = -\frac{1}{2}\Delta x$ and the outflow boundary $x = N + \frac{1}{2}\Delta x$. The numerical evaluation of the boundary states

$$U^{in}_{k+1} \equiv U(-\tfrac{1}{2}\Delta x - 0, (k + 1)\Delta t),$$

$$U^{out}_{k+1} \equiv U((N + \tfrac{1}{2})\Delta x + 0, (k + 1)\Delta t),$$

are done as follows. Since we are dealing with supersonic inflow, the state U^{in} at the inflow is a given boundary datum and so we set $U^{in}_{k+1} \equiv U^{in}$. The downstream boundary condition should be prescribed only if the flow is subsonic there, and then only one fluid variable should be prescribed there. In the present study, we prescribe the outflow density $\rho = \rho^{out}$. Since we do not know *a priori* whether the outflow is supersonic or subsonic at any given moment, a desirable scheme would be to implement the boundary condition $\rho = \rho^{out}$ only when the outflow is subsonic. This is achieved by calculating the state U^{out}_{k+1} according to the following[10]:

U^{out}_{k+1} is connected to $U^{+}_{N,k+1}$ by either a 1-shock wave or

a 1-rarefaction wave and $\rho(U^{out}_{k+1}) = \rho_{out}$.

This determined the state U^{out}_{k+1} uniquely since $U^{+}_{N,k+1}$ is an interior state whose calculation has been described in the first half of this section. The outflow condition may be replaced by prescribing the pressure p, the velocity u or the temperature T.

4. WAVE ITEGRATIONS I

The propagation of elementary waves for (1.2) through a steady wave is rather complicated and no exact analytical description exist. We now describe an approximating procedure with second order of accuracy[2] and then use it to derive the estimate (3.1). Consider the region

$$\Omega \equiv \{(x, t): 0 \le x \le \Delta x, \quad 0 \le t \le \Delta t\}.$$

Suppose that at $t = 0$, an approximate solution $U(x, t)$, consists of a steady wave for $0 < x < \Delta x$ and elementary waves for (1.2) issued from $x = 0$ and $x = \Delta x$ and propagating into the region Ω. The end states of the steady wave at $x = 0$ and $x = \Delta x$ are denoted by U_- and U_+. The average of the steady wave is $\bar{U}$. The elementary waves issued from $x = 0$ (or $x = \Delta x$) are the solution of the Riemann problems (U_l, U_-) and (U_+, U_r), respectively (Fig. 2). Our purpose is to estimate the average value of $U(x, \Delta t)$ over $0 \le x \le \Delta x$. For definiteness and simplicity, we assume that the solution of (U_l, U_-) consists of a contact discontinuity (U_l, U_1)

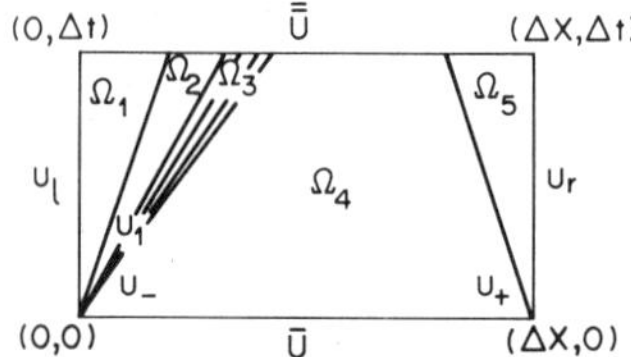

Fig. 2. Propagation of elementary waves at (0, 0) and $(\Delta x, 0)$ through the steady wave (U_-, U_+).

and a 3-rarefaction wave (U_1, U_-), and that the solution of $U_+, U_r)$ is a 1-shock wave. These elementary waves divide the region Ω into subregions Ω_i, $i = 1, 2, \ldots, 5$ (Fig. 2). The approximating procedure is that in Ω_1 and Ω_2, $U(x, t)$ is a steady wave with the given values U_l and U_1, respectively at $x = 0$. In Ω_4, $U(x, t)$ is the steady wave (U_-, U_+). Similarly, in Ω_5, $U(x, t)$ is a steady wave with the given state U_r at $x = \Delta x$. In the rarefaction region Ω_3, $U(x, t)$ is a steady wave along the ray $\xi = $ constant, $\xi = x/t \in (\lambda_3(U_1), \lambda_3(U_-))$, and coincide with the rarefaction wave (U_1, U_-) at $(0, 0)$.

As usual, we integrate over Ω to estimate the average value of $U(x, \Delta t)$:

$$\bar{\bar{U}} - \bar{U} \equiv \frac{1}{\Delta x} \left[\int_0^{\Delta x} U(x, \Delta t)\, dx - \int_0^{\Delta x} U(x, 0)\, dx \right]$$

$$= \frac{\Delta t}{\Delta x} [F(U_l) - F(U_r)] + \frac{1}{\Delta x} \sum_{i=1}^{5} \int \int_{\Omega_i} G(x, U)\, dx\, dt. \quad (4.1)$$

Since $U|_{\Omega_1}$ is a steady wave, we have

$$\int \int_{\Omega_1} G(x, U)\, dx = \int \int_{\Omega_1} F_x(U)\, dx = \int_{\Omega_1 \cap \Omega_2} F(U(x - 0, t)\, dt - F(U_l)\Delta t,$$

where we have used the fact that $U = U_l$ along $t = 0$ because U is steady there. Similar estimates also hold for the integration of $G(x, U)$ over Ω_2, Ω_4 and Ω_5 as U is also steady in these regions. Consequently we have

$$\int \int_{\Omega} G(x, U)\, dx\, dt = \int \int_{\Omega_3} G(x, U)\, dx\, dt$$

$$+ \int_{\Omega_1 \cap \Omega_2} [F(U(x - 0, t)) - F(U(x + 0, t))]\, dt$$

$$+ \int_{\Omega_2 \cap \Omega_3} F(U(x, t))\, dt - \int_{\Omega_3 \cap \Omega_4} F(U(x, t))\, dt$$

$$+ \int_{\Omega_4 \cap \Omega_5} [F(U(x - 0, t)) - F(U(x + 0, t))]\, dt$$

$$+ [F(U_r) - F(U_l)]\Delta t. \quad (4.2)$$

Let α_i, $i = 1, 2, 3$, be the strength of elementary waves:

$$\alpha_1 \equiv \|U_1 - U_l\|,$$

$$\alpha_2 \equiv \|U_- - U_1\|,$$

$$\alpha_3 \equiv \|U_r - U_+\|, \quad \alpha \equiv \alpha_1 + \alpha_2 + \alpha_3.$$

Since $F(U(x \pm 0, t))$ satisfy (1.3) along $\Omega_1 \cap \Omega_2$ and also along $\Omega_4 \cap \Omega_5$, we have from elementary theory of ordinary differential equations that

$$\int_{\Omega_1 \cap \Omega_2} [F(U(x - 0, t)) - F(U(x + 0, t))]\, dt = [F(U_l) - F(U_1)]\Delta t + O(1)\alpha_1(\Delta t)^2,$$

$$\int_{\Omega_4 \cap \Omega_5} [F(U(x - 0, t)) - F(x + 0, t))]\, dt = F(U_+) - F(U_r)]\Delta t + O(1)\alpha_3(\Delta t)^2.$$

Similarly, we have

$$\int_{\Omega_2 \cap \Omega_3} F(U)\, dt - \int_{\Omega_3 \cap \Omega_4} F(U)\, dt = [F(U_1) - F(U_-)]\Delta t + O(1)\alpha_2(\Delta t)^2.$$

The area of the region Ω_3 occupied by the rarefaction wave (U_1, U_-) is

$$\tfrac{1}{2}(\lambda_3(U_-) - \lambda_3(U_1))(\Delta t)^2 = O(1)\alpha_2(\Delta t)^2,$$

and thus

$$\int\int_{\Omega_3} G(x, U)\ dx\ dt = O(1)\alpha_2(\Delta t)^2.$$

The above estimates and (4.2) yield

$$\int\int_\Omega G(x, U)\ dx\ dt = [F(U_+) - F(U_-)]\Delta t + O(1)\alpha(\Delta t)^2,$$

and so from (4.1)

$$\bar{\bar{U}} = \bar{U} + \frac{\Delta t}{\Delta x}[F(U_l) - F(U_r) + F(U_+) - F(U_-)] + \text{error},$$

$$\text{error} = O(1)\alpha(\Delta t)^2, \tag{4.3}$$

which is formula (3.1) plus the error term. Assuming that the total strength of waves is finite, then for each time level the total error is $O(1)(\Delta t)^2$. And, for the finite region $0 \le t \le T$, there are $T(\Delta t)^{-1}$ time levels and the total amount of errors would then be $O(1)T\Delta t$, which tends to zero as the mesh size Δt tends to zero and the scheme is consistent. Of course, there are other errors which have been committed. One is the approximation procedure imposed at the beginning of this section, on the propagation of waves, and the other is the averaging procedure on $t = k\Delta t$. The former has been shown to contain second-order as above (cf. [2]); the latter is again a consistent procedure, as in Godunov scheme for conservation laws[4].

5. WAVE INTEGRATIONS II

A simpler and revealing way of deriving the basic estimate (4.3) is presented in this section without resorting to the approximating procedure described at the beginning of the last section. As in the last section, consider the region $\Omega = \{(x, t): 0 \le x \le \Delta x, 0 \le t \le \Delta t\}$ and an approximate solution $U(x, t)$ defined in Ω. However, we now approximate the cross section of the duct by a step function with a discontinuity at $x = \Delta x/2$, so that the steady wave (U_-, U_+) is now concentrated at the line $x = \Delta x/2$ [Figs. 3(a,b)].

In the region $\Omega- \equiv \{(x, t): 0 < x < \Delta x/2, 0 < t < \Delta t\}$ and also in $\Omega+ \equiv \{(x, t): \Delta x/2 < x < \Delta x, 0 < t < \Delta t\}$, the duct is uniform and so $U(x, t)|_{\Omega_\pm}$ solves the conservation laws (1.2). Thus integrating $U(x, t)$ over $\Omega+$ and $\Omega-$ we obtain

$$\int_0^{\Delta x/2} U(x, \Delta t)\ dx = \frac{\Delta x}{2}U_- + [F(U_l) - F(U_-)]\Delta t,$$

$$\int_{\Delta x/2}^{\Delta x} U(x, \Delta t)\ dx = \frac{\Delta x}{2}U_+ + [F(U_+) - F(U_r)]\Delta t,$$

and so

$$\bar{\bar{U}} - \bar{U} \equiv \frac{1}{\Delta x}\left[\int_0^{\Delta x/2} U(x, \Delta t)\ dx + \int_{\Delta x/2}^{\Delta x} U(x, \Delta t)\ dx\right] - \frac{1}{2}(U_- + U_+)$$

$$= \frac{\Delta t}{\Delta x}[F(U_l) - F(U_r) + F(U_+) - F(U_-)],$$

which is the same as estimate (4.3) with no error terms. The numerical error caused by the

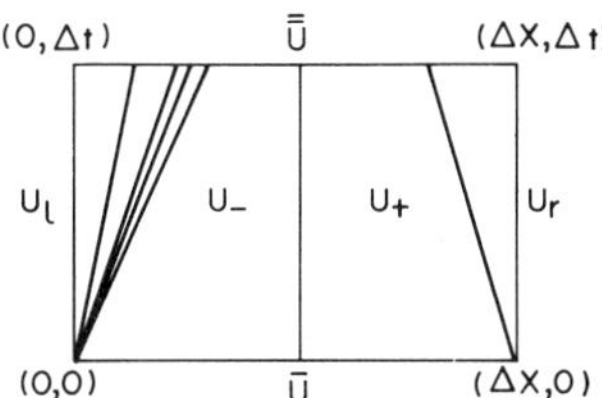

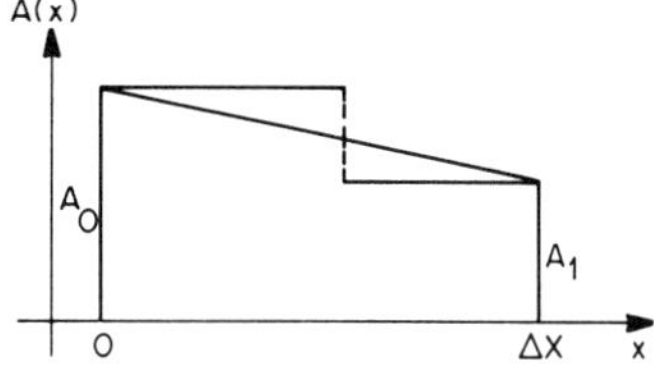

Fig. 3. (a) Elementary wave pattern when the cross section of the duct is
a step function. (b) Approximation of the duct shape by a step function.

approximation of the cross section of the duct by a step function is of the same order of the error that occurs in (4.3).

6. ADMISSIBILITY

For an admissible solution of the Euler eqs. (1.2) the thermodynamics entropy S is non-decreasing along each fluid element:

$$\frac{\partial S}{\partial t} + u \frac{\partial S}{\partial x} \geq 0. \tag{6.1}$$

Direct calculations using the thermodynamics relation $de = T\, dS - p\, d(1/\rho)$ also shows that (6.1) holds for admissible solutions of (1.1). We now show that, except for a second-order numerical error, our approximate solutions also satisfy (6.1). First we point out that along a steady wave S is constant. Consequently, in each zone $k\Delta t < t < (k+1)\Delta t$, (6.1) is satisfied exactly for an approximation solution. This is clear when the formulation of Sec. 5 is used because (6.1) holds for a solution of the Riemann problem.

At the time $k\Delta t$, the averaging process introduces a change in the entropy production. The amount of entropy averaged over an interval $(a, b) \equiv ((h - \frac{1}{2})\Delta x, (h + \frac{1}{2})\Delta x)$ at time $k\Delta t + 0$ and $k\Delta t - 0$ are, respectively,

$$\frac{1}{b-a} \int_a^b S(U(x, kt + 0))\, dt = S\left(\frac{1}{b-a} \int_a^b U(x, kt - 0)\, dt\right)$$

and

$$\frac{1}{b-a} \int_a^b S(U(x, kt - 0))\, dx,$$

where we have noted that $U(x, k\Delta t + 0)$, $a < x < b$, is the constant state

$$\frac{1}{b-a} \int_a^b U(x, k\Delta t - 0)\, dx.$$

For the constitutive relation (2.1) we have $S = k + \ln e + (1 - \gamma) \ln \rho$ and so $-S$ is a convex function of $U = (\rho, \rho u, \rho E)$. Thus

$$\frac{1}{b-a} \int_a^b S(U(x, k\Delta t - 0))\, dt \leq s\left(\frac{1}{b-a} \int_a^b U(x, k\Delta t - 0)\, dx\right),$$

and so the entropy averaged over any mesh interval $((h - \frac{1}{2})\Delta x, (h + \frac{1}{2})\Delta x)$ increases from $t = k\Delta t - 0$ to $t = k\Delta t + 0$.

Finally, we show that the average of the entropy between two particle paths is a nondecreasing function of time. This is, in fact, the consequence of the above observations on the

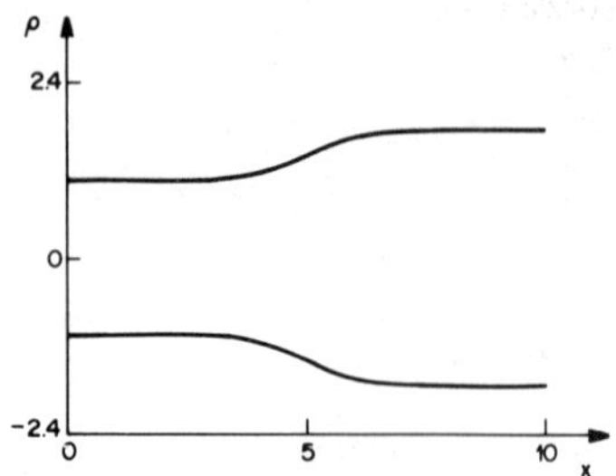

Fig. 4. The shape of the diverging duct.

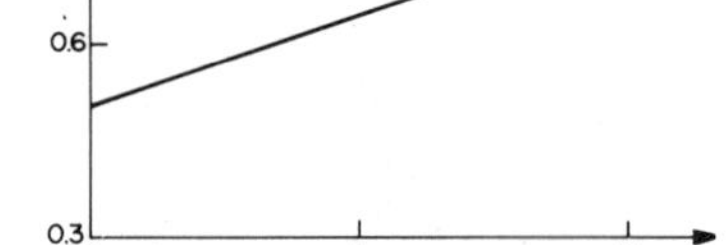

Fig. 5. Linear initial density profile.

entropy production. To apply the above arguments in a straightforward way we have to assume that the particle paths pass through the mesh points $\{((h \pm \frac{1}{2})\Delta x, k\Delta t)\}$, h integers, k any positive integer, which is, of course, in general not the case. In other words, there is a slight numerical error for the approximate solution to violate (6.1); details are omitted. This completes the proof of the admissibility of our scheme.

7. NUMERICAL RESULTS

Numerical calculations are carried out for a diverging duct and also for a converging–diverging duct. The ratio of specific heats is taken to be $\gamma = 1.4$. We first describe the numerical results for a diverging duct. The length of the duct is 10 m and the cross section of the duct 13 m (Fig. 4):

$$A(x) = 1.398 + 0.347 \tanh(0.8x - 4), \quad 0 < x < 10.$$

Forty mesh points are used; $\Delta x = 0.25$. The inflow boundary condition is $(p_{\text{in}}, \rho_{\text{in}}, u_{\text{in}}) = (0.3809, 0.502, 1.299)$ and the out flow density is $\rho_{\text{out}} = 0.776$. These boundary data can be connected by a steady wave solution with a standing shock wave located at $x = 4.816$ (the solid line in Fig. 5)[5]. The initial data is a linear profile connecting U_{in} and $(P, \rho, u) = (0.745, 0.776, 0.505)$. The Riemann problem and the steady wave equations (1.3) respectively are solved within the numerical error 10^{-6} for the pressure variable and $(M_2)^2$. In our calculation, these problems usually require one or two iterations, and never exceed four iterations for the Riemann problem or five iterations for the steady-wave problem. Our scheme is compared to a *fractional Godunov scheme*; it uses the Godunov scheme to solve (1.2) to obtain $\bar{U}$[4] and then alternates with solving the ordinary differential equations

$$U_t = G(t, \bar{U}).$$

Our scheme compares favorably with the fractional Godunov scheme in that it yields a much sharper shock wave and less dissipation elsewhere. Figure 6 depicts the density profile at time $t = 2.67$ (50 time steps) using our scheme. It shows that a stable wave profile is already being formed. The solid line in Fig. 6 represents the asymptotic steady wave. The density profile after 50 time steps using the fractional Godunov scheme is depicted in Fig. 7.

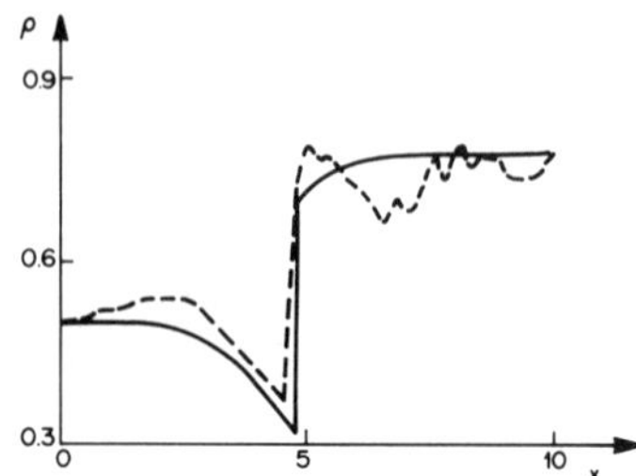

Fig. 6. The density profile (dotted line) at $t = 2.67$ (50 time steps), conservative piecewise steady (CPS) scheme, and the asymptotic profile (solid line).

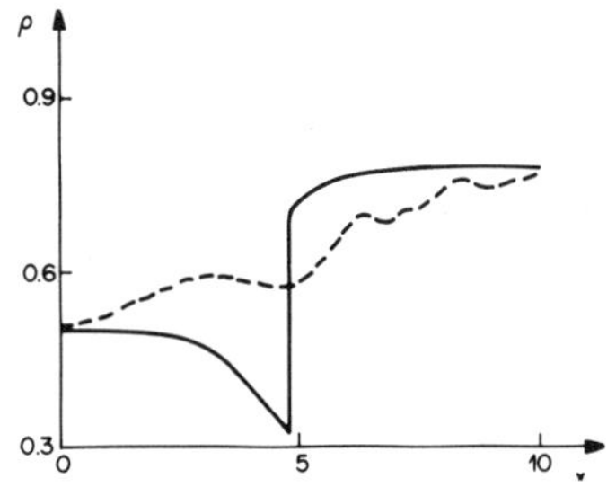

Fig. 7. The density profile at $t = 2.69$ (50 time steps) using the fractional Godunov scheme.

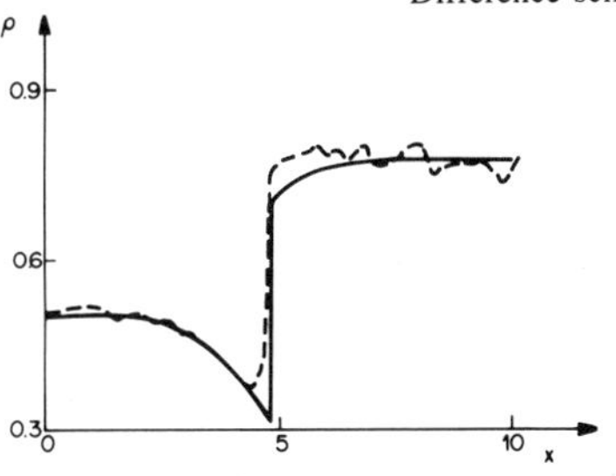

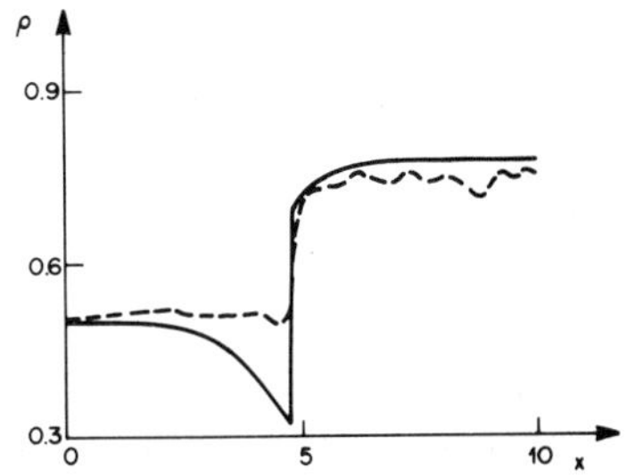

Fig. 8. The density profile at $t = 15.18$ (250 time steps) using the CPS scheme.

Fig. 9. The density profile at $t = 13.42$ (250 time steps) using the fractional Godunov scheme.

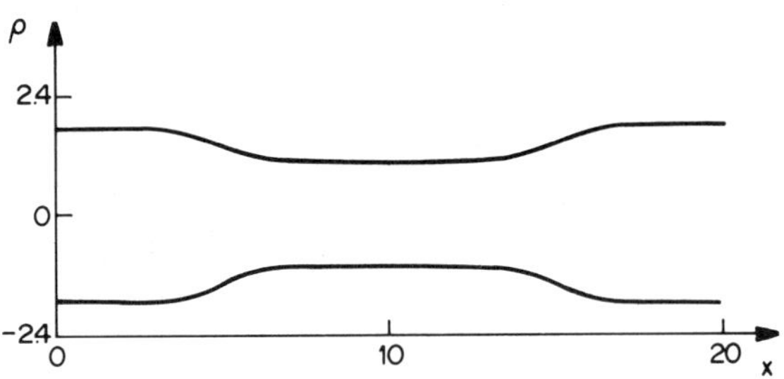

Fig. 10. The shape of the converging–diverging duct.

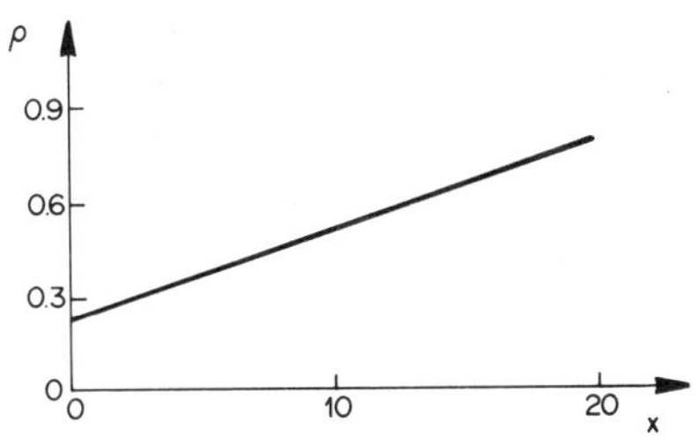

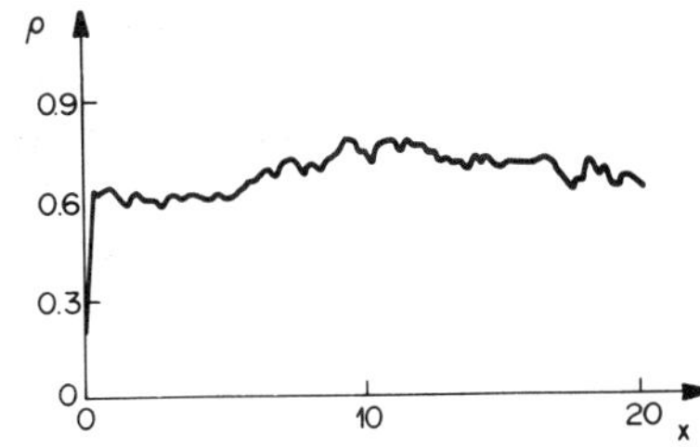

Fig. 11. A linear initial density profile.

Fig. 12. The density profile at 1800 time steps with the linear profile (Fig. 11) using the CPS scheme.

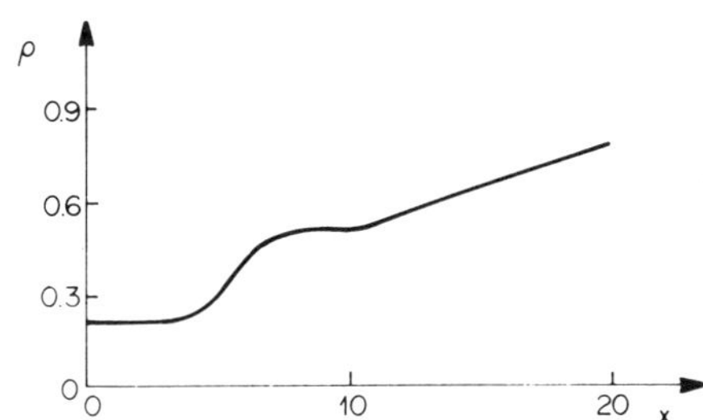

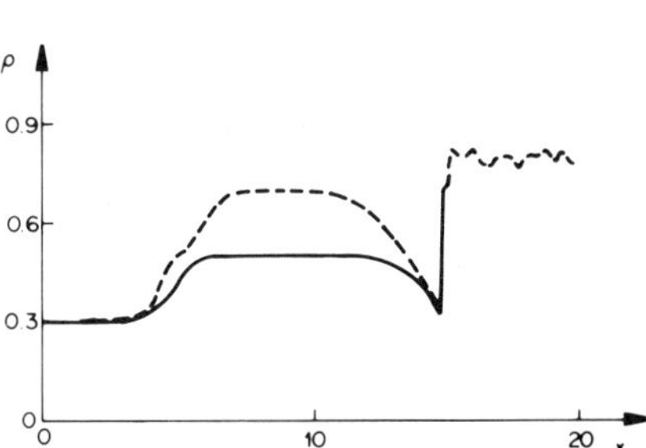

Fig. 13. A partly linear and partly steady initial density profile.

Fig. 14. The density profile at 900 time steps with the initial profile depicted in Fig. 13 using the CPS scheme.

After 250 time steps ($t = 15.18$), our scheme yields a wave profile close to the asymptotic steady profile (Fig. 8). The density profile at time step 250 using the fractional Godunov scheme is shown in Fig. 9.

It is noted that our scheme (Fig. 8 yields a sharper shock wave and smaller oscillations than the fractional Godunov scheme (Fig. 9). While it takes 50 time steps for our scheme to produce a clear stable wave pattern (Fig. 6) it takes 200 time steps for the fractional Godunov scheme to produce a comparable result. In Figs. 5–9 the transient solutions in dotted lines are superimposed with the exact asymptotic steady profile (solid lines).

The shape of the converging–diverging duct, Fig. 10 for our calculation, is the reflection of the above diverging duct (Fig. 4). There are 80 meshes; $\Delta x = 0.25$. We extend the steady wave for the diverging duct in the above case to the converging portion of the duct by solving (1.3) to obtain an upstream state $(\bar{p}_{in}, \bar{\rho}_{in}, \bar{u}_{in}) = (0.12071, 0.22110, 1.78255)$. Thus $\bar{U}_{in}$ can be connected to U_{out} by a steady wave with a standing shock wave at the diverging portion of the duct. The state $\bar{U}_{in}$ can also be connected to a downstream state with the given outflow density $\rho_{out} = 0.776$ by a shock wave with a negative speed at the inflow boundary followed by a subsonic steady wave. This is indeed the asymptotic wave profile which a linear initial data connecting $\bar{U}_{in}$ and $\bar{U}_{out}$ approaches. The initial density profile and the density profile at 1800 time steps are depicted in Figs. 11 and 12, respectively. Note that in Fig. 12 the density profile behind the shock wave at the inflow is close to a constant. This is reasonable because the subsonic steady wave in the exact asymptotic state has a small velocity and a near-constant density profile.

To produce an asymptotic state with the given inflow state $\bar{U}_{in}$ and outflow density ρ_{out} (solid line in Fig. 14) we choose an initial profile which is steady for the converging portion of the duct and the same linear profile for the diverging portion of the duct as before (Fig. 13).

The density profile after 900 time steps (in dotted lines) and the exact asymptotic steady density profile (in solid lines) are depicted in Fig. 14.

Acknowledgement—Research supported in part by the Department of Mathematics, University of Maryland under the exchange agreement with the Peking University.

Research (T-P.L.) supported in part by the NSF.

Computer time needed to carry out the present work was provided by the Computer Science Center of the University of Maryland.

REFERENCES

1. R. Courant and K. O. Friedrichs, *Supersonic Flow and Shock Waves*. Wiley-Interscience, New York (1948).
2. T.-P. Liu, Nonlinear stability and instability of transonic gas flow through a nozzle. *Comm. Math. Phys.* **83**, 243–260 (1982).
3. T.-P. Liu, Transonic gas flows along a duct of varying area. *Arch. Rat. Mech. Anal.* **80**, 1–18 (1982).
4. C. K. Godunov, *Numerical Methods of Multidimension Problems in Gas Dynamics*. Moscow (1976).
5. G. R. Shubin, A. B. Stephens and H. M. Glaz, Steady shock tracking and Newton's method applied to one-dimensional duct flow. *J. Comp. Phys.* **39**, 364–374 (1981).
6. P. Colella, A direct Eulerian MUSCL scheme for gas dynamics. Lawrence Berkeley Lab. Report LBL-14104 (1982).
7. B. van Leer, On the relation between the upwind-differencing schemes of Godunov, Enquist-Osler, and Roe. ICASE Report No. 81-11 (1981).
8. L. C. Huang, Pseudo-unsteady difference schemes for discontinuous solutions of steady-state, one-dimensional fluid dynamics problems. *J. Comp. Phys.* **42**, 195–211 (1981).
9. Computational Shock Wave Group, Peking Univ., The propagation of shock waves in a duct of variable cross-section and the formation and development of some new shocks. *Scientia Sinica* **XVIII**(6), 715–734 (1975).
10. H. Glaz and T.-P. Liu, The asymptotic analysis of wave interactions and numerical calculations of transonic nozzle flow **5**, 111–146 (1984).
11. S. K. Fok, Extention of Glimm's method to the problem of gas flow in a duct of variable cross section. Lawrence Berkeley Lab. Report LBL-12322 (1980).
12. J. Glimm, G. Marshall and B. Plohr, A generalized Riemann problem for quasi-one-dimensional gas flows (preprint).
13. G. A. Sod, Survey of finite difference methods. *J. Comp. Phys.* **27**, 1–31 (1978).
14. H. W. Liepmann and A. Roshko, *Elements of Gas Dynamics*, Chap. 5. John-Wiley, New York (1956).

Comp. & Maths. with Appls. Vol. 12A, Nos. 4/5, pp. 389–412, 1986
Printed in Great Britain.

0886-9553/86 $3.00 + .00
© 1986 Pergamon Press Ltd.

DISCRETIZATION OF IDEAL FLUID DYNAMICS IN THE EULERIAN REPRESENTATION

Boris A. Kupershmidt
The University of Tennessee Space Institute, Tullahoma, TN 37388, U.S.A.

and

Center for Nonlinear Studies, Los Alamos National Laboratory, Los Alamos, NM 87545, U.S.A.

Abstract—A Hamiltonian discretization of one-dimensional compressible fluid dynamics is made possible by analyzing the properties of semidirect product Lie algebras associated to representations of the Lie algebra of vector fields in $\mathbf{R}^n$.

1. INTRODUCTION

Analytical models of motion operate with continuous notions; computational models work with the language of discrete. The descriptive abilities of these two languages are incomparable: the differential calculus is old and powerful; the discrete calculus is young and weak. And although on fundamental level the nature is discrete, the discrete description of the world, directly from the first principles, is not available at the present time; instead, one is forced *to discretize the continuous models themselves.*

The question then arises of how such discretizations are to be made. (To have a genuine dynamics, we discretize space and keep time continuous.) Naturally, we expect the behaviour of a discrete model being similar to the behaviour of the corresponding continuous model which is being discretized; this is, however, a requirement on the properties of the final product and it provides no clues of how such a discretization should be performed. The only alternative is to make use of the mathematical properties of the continuous model, and to try to discretize them. What one should mean by "mathematical properties" of a given dynamical system? Certainly, conservation laws and symmetries should be counted in, as well as various other particulars. More generally though, we should look at the relations of our system with other dynamical systems; in other words: a) at the place our system occupies among others; and b) at the maps (morphisms) that connect it to other systems.

To make our discussion a little bit more concrete, let us concentrate on the class consisting of continuous conservative dynamical systems; that is to say, ideal fluids interacting selfconsistently with external fields.

Two representations are useful in fluid dynamics: Lagrangian and Eulerian. The discretization problem in the Lagrangian representation turns out to be not too difficult and is effectively solved in Holm, Kupershmidt, and Levermore[1,2] (except for superfluids, quantum fluids, and in all situations where the Lagrangian representation is not feasible), and we shall not discuss it here. In the Eulerian representation, the mathematical properties of all known ideal fluid dynamical systems are these: a) all such systems are Hamiltonian, and corresponding Hamiltonian structures are given in terms of 2-cocycles on appropriate differential Lie algebras of semidirect product type (Dzyaloshinskii and Volovick[3], Holm and Kupershmidt[4,5,7,9,10], Gibbons, Holm, and Kupershmidt[6,8]); and b) to each such system, one associates other systems in the space of Clebsch potentials together with maps called Clebsch representations, which are quadratic and canonical, and which have a natural Lie algebraic interpretation (Holm and Kupershmidt[7], Kupershmidt[11, Ch. VIII, Sec. 4]). Since the energy of every such system is its Hamiltonian function, there is no difficulty in discretizing the Hamiltonian. The whole problem of discretizing is then reduced to the problem of discretizing the Hamiltonian structures of such systems, and this is the topic addressed in the paper.

The Hamiltonian structures of fluid dynamical systems are the natural ones associated to dual spaces of differential Lie algebras (we can disregard at will 2-cocycles since they are represented by constant-coefficient operators and there is no difficulty in discretizing such) of semidirect product type: $D_n \propto V$, where D_n is the Lie algebra of vector fields on $\mathbf{R}^n$ and V is the direct sum of various tensor fields on $\mathbf{R}^n$; D_n acts on V by the Lie derivative. (In practice, the situation could be more complex, e.g. for superfluids and Yang–Mills fluids, but we disregard such possibility and concentrate instead on the typical cases.) It is natural to try to discretize D_n, V, and the action of D_n on V in order to preserve the most important feature of the Hamiltonian structure of a given fluid system: its Lie algebraic character. The first result of this paper is that this is impossible (Theorem 1.1). More precisely, set $K_n = C^\infty(\mathbf{R}^n)$, and denote $\partial_i = \partial/\partial x_i : K_n \to K_n$, $i = 1, \ldots, n$, the natural derivations. For $n = 1$, we denote ∂_1 by ∂. According to Ritt[12], there exist precisely two one-dimensional Lie algebras over $K_1 : D_1$ and the Abelian one.

Theorem 1.1

Let $\overline{K}$ be a commutative ring over which $\mathbf{Z}$ acts by automorphisms. Then there exists only one one-dimensional operator Lie algebra over K_1: the Abelian one.

It follows that the Lie algebraic connections should not be expected from discretizations of fluid dynamical systems. The problem of discretization is, hence, an extremely complex one. In retrospect though, this conclusion might have been expected. Indeed, upon browsing through the numerical analysis literature of the last few dozen years, one is struck by the total absence of elegant formulae, and not at all due to the venerable age of the subject. [Hardy believed that there were no new simple formulae to be discovered (in reference to Ramanujan) because they all had been found already by the end of the XIX century.] The inference that no simple discretizations of fluid dynamics exist is, thus, hardly escapable.

The Lie algebraic route [a)] to Hamiltonian discretizations being blocked, the only one left is to make sure use of the morphisms [b)] known in fluid dynamics. Not surprisingly, this route is not available either (except for $D_1 \propto K_1$, about which more will be said later) since the quadratic Clebsch maps cease to be canonical after being discretized. The whole problem is, therefore, a giant puzzle, with no general clues as how to proceed. It is natural then to analyze in greatest possible detail the first nontrivial case $n = 1$, and this is exactly what we shall do. This analysis leads, in Sec. 9, to a solution of the discretization problem (for $n = 1$), in the following form. We shall regularize the Hamiltonian matrix $B(N)$, corresponding to the Lie algebra $D_1 \propto (\oplus_{i=1}^N K_1)$, by adding to it a small (proportional to ϵ) nontrivial 2-cocycle such that the new matrix will become constant-coefficient after an appropriate change of variables has been made; the corresponding discrete Hamiltonian matrix will be then algebraic and singular in ϵ. (The singularity disappears upon passing to the continuous limit.)

The plan of the paper is as follows. In Sec. 2 we prove Theorem 1.1. In Sec. 3 we classify scalar Hamiltonian operators of nearest neighbor type, on one-dimensional lattice ($\mathbf{Z}$), and then find a nonlinear discretization of the Hamiltonian matrix $B(D_1) = B(0)$. In Sec. 4 we address the question of periodic systems and classify scalar Hamiltonian operators and one-dimensional Lie algebras over $\mathbf{Z}_2$ and $\mathbf{Z}_3$. In Sec. 5 we turn to one-dimensional hydrodynamics. We discretize barotropic fluids and examine the adiabatic case. We begin to study Lie algebras underlying fluid dynamics (in arbitrary number of dimension) in Sec. 6, analyze various canonical equivalences of the corresponding Hamiltonian structures in Sec. 7, and compute 2-cocycles on these Lie algebras (for one-dimensional case) in Sec. 8. One of such 2-cocycles, considered as a perturbation of the Hamiltonian structure of adiabatic one-dimensional dynamics, makes it possible in Sec. 9 to derive a desired discretization.

2. ONE-DIMENSIONAL LIE ALGEBRAS OVER Z

In this section we prove Theorem 1.1. Let $\overline{K}$ be a commutative ring and $\Delta : \overline{K} \to \overline{K}$ be an automorphism. We seek a bilinear operation $\overline{K} \times \overline{K} \to \overline{K}$ of the form

$$[X, Y] = \sum_{i,j} a_{ij} X^{(i)} Y^{(j)}, \quad \text{finite sum}, \quad X, Y \in \overline{K}, a_{ij} \in \overline{k}, \tag{2.1}$$

where $\bar{k} = \{a \in \overline{K} | \Delta(a) = a\}$ is the subring of constants, and $X^{(i)} = \Delta^i(X)$, such that

$$[X, Y] = -[Y, X], \tag{2.2}$$

$$[X, [Y, Z]] + [Y, [Z, X]] + [Z, [X, Y]] = 0, \tag{2.3}$$

for any $X, Y, Z, \in \overline{K}$.

[An informal model: $\overline{K} = \{f : \mathbf{Z} \to \mathbf{C}\}$, $[\Delta(f)](n) = f(n + 1)$, $\bar{k} \approx \mathbf{C}$.]

We want to show that all a_{ij}s in (2.1) vanish. Assume that this is not so. Let

$$A(x, y) = \Sigma a_{ij} x^i y^j \in \bar{k}[x, x^{-1}, y, y^{-1}] \tag{2.4}$$

be the corresponding generating function. Since

$$[[X, Y], Z] = \Sigma a_{pq} \Delta^p([X, Y]) \Delta^q(Z) = \Sigma a_{pq} a_{ij} X^{(i+p)} Y^{(j+p)} Z^{(q)},$$

the equalities (2.2) and (2.3) are equivalent to

$$A(y, x) = -A(x, y), \tag{2.5}$$

$$A(x, y)A(xy, z) + A(y, z)A(yz, x) + A(z, x)A(zx, y) = 0. \tag{2.6}$$

Let $N \in \mathbf{Z}$ be the maximal number such that

$$A(x, y) = (xy)^N B(x, y), \quad B(x, y) \in \bar{k}[x, y]. \tag{2.7}$$

Rewriting (2.5), (2.6) in terms of $B(x, y)$, we obtain

$$B(y, x) = -B(x, y), \tag{2.8}$$

$$(xy)^N B(x, y)B(xy, z) + (yz)^N B(y, z)B(yz, x) + (zx)^N B(z, x)B(zx, y) = 0. \tag{2.9}$$

Since N in (2.7) is maximal and we assumed $B \neq 0$, it follows that

$$P(x) = B(x, 0) = -B(0, x) \neq 0. \tag{2.10}$$

We deduce a contradiction by showing that $P(x) = 0$.

Using

$$(B(x, y) - [P(x) - P(y)]) \in xy\bar{k}[x, y], \tag{2.11}$$

we consider (2.9) for three cases: $N = 0$, $N < 0$, and $N > 0$.

Let first $N = 0$. Picking out from (2.9) terms not containing xy, we get

$$[P(x) - P(y)][-P(z)] + B(y, z)[P(yz) - P(x)]$$
$$+ B(z, x)[P(zx) - P(y)] = 0 \ (\text{mod } xy). \tag{2.12}$$

Picking out from (2.12) terms not containing x, and noticing that $P(0) = B(0, 0) = 0$, we obtain

$$0 = P(y)P(z) + B(y, z)P(yz) + P(z)[-P(y)] = B(y, z)P(yz), \tag{2.13}$$

so that either $B = 0$ or $P = 0$, and neither outcome is agreeable.

Let now $N < 0$. Dividing out (2.9) by $(xyz)^N$ and picking out from the result terms not containing x, we obtain

$$0 = z^{-N}[-P(y)][-P(z)] + y^{-N}P(z)[-P(y)] = (z^{-N} - y^{-N})P(y)P(z), \qquad (2.14)$$

which implies $P = 0$.

Finally, let $N > 0$. Picking out from (2.9) terms not containing x, we get

$$0 = (yz)^N B(y, z)P(yz), \qquad (2.15)$$

so that again $B = 0$ or $P = 0$. This finishes the proof.

Remark 2.1. Two-dimensional differential Lie algebras over K_1 were classified by Ritt[12]; the list of those can be found, e.g. in Cassidy[13, pp. 268, 269] (see also formula (1) in Kupershmidt[14]). For the discrete case there is no classification available as yet; the following formula provides a one-parameter family $L(\alpha)$ of two-dimensional Lie algebras over $\overline{K}$:

$$\left[\begin{pmatrix} X_0 \\ X_1 \end{pmatrix}, \begin{pmatrix} Y_0 \\ Y_1 \end{pmatrix} \right] = \begin{pmatrix} 0 \\ X_1(\Delta + \alpha)(Y_0) - Y_1(\Delta + \alpha)(X_0) \end{pmatrix}, \quad \alpha \in \overline{k}. \qquad (2.16)$$

The corresponding Hamiltonian matrix $B(\alpha)$ on the dual space to the Lie algebra $L(\alpha)$ (Kupershmidt[11,15]) gives rise to the following equations of motion

$$\dot{q}_0 = -(\alpha + \Delta^{-1})\left(q_1 \frac{\delta H}{\delta q_1} \right), \quad \dot{q}_1 = q_1(\Delta + \alpha)\left(\frac{\delta H}{\delta q_0} \right). \qquad (2.17)$$

For $\alpha = -1$, these equations describe the first Hamiltonian structure of the infinite Toda lattice (see Kupershmidt[13, p. 89; 11]). Notice that the Lie algebra $L(-1)$ has a very simple interpretation: Let $R = \{\Sigma_{i=0}^{M} f_i \zeta^i | 0 \leq M < \infty; f_i \in \overline{K}; \zeta^s f = \Delta^s(f)\zeta^s, s \in \mathbf{Z}_+, f \in \overline{K}\}$ be the associative ring of polynomial operators over $\overline{K}$, and let $(R)^{\mathrm{Lie}}$ be the corresponding Lie algebra. Let $I_M = \{\Sigma f_i \zeta^i | f_i = 0 \text{ for } i < M\}$ be the ideal in R. Then $(R_M)^{\mathrm{Lie}} = (R/I_M)^{\mathrm{Lie}}$ is a M-dimensional Lie algebra which is non-Abelian for $M > 1$. In particular, $(R_2)^{\mathrm{Lie}} \approx L(-1)$.

3. CLASSIFICATION OF SCALAR HAMILTONIAN OPERATORS OF NEAREST NEIGHBOR TYPE

The Hamiltonian operator associated to the Lie algebra D_1,

$$B(D_1) = u\partial + \partial u, \qquad (3.1)$$

cannot be linearly discretized (Theorem 1.1). Nevertheless, it can be discretized nonlinearly, and in essentially unique manner. This will follow from the description given below of all Hamiltonian operators of the form

$$B = \varphi\Delta - \Delta^{-1}\varphi, \qquad (3.2)$$

with appropriate $\varphi \in \overline{C} = \overline{K}[u^{(s)}], s \in \mathbf{Z}$. (See Kupershmidt[16,11] for the basics of the discrete calculus of variation.) We want to find all φ such that the skew-symmetric operator B in (3.2) is Hamiltonian; this means that the associated Poisson bracket $\{H, F\} = X_H(F), H, F \in \overline{C}$, satisfies the Jacobi identity modulo $\mathrm{Im}(\Delta - 1)$, where X_H stands for an evolution derivation of $\overline{C}$ (i.e. commuting with Δ and $\overline{K}$) given on the generator u by the equation

$$\dot{u} = (\varphi\Delta - \Delta^{-1}\varphi)\left(\frac{\delta H}{\delta u} \right), \qquad (3.3)$$

where

$$\frac{\delta H}{\delta u} = \sum_{s \in \mathbf{Z}} \Delta^{-s}\left(\frac{\partial H}{\partial u^{(s)}}\right). \tag{3.4}$$

In the familiar notations of numerical analysis, the equation (3.3) reads

$$\dot{u}_n = \varphi_n \frac{\partial H}{\partial u_{n+1}} - \varphi_{n-1} \frac{\partial H}{\partial u_{n-1}}, \quad n \in \mathbf{Z}. \tag{3.5}$$

To classify Hamiltonian operators B of the form (3.2), we use the following criterion for a skew-symmetric operator B to be Hamiltonian (see Kupershmidt and Manin[17], Kupershmidt[18]):

$$B \frac{\delta}{\delta u} [B(X)'Y] = D(BY)B(X) - D(BX)B(Y), \tag{3.6}$$

where X and Y are arbitrary vectors with indeterminate entries X_i, Y_i, $i = 1, \ldots, N$, where N is the number of independent generators of $\overline{C} = \overline{K}[u_i^{(s)}]$ (in our case, $N = 1$), and $D(Z)$, for a vector Z, is the Fréchet derivative operator:

$$[D(Z)]_{ij} = \sum_s \frac{\partial Z_i}{\partial u_j^{(s)}} \Delta^s. \tag{3.7}$$

The main result of this section is this:

THEOREM 3.1

The operator B in (3.2) is Hamiltonian if and only if it can be written in the form

$$B = a(u)(\Delta - \Delta^{-1})a(u), \tag{3.8}$$

with some $a(u) \in \overline{K}[u]$, where $u = u^{(0)}$.

Proof. We have

$$B(X)Y = [\varphi X^{(1)} - \varphi^{(-1)}X^{(-1)}]Y \sim \varphi[X^{(1)}Y - XY^{(1)}],$$

where $\sim$ means equality modulo $\mathrm{Im}(\Delta - 1)$. Hence,

$$\frac{\delta}{\delta u} [B(X)Y] = \sum_{k=n}^{m} \Delta^{-k}\{\varphi_k[X^{(1)}Y - XY^{(1)}]\},$$

where we assume

$$\varphi = \varphi(u^{(n)}, \ldots, u^{(m)}), \quad n \leq m, \tag{3.9}$$

and denote

$$\varphi_k = \frac{\partial \varphi}{\partial u^{(k)}}. \tag{3.10}$$

Hence, for the left-hand side of (3.6) we obtain

$$(\varphi \Delta - \Delta \varphi) \sum_{k=n}^{m} \Delta^{-k} \varphi_k [X^{(1)}Y - XY^{(1)}]. \tag{3.11L}$$

Now, for the right-hand side of (3.6) we consequently get

$$B(Y) = (\varphi\Delta - \Delta^{-1}\varphi)(Y),$$

$$D(BY) = Y^{(1)}D(\varphi) - \Delta^{-1}YD(\varphi) = (Y^{(1)} - \Delta^{-1}Y)\sum_{k=n}^{m}\varphi_k\Delta^k,$$

so that the right-hand side of (3.6) equals

$$(Y^{(1)} - \Delta^{-1}Y)\sum_{k=n}^{m}\varphi_k\Delta^k(\varphi\Delta - \Delta^{-1}\varphi)(X)$$

$$- (X^{(1)} - \Delta^{-1}X)\sum_{k=n}^{m}\varphi_k\Delta^k(\varphi\Delta - \Delta^{-1}\varphi)(Y). \quad (3.11R)$$

Let us compare terms containing $X^{(s)}$ with highest possible s, from both sides of (3.11). We get

$$\varphi\Delta^{1-n}\varphi_n(X^{(1)}Y) = \varphi\varphi_n^{(1-n)}X^{(2-n)}Y^{(1-n)}, \quad (3.12L)$$

$$Y^{(1)}\varphi_m\varphi^{(m)}X^{(m+1)}, \quad (3.12R1)$$

$$-X^{(1)}\sum_{k=n}^{m}\varphi_k\Delta^k(\varphi\Delta - \Delta^{-1}\varphi)(Y). \quad (3.12R2)$$

Since (3.12L) is a monomial and (3.12R2) is not and neither is (3.12R1) + (3.12R2) (even when $n = m$), we have to equate (3.12L) with (3.12R1):

$$\varphi\varphi_n^{(1-n)}X^{(2-n)}Y^{(1-n)} = \varphi_m\varphi^{(m)}X^{(m+1)}Y^{(1)}. \quad (3.13)$$

It follows that

$$n = 0, \quad m = 1, \quad (3.14)$$

$$\varphi\varphi_0^{(1)} = \varphi_1\varphi^{(1)}, \quad \varphi = \varphi(u, u^{(1)}). \quad (3.15)$$

Since $\varphi_0^{(1)} = [\varphi^{(1)}]_1$, (3.15) is equivalent to

$$(\ln \varphi)_1 = [\ln \varphi^{(1)}]_1. \quad (3.16)$$

Denote $\psi = \ln \varphi$. Writing (3.16) in long hand as

$$\frac{\partial\psi(u, u^{(1)})}{\partial u^{(1)}} = \frac{\partial\psi(u^{(1)}, u^{(2)})}{\partial u^{(1)}}, \quad (3.17)$$

we see that $\partial^2\psi/\partial u\partial u^{(1)} = 0$, so that $\psi = A(u) + B(u^{(1)})$, and (3.17) yields $\partial A(u)/\partial u = \partial B(u)/\partial u$. Hence, we can take $B(u) = A(u)$ without any loss of generality, so that

$$\varphi = a(u)a(u^{(1)}). \quad (3.18)$$

Hence,

$$B = \varphi\Delta - \Delta^{-1}\varphi = a(u)(\Delta - \Delta^{-1})a(u), \quad (3.19)$$

which is (3.8). It remains to notice that (3.19) is always Hamiltonian no matter what $a(u)$ is. The easiest way to see this is to observe that after a change of variables $U = U(u)$ such that

$$\frac{dU}{du} = \frac{1}{a(u)}, \quad (3.20)$$

the matrix (3.18) becomes

$$B = \Delta - \Delta^{-1}, \tag{3.21}$$

which is Hamiltonian since it is constant coefficient. ∎

The matrix (3.19) has a continuum limit. To perform the limiting procedure, one substitutes $\exp(\lambda s\partial)$ instead of Δ^s, and keeps only the first nontrivial order (in λ) terms. For (3.19), we obtain

$$B = a2\partial a = a^2\partial + \partial a^2. \tag{3.22}$$

Hence, to get $B(D_1)$ in (3.1) one has to take $a = \sqrt{u}$. The result of this reasoning we collect into the following:

Proposition 3.2. The matrix

$$B = \sqrt{u}(\Delta - \Delta^{-1})\sqrt{u} \tag{3.23}$$

is Hamiltonian, has $B(D_1)$ as its continuous limit, and is unique with such properties among all the Hamiltonian matrices of the nearest neighbor interaction type (3.2).

Remark 3.3. The change of variables

$$U = 2u^{-1/2} \tag{3.24}$$

transforms $B = \sqrt{u}(\Delta - \Delta^{-1})\sqrt{u}$ into $\Delta - \Delta^{-1}$, and $B(D_1) = u\partial + \partial u$ into 2∂, the latter being the continuous limit of $\Delta - \Delta^{-1}$.

4. PERIODIC PROBLEM

Oftentimes, the physical model is periodic in space. Its discretization then must be performed not over $\mathbf{Z}$ but over $\mathbf{Z}_n$, where n is the number of points we choose for our discrete space to have. In contrast to the case of $\mathbf{Z}$, one-dimensional Lie algebras over $\mathbf{Z}_n$, apparently, abound. In this section we consider $\mathbf{Z}_2$ and $\mathbf{Z}_3$. After we have classified all scalar Hamiltonian operators, it will be easy to extract those of Lie algebraic type.

We begin with $\mathbf{Z}_2$. A skew-symmetric operator has the form $f\Delta - \Delta^{-1}f = f\Delta - \Delta f = (f - f^{(1)})\Delta$, since $\Delta^2 = 1$. In other words, we look at the operators of the form

$$B = \varphi\Delta, \quad \varphi = \varphi(u, u^{(1)}) = -\varphi(u^{(1)}, u) = -\varphi^{(1)}. \tag{4.1}$$

To simplify notation, we shall write $\varphi = \varphi(0, 1)$ instead of $\varphi(u, u^{(1)})$, $\varphi(1, 0)_0 = \varphi_1(1, 0) = (\partial/\partial u)\varphi(u^{(1)}, u) = \Delta[(\partial/\partial u^{(1)})\varphi]$, etc.

Using the criterion (3.6), we get

$$B(X)Y = \varphi X^{(1)}Y,$$

$$\frac{\delta}{\delta u}[B(X)Y] = \varphi_0 X^{(1)}Y + \varphi(1, 0)_0 XY^{(1)},$$

$$B\frac{\delta}{\delta u}[B(X)Y] = \varphi[\varphi(1, 0)_1 XY^{(1)} + \varphi_1 X^{(1)}Y]; \tag{4.2L}$$

$$B(Y) = \varphi Y^{(1)},$$

$$D(BY) = Y^{(1)}(\varphi_0 + \varphi_1\Delta),$$

$$D(BY)B(X) - D(BX)B(Y) = Y^{(1)}(\varphi_0 + \varphi_1\Delta)(\varphi X^{(1)}) - X^{(1)}(\varphi_0 + \varphi_1\Delta)(\varphi Y^{(1)})$$
$$= \varphi_1[Y^{(1)}X - X^{(1)}Y]\varphi^{(1)}. \tag{4.2R}$$

Equating $XY^{(1)}$ − and $X^{(1)}Y$ − terms in (4.2), we obtain

$$\varphi\varphi(1, 0)_1 = \varphi_1\varphi^{(1)}, \tag{4.3}$$

$$\varphi\varphi_1 = -\varphi_1\varphi^{(1)}. \tag{4.4}$$

By (4.1), $\varphi^{(1)} = -\varphi$. Hence,

$$\varphi(1, 0)_1 = -\varphi_1. \tag{4.5}$$

Therefore, both (4.3) and (4.4) are identically satisfied. Thus, we have proved the following Theorem:

THEOREM 4.1

Every scalar skew-symmetric operator $B = \varphi\Delta$ over $\mathbf{Z}_2$ is Hamiltonian.

COROLLARY 4.2.

There exists just one (up to a constant multiple) one-dimensional Lie algebra structure over $\mathbf{Z}_2$, given by the formula

$$[X, Y] = X\Delta(Y) - Y\Delta(X). \tag{4.6}$$

Proof. We look at those φ that are linear in u, $u^{(1)}$. Since φ is skew-symmetric, $\varphi^{(1)} = -\varphi$, we must have

$$\varphi = \alpha(u - u^{(1)}), \quad \alpha \in \bar{k}. \tag{4.7}$$

To recover the commutator of the associated Lie algebra, we write

$$u[X, Y] \sim B(X)Y = \alpha(u - u^{(1)})X^{(1)}Y \sim \alpha u[X^{(1)}Y - XY^{(1)}],$$

and (4.6) follows.   ∎

Remark 4.3. Due to skew-symmetry of φ, additive constants can't be present in φ. In other words, the Lie algebra (4.6) has no nontrivial 2-cocycles on it.

We now turn to $\mathbf{Z}_3$. Since $\Delta^3 = 1$, an arbitrary scalar skew-symmetric operator has the form

$$B = \varphi\Delta - \Delta^{-1}\varphi, \quad \varphi = \varphi(u, u^{(1)}, u^{(2)}). \tag{4.8}$$

Working out the criterion (3.6), we obtain

$$B(X)Y = Y[\varphi X^{(1)} - \varphi^{(1)}X^{(1)}] \sim \varphi[YX^{(1)} - XY^{(1)}] = \varphi Z,$$

where we denote

$$Z = YX^{(1)} - XY^{(1)}; \tag{4.9}$$

$$\frac{\delta}{\delta u}[B(X)Y] = \varphi_0 Z + \varphi_1^{(2)}Z^{(2)} + \varphi_2^{(1)}Z^{(1)},$$

$$B\frac{\delta}{\delta u}[B(X)Y] = \varphi[\varphi_0^{(1)}Z^{(1)} + \varphi_1 Z + \varphi_2^{(2)}Z^{(2)}] - \varphi^{(2)}[\varphi_0^{(2)}Z^{(2)} + \varphi_1^{(1)}Z^{(1)} + \varphi_2 Z]$$

$$= Z[\varphi\varphi_1 - \varphi^{(2)}\varphi_2] + Z^{(1)}[\varphi\varphi_0^{(1)} - \varphi^{(2)}\varphi_1^{(1)}] + Z^{(2)}[\varphi\varphi_2^{(2)} - \varphi^{(2)}\varphi_0^{(2)}]; \tag{4.10L}$$

$$B(Y) = \varphi Y^{(1)} - \Delta^{-1}(\varphi Y),$$

$$D(BY) = (Y^{(1)} - \Delta^{-1}Y)D(\varphi) = (Y^{(1)} - \Delta^2 Y)(\varphi_0 + \varphi_1\Delta + \varphi_2\Delta^2),$$

$$D(BY)B(X) - D(BX)B(Y) = (Y^{(1)} - \Delta^2 Y)$$

$$\times (\varphi_0 + \varphi_1\Delta + \varphi_2\Delta^2)(\varphi\Delta - \Delta^2\varphi)(X) - \longleftrightarrow = \tag{4.10R}$$

[where "$-\leftrightarrow$" means: "minus the same expression with X and Y interchanged"]

$$= (Y^{(1)} - \Delta^2 Y)[\varphi_2\varphi^{(2)} - \varphi_1\varphi + (\varphi_0\varphi - \varphi_2\varphi^{(1)})\Delta + (\varphi_1\varphi^{(1)} - \varphi_0\varphi^{(2)})\Delta^2](X) - \longleftrightarrow$$

$$= Y^{(1)}[aX + bX^{(1)} + cX^{(2)}] - \longleftrightarrow + (-Y^{(2)})[a^{(2)}X^{(2)} + b^{(2)}X + c^{(2)}X^{(1)}] - \longleftrightarrow$$

{where we denote

$$a = \varphi_2\varphi^{(2)} - \varphi_1\varphi, \quad b = \varphi_0\varphi - \varphi_2\varphi^{(1)}, \quad c = \varphi_1\varphi^{(1)} - \varphi_0\varphi^{(2)}\}$$

$$= Z(-a) + Z^{(1)}[c + c^{(2)}] - Z^{(2)}b^{(2)}. \tag{4.11}$$

Comparing (4.10L) with (4.10R) and using (4.11), we see that only $Z^{(1)}$-terms provide a nontrivial condition, namely,

$$\varphi\varphi_0^{(1)} - \varphi^{(2)}\varphi_1^{(1)} = c^{(2)} + c = (\varphi_1^{(2)}\varphi - \varphi_0^{(2)}\varphi^{(1)}) + (\varphi_1\varphi^{(1)} - \varphi_0\varphi^{(2)}), \tag{4.12}$$

which can be rewritten in the form

$$0 = \varphi[\varphi_0^{(1)} - \varphi_1^{(2)}] + \varphi^{(1)}[\varphi_0^{(2)} - \varphi_1] + \varphi^{(2)}[\varphi_0 - \varphi_1^{(1)}],$$

or, better still,

$$(1 + \Delta + \Delta^2)[\varphi(\varphi_0^{(1)} - \varphi_1^{(2)})] = 0. \tag{4.13}$$

Thus, we have proved

THEOREM 4.4

An operator $\varphi\Delta - \Delta^{-1}\varphi$ over $\mathbf{Z}_3$ is Hamiltonian if and only if $(1 + \Delta + \Delta^2)[\varphi(\varphi_0^{(1)} - \varphi_1^{(2)})] = 0$.

Remark 4.5. By Theorem 3.8, an operator $a(u)(\Delta - \Delta^{-1})a(u)$ is Hamiltonian over $\mathbf{Z}$. Considering only those objects in the Hamiltonian formalism over $\mathbf{Z}$ which are invariant with respect to Δ^n, for a fixed $n \in \mathbf{N}$, we can pass to the quotient which is the Hamiltonian formalism over $\mathbf{Z}_n$. Hence, the operator $a(u)(\Delta - \Delta^{-1})a(u)$ is Hamiltonian over $\mathbf{Z}_n$ as well. For $n = 2$, $\Delta^{-1} = \Delta$, and $a(\Delta - \Delta^{-1})a = 0$. Let us see that, for $n = 3$, $\varphi = a(0)a(1)$ satisfies our criterion (4.13). We have,

$$\varphi = a(0)a(1), \quad \varphi_0^{(1)} = a'(1)a(2), \quad \varphi_1^{(2)} = a(2)a'(0),$$

$$\varphi(\varphi_0^{(1)} - \varphi_1^{(2)}) = a(0)a(1)a(2)[a'(1) - a'(0)],$$

and $(1 + \Delta + \Delta^2)[a'(1) - a'(0)] = 0$, so that (4.13) is indeed satisfied.

We now can classify one-dimensional Lie algebras over $\mathbf{Z}_3$. Let φ be linear,

$$\varphi = c_0 u + c_1 u^{(1)} + c_2 u^{(2)}, \quad c_i \in \bar{k}. \tag{4.14}$$

Then $\varphi_0^{(1)} - \varphi_1^{(2)} = c_0 - c_1$, and (4.13) becomes

$$(c_0 - c_1)(c_0 + c_1 + c_2) = 0. \tag{4.15}$$

Thus, we have two lines in $\bar{k}\mathrm{P}^2$ describing the variety of one-dimensional Lie algebras over $\mathbf{Z}_3$. It seems likely that the dimension of the space of one-dimensional Lie algebras over $\mathbf{Z}_n$ tends to infinity as n grows.

Remark 4.6. Formula (4.15) had been obtained earlier by Dale Peterson and myself, by a direct calculation.

We conclude by computing all 2-cycles on Lie algebras defined by (4.15). If

$$\varphi = c_0 u + c_1 u^{(1)} + c_2 u^{(2)} + \alpha, \quad c_i, \alpha \in \bar{k}, \tag{4.16}$$

then (4.13) yields

$$(c_0 - c_1)(c_0 + c_1 + c_2) = 0, \quad (c_0 - c_1)\alpha = 0. \tag{4.17}$$

Thus, there are no nontrivial 2-cocycles for $c_0 \neq c_1$. When $c_0 = c_1$, the space of nontrivial 2-cocycles is one-dimensional and is generated by $\Delta - \Delta^{-1}$.

5. ONE-DIMENSIONAL BAROTROPIC AND ADIABATIC FLUIDS

Adiabatic fluid equations in one space dimension are

$$-\dot{M} = \partial(M^2/\rho + P),$$

$$-\dot{\rho} = \partial(M), \quad -\dot{\eta} = \rho^{-1}M\partial(\eta), \tag{5.1}$$

where M is momentum density, ρ is mass density, η is specific entropy, P is fluid pressure: $P = \rho^2 e_{,\rho}$, where $e = e(\rho, \eta)$ is specific internal energy. The system (5.1) is a Hamiltonian system: it can be written in the form (see Holm and Kupershmidt[7])

$$-\begin{pmatrix} M \\ \rho \\ \sigma \end{pmatrix}_t = \begin{pmatrix} M\partial + \partial M & \rho\partial & \sigma\partial \\ \partial\rho & 0 & 0 \\ \partial\sigma & 0 & 0 \end{pmatrix}\begin{pmatrix} \delta H/\delta M \\ \delta H/\delta\rho \\ \delta H/\delta\sigma \end{pmatrix}, \tag{5.2}$$

where $\sigma = \rho\eta$ is entropy per unit volume, and

$$H = M^2/2\rho + \rho e \tag{5.3}$$

is the total energy density. In the case $e = e(\rho)$, the entropy η can be taken to be zero; the resulting barotropic fluid is again a Hamiltonian system, of the form

$$-\begin{pmatrix} M \\ \rho \end{pmatrix}_t = \begin{pmatrix} M\partial + \partial M & \rho\partial \\ \partial\rho & 0 \end{pmatrix}\begin{pmatrix} \delta H/\delta M \\ \delta H/\delta\rho \end{pmatrix}. \tag{5.4}$$

The 3 by 3 (respectively, 2 by 2) matrix in the right-hand side of the equation (5.2) (respectively, (5.4)) is called the associated Hamiltonian matrix. As has been explained in Sec. 1, we strive to discretize exactly such matrices.

We now show that the Hamiltonian matrix in (5.4) is equivalent (after a change of variables) to a constant-coefficient one. It would follow, in particular, that the matrix in (5.4) can be easily discretized.

Let

$$b = \begin{pmatrix} 0 & s \\ -s^\dagger & 0 \end{pmatrix}$$

be a Hamiltonian matrix in variables q and p, where s is a constant-coefficient operator and $s^\dagger$ is the adjoint to s. Let

$$M = pq, \quad \rho = p, \tag{5.5}$$

be an invertible change of variables. Then the Hamiltonian matrix B in the variables M and ρ

satisfies (see Holm and Kupershmidt[7, Sec. 3])

$$B = JbJ^\dagger, \tag{5.6}$$

where J is the Jacobian of the map (5.5). In our case,

$$J = \begin{pmatrix} p & q \\ 0 & 1 \end{pmatrix}, \quad J^\dagger = \begin{pmatrix} p & 0 \\ q & 1 \end{pmatrix},$$

so that

$$B = \begin{pmatrix} \rho s \dfrac{M}{\rho} - \dfrac{M}{\rho} s^\dagger \rho & \rho s \\ -s^\dagger \rho & 0 \end{pmatrix}. \tag{5.7}$$

For $s = \partial$, we obtain $\rho\partial(M/\rho) - (M/\rho)(-\partial)\rho = M\partial + \partial M$, so that (5.4) results. In particular, for $s = (\Delta - \Delta^{-1})/2$, (5.7) becomes a discrete version of (5.4):

$$B = \begin{pmatrix} \rho\, \dfrac{\Delta - \Delta^{-1}}{2}\, \dfrac{M}{\rho} + \dfrac{M}{\rho}\, \dfrac{\Delta - \Delta^{-1}}{2}\, \rho & \rho\, \dfrac{\Delta - \Delta^{-1}}{2} \\[2ex] \dfrac{\Delta - \Delta^{-1}}{2}\, \rho & 0 \end{pmatrix}, \tag{5.8}$$

which provides a discretization of barotropic fluid dynamics in one space dimension.

Let us turn to the matrix (5.2). To show that it is not equivalent to a constant-coefficient one we may try to employ the following obvious sufficient criterion: if B is a matrix of differential operators without constant terms and the dimension of the ker B is less than the size of B, then B is not equivalent to a constant-coefficient matrix; here ker B consists of all those nontrivial H for which the corresponding motion equations vanish. It is easy to see that ker $B = \{c^1\rho + c_2\sigma + c_3(\rho^2 + \sigma^2)^{1/2}|c_i \in k = \ker \partial\}$ for (5.2), so that dim ker $B = 3 = $ size of B, and we can't deduce a definite result; however, it's almost obvious that this matrix can't be reduced to a constant-coefficient form. (Note that ker $B = \{c_1 M/\rho + c_2\rho|c_i \in k\}$ for (5.4), and this is the origin of the formula (5.5): in general, the best coordinates are found in ker B.)

It is now a bit more clear why discretization of adiabatic fluid dynamics is so much more difficult than of the barotropic one. As an additional illustration of difficulties involved, let me show that the direct generalization to the adiabatic case of the discrete barotropic matrix (5.8) is not possible. Consider the matrix

$$B = \begin{pmatrix} a\Delta - \Delta^{-1}a & \rho(\Delta - \Delta^{-1}) & \sigma(\Delta - \Delta^{-1}) \\ (\Delta - \Delta^{-1})\rho & 0 & 0 \\ (\Delta - \Delta^{-1})\sigma & 0 & 0 \end{pmatrix}, \tag{5.9}$$

where a is an arbitrary function of $\{M^{(s)}, \rho^{(s)}, \sigma^{(s)}|s \in \mathbf{Z}\}$. Let us see that no matter what a is, the matrix (5.9) is not Hamiltonian. We use criterion (3.6) for $X = (X_1, X_2, X_3)$, $Y = (Y_1, Y_2, Y_3)$, and pay attention only to the second and third elements of both vectors in (3.6). We have, by denoting $\rho_2 = \rho$, $\rho_3 = \sigma$, and letting i to be 2 or 3:

$$\left(B\, \frac{\delta}{\delta u}\, [Y'B(X)]\right)_i = (\Delta - \Delta^{-1})(\rho_i\Omega), \tag{5.10L}$$

where $\Omega = (\delta/\delta M)[Y'B(X)]$; also,

$$
B(Y) = \begin{pmatrix} \cdots \\ (\Delta - \Delta^{-1})(\rho_2 Y_1) \\ (\Delta - \Delta^{-1})(\rho_3 Y_1) \end{pmatrix},
$$

$$
D(BY) = \begin{pmatrix} \cdots & \cdots & & \cdots \\ 0 & (\Delta - \Delta^{-1})Y_1 & & 0 \\ 0 & & 0 & (\Delta - \Delta^{-1})Y_1 \end{pmatrix},
$$

$$
(D(BY)B(X) - \longleftrightarrow)_i = (\Delta - \Delta^{-1})[Y_1(\Delta - \Delta^{-1})(\rho_i X_1) - \longleftrightarrow]. \tag{5.10R}
$$

Equating (5.10L) with (5.10R), we first get rid of $\Delta - \Delta^{-1}$; then, to eliminate Ω, we multiply the result by $\rho_{\bar{i}}$, where $\bar{i} = 5 - i$, to arrive at the equality

$$
\sigma[Y_1(\Delta - \Delta^{-1})(\rho X_1) - \longleftrightarrow] = \rho[Y_1(\Delta - \Delta^{-1})(\sigma X_1) - \longleftrightarrow], \tag{5.11}
$$

which is clearly impossible: the left-hand side involves ρ only as $\rho^{(1)}$ and $\rho^{(-1)}$ while the right-hand side has only $\rho = \rho^{(0)}$.

Remark. If more variables $\sigma_2, \ldots, \sigma_n$, similar to $\sigma = \sigma_1$, are added to the matrix (5.2), the resulting matrix B has ker $B = \{\Sigma\, c_i \sigma_i + c_0 \rho + c\sqrt{\rho^2 + \sigma^2}\}$, so that dim Ker $B = n + 2 = $ size of B; but B is certainly not equivalent to a constant-coefficient matrix. Analogously, if more σ's are added to the matrix (5.9), the resulting matrix will be not Hamiltonian.

6. SEMIDIRECT PRODUCTS ASSOCIATED TO ONE-DIMENSIONAL REPRESENTATIONS OF D_n

In this section we begin a systematic study of Lie algebras underlying the Hamiltonian structures of fluid dynamics.

LEMMA 6.1

Let $X, Y \in D_n$. Then

$$
\text{div}([X, Y]) = X(\text{div}(Y)) - Y(\text{div}(X)). \tag{6.1}
$$

Proof. Let $X = \Sigma\, X_j \partial_j$, $Y = \Sigma\, Y_i \partial_i$. Then

$$
[X, Y] = \sum X_j Y_{i,j} \partial_i - \sum Y_i X_{j,i} \partial_j
$$

so that

$$
\begin{aligned}
\text{div}([X, Y]) &= \sum (X_j Y_{i,j})_{,\,i} - \sum (Y_i X_{j,i})_{,\,j} \\
&= \sum (X_{j,i} Y_{i,j} + X_j Y_{i,ij} - Y_{i,j} X_{j,i} - Y_i X_{j,ji}) \\
&= \sum X_j \partial_j(\text{div}(Y)) - \sum Y_i \partial_i(\text{div}(X)). \quad\blacksquare
\end{aligned}
$$

Remark 6.2. The proof above becomes more transparent in the geometric situation: Suppose M is a manifold, w is a volume form on M, and X and Y are vector fields on M. Then

$$
\begin{aligned}
\text{div}([X, Y])w &= [X, Y](w) = X(Y(w)) - Y(X(w)) \\
&= X(\text{div}(Y)w) - Y(\text{div}(X)w) = X(\text{div}(Y))w + \text{div}(Y)\text{div}(X)w \\
&\quad - Y(\text{div}(X))w - \text{div}(X)\text{div}(Y)w \\
&= [X(\text{div}(Y)) - Y(\text{div}(X))]w.
\end{aligned}
$$

THEOREM 6.3

Let $\lambda \in k = \cap_i \ker \partial_i|_{K_n}$. Then the map $\sigma_\lambda: D_n \to \mathrm{Diff}_1(K_n)$ is a representation of Lie algebras, where

$$\sigma_\lambda: X \mapsto X + \lambda \, \mathrm{div}(X), \qquad (6.2)$$

and $\mathrm{Diff}_1(K_n)$ is the Lie algebra of differential operators on K_n, of order ≤ 1.

Proof. Let $X, Y \in D_n$. Then

$$\sigma_\lambda([X, Y]) = [X, Y] + \lambda \, \mathrm{div}([X, Y]) \; [\text{by (6.1)}]$$
$$= [X, Y] + \lambda[X(\mathrm{div}(Y)) - Y(\mathrm{div}(X))]$$
$$= [X + \lambda \, \mathrm{div}(X), Y + \lambda \, \mathrm{div}(Y)] = [\sigma_\lambda(X), \sigma_\lambda(Y)]. \qquad \blacksquare$$

Remark 6.4. Representations σ_λ are one-dimensional. When external fields interact with fluids, one also needs to consider higher-dimensional D_n-modules consisting of natural tensor fields on $\mathbf{R}^n$ (see Holm and Kupershmidt[7]). Various other representations of D_n exist, some of which are important in fluid dynamics. For example, suppose we consider flows in $\mathbf{R}^n$ depending only upon one of coordinates (more generally: only upon some of coordinates, may be curvelinear); *a priori*, there are no reasons why such flows should be Hamiltonian, but they are. The corresponding representations of D_1 (in the case of one-coordinate dependence) are described by a pair of constant (i.e. with elements in k) matrices A, C:

$$X \mapsto XA\partial + X'C, \quad X' = \partial(X), \quad X \in D_1, \qquad (6.3)$$

satisfying the equations

$$(A - 1)C = (C - A)(A - 1) = 0. \qquad (6.4)$$

In particular,

$$A = 1, \quad C \text{ arbitrary}, \qquad (6.5)$$

satisfies (6.4) and generalizes (6.2) (when C is not diagonal). Representations of D_n useful outside fluid dynamics (e.g. in the theory of Virasoro algebras) are of the form (in the case $n = 1$)

$$X \mapsto XA\partial + X'C + XF + x^{-1}XG, \qquad (6.6)$$

where A, C, F and G are constant matrices, and $x \in K_1$ (or $K_1[x]$) is such that $\partial(x) = 1$.

We now can elucidate the origin of the matrices (5.2) and (5.4). Recall that if $\sigma: L \to \mathrm{End} \, V$ is a representation of a Lie algebra L on a vector space V, then the semidirect product $L \times_\sigma V = L \ltimes V$ is a new Lie algebra with the commutator

$$\left[\begin{pmatrix} X \\ u \end{pmatrix}, \begin{pmatrix} Y \\ v \end{pmatrix} \right] = \begin{pmatrix} [X, Y] \\ \sigma(X)(v) - \sigma(Y)(u) \end{pmatrix}, \quad X, Y \in L, u, v \in V. \qquad (6.7)$$

Denote V_λ a free one-dimensional D_n- and K_n-module, on which D_n acts by formula (5.6). We set $V = V_{\underline{\lambda}} = \oplus_j V_{\lambda_j}$ for $\underline{\lambda} = (\lambda_1, \ldots, \lambda_m)$, and denote

$$L_{\underline{\lambda}}(n) = D_n \ltimes V_{\underline{\lambda}}. \qquad (6.8)$$

For $n = 1$, we write $L_{\underline{\lambda}}$ instead of $L_{\underline{\lambda}}(1)$. Let us compute the natural Hamiltonian structure associated to the Lie algebra $L_{\underline{\lambda}}(n)$ (see Kupershmidt[11,15]). Let M_i be the coordinate dual to

∂_i, and ρ_j be the coordinate dual to $1 \in V_{\lambda_j}$. Then

$$(X, u)'B\binom{Y}{v} \sim \sum M_i[X, Y]_i + \sum \rho_j[\sigma(X)(v)$$

$$- \sigma(Y)(u)]_j = \sum M_i(X_k Y_{i,k} - Y_k X_{i,k})$$

$$+ \sum \rho_j(X_k v_{j,k} + \lambda_j X_{k,k} v_j - Y_k u_{j,k} - \lambda_j Y_{k,k} u_j)$$

$$\sim \sum X_k(M_i \partial_k + \partial_i M_k)(Y_k) + \sum X_k(\rho_j \partial_k - \lambda_j \partial_k \rho_j)(v_j)$$

$$+ \sum u_j(\partial_k \rho_j - \lambda_j \rho_j \partial_k)(Y_k). \tag{6.9}$$

Hence, the corresponding Hamiltonian systems are of the form

$$\dot{M}_k = \Sigma(M_i \partial_k + \partial_i M_k)(\delta H/\delta M_i) + \Sigma(\rho_j \partial_k - \lambda_j \partial_k \rho_j)(\delta H/\delta \rho_j),$$

$$\dot{\rho}_j = \Sigma(\partial_k \rho_j - \lambda_j \rho_j \partial_k)(\delta H/\delta M_k). \tag{6.10}$$

This general form covers n-dimensional fluid dynamics; for $n = 1$, $\underline{\lambda} = (0, 0)$ results in (5.2) while $\underline{\lambda} = (0)$ results in (5.4).

Remark 6.5. In this paper we study only the usual hydrodynamics, i.e., one-fluid system. More complex two-fluid systems, e.g. superfluids and quantum fluids (see Holm and Kupershmidt[4]), have two copies of D_n (one copy for each fluid) acting by derivations in corresponding semidirect products. In the language of this section, a convenient model for such actions is this (for $n = 1$). Consider $L_C = D_1 \propto V$, where the action of D_1 is given by (6.3), (6.5). Then the only action of D_1, of the same form (6.3) and (6.5), on L_C by derivations, is given by the matrix $\overline{C}$ of the form

$$\overline{C} = \begin{pmatrix} -1 & 0 \cdots 0 \\ 0 & \\ \vdots & C \\ 0 & \end{pmatrix}.$$

The corresponding semidirect product of D_1 and L_C has a nonabelian piece in it, L_C, and has two copies of D_1 present.

Denote $B_\lambda(n)$ the Hamiltonian matrix in (6.10) associated to the Lie algebra $L_\lambda(n)$; for $n = 1$, we write simple B_λ. As we shall see, there exist numerous hidden isomorphisms between $B_\lambda(n)$'s for various λ's. Our first step is to make sure that those isomorphisms are not due to isomorphisms of Lie algebras $L_\lambda(n)$'s themselves.

Theorem 6.6

If $\Theta : L_\lambda(n) \to L_\mu(n)$ is a Lie algebra isomorphism then $\lambda_j = [w(\underline{\mu})]_j$, $j = 1, \ldots, m$, for some permutation $w \in S_m$, and Θ has the form, $\Theta = 1 \oplus \overline{\Theta}$, where $\overline{\Theta}$ is the direct sum of linear constant-coefficient invertible transformations $\overline{\Theta}_\lambda$ on $\oplus_{j:\lambda_j = \lambda} V_{\lambda_j}$. We break the proof into several Lemmas.

Lemma 6.7

Let $l_\lambda(n)$ be the n-dimensional Abelian subalgebra $\oplus_j V_{\lambda_j}$ in $L_\lambda(n)$. If R is a one-dimensional Abelian subalgebra of $L_\lambda(n)$ then $R \subset l_\lambda(n)$. In particular, any n-dimensional Abelian subalgebra in $L_\lambda(n)$ is $l_\lambda(n)$ itself.

Proof. Let $\binom{X}{u} \in R$. Then $f\binom{X}{u} = \binom{fX}{fu} \in R$, and since R is Abelian, $\binom{0}{0} = [\binom{X}{u}, \binom{fX}{fu}] = \binom{X(f)X}{\cdots}$. Since f is arbitrary, X must be zero. ∎

LEMMA 6.8

Let $\alpha: K_n^m \to D_n$ be a linear differential operator. If $\mathrm{Im}(\alpha)$ is an Abelian subalgebra then $\alpha = 0$.

Proof. Let $u \in K_n^m$ be such that $\alpha(u) \neq 0$. We will arrive at a contradiction as follows. Let $\alpha(su) = \Sigma s^{(\sigma)} X_\sigma$, where $s \in K_n$, $X_\sigma \in D_n$, $\sigma \in \mathbf{Z}_+^n$ is a multi-index, $s^{(\sigma)} = \partial_1^{\sigma_1} \cdots \partial_n^{\sigma_n}(s)$ for $\sigma = (\sigma_1, \ldots, \sigma_n)$. For $s, g \in K_n$, we have

$$0 = [\alpha(su), \alpha(gu)] = \left[\sum s^{(\sigma)} X_\sigma, \sum g^{(\nu)} X_\nu \right]$$

$$= \sum s^{(\sigma)} g^{(\nu)} [X_\sigma, X_\nu] + \sum s^{(\sigma)} X_\sigma(g^{(\nu)}) X_\nu - \sum g^{(\nu)} X_\nu(s^{(\sigma)}) X_\sigma.$$

Denote $|\sigma| = \sigma_1 + \cdots + \sigma_n$, and let k be the maximal $|\sigma|$ for which $X_\sigma \neq 0$. Then the highest number of derivations acting on g happens in the term $\Sigma_{|\nu|=k} s^{(\sigma)} X_\sigma(g^{(\nu)}) X_\nu$. Since s and g are arbitrary, we conclude that $X_\nu = 0$, whenever $|\nu| = \mathrm{k}$. Hence, all X_σ's vanish. Thus, $\alpha(u) = \alpha(1u) = 0$. $\blacksquare$

Let $\Theta: L_\lambda(n) \to L_\mu(n)$ be a linear differential operator and an isomorphism of Lie algebras. Denote by $\overline{\Theta}$ the restriction of Θ on $l_\lambda(n)$.

LEMMA 6.9

$\overline{\Theta}(l_\lambda(n)) \subset l_\mu(n)$.

Proof. $\overline{\Theta}(l_\lambda(n))$ is Abelian. Therefore, by Lemma 6.8, $\Theta(l_\lambda(n))$ intersects $D_n \subset L_\mu(n)$ trivially. $\blacksquare$

LEMMA 6.10

Let $\alpha: D_n \to D_n$ be a linear differential operator and an isomorphism of Lie algebras. Then $\alpha = id$.

Proof. Let $X \in D_n$, $s, g \in K_n$. Let $\alpha(sX) = \Sigma s^{(\sigma)} X_\sigma$. Then

$$\alpha([sX, gX]) = \alpha([sX(g) - gX(s)]X) = \sum [sX(g) - gX(s)]^{(\sigma)} X_\sigma, \qquad (6.11\mathrm{L})$$

$$[\alpha(sX), \alpha(gX)] = \left[\sum s^{(\sigma)} X_\sigma, \sum g^{(\nu)} X_\nu \right] = \sum s^{(\sigma)} g^{(\nu)} [X_\sigma, X_\nu]$$

$$+ \sum s^{(\sigma)} X_\sigma(g^{(\nu)}) X_\nu - \sum g^{(\nu)} X_\nu(s^{(\sigma)}) X_\sigma. \qquad (6.11\mathrm{R})$$

As in the proof of Lemma 6.8, comparing the highest-order derivatives acting on g in (6.11), we find that

$$\sum sX(g^{(\sigma)}) X_\nu = \sum s^{(\sigma)} X_\sigma(g^{(\nu)}) X_\nu,$$

so that $sX = \Sigma s^{(\sigma)} X_\sigma$. In other words, $\alpha = id$. $\blacksquare$

LEMMA 6.11

$\Theta = 1 \oplus \overline{\Theta}$.

Proof. Since Θ^{-1} is also an isomorphism, from Lemma 6.9 it follows that $\Theta(D_n) \subset D_n$. Therefore, $\Theta|_{D_n} = id$ by Lemma 6.10. $\blacksquare$

LEMMA 6.12

Let $\sigma_i: L \to \mathrm{End}\, V_i$ be two representations, $i = 1, 2$, and let $\alpha: V_1 \to V_2$ be a homomorphism. Then $1 \oplus \alpha: L \times_\sigma V_1 \to L \times_\sigma V_2$ is a homomorphism of Lie algebras if and only if α intertwines the representations σ_1 and σ_2, i.e.

$$\sigma_2(X)\alpha = \alpha\sigma_1(X), \quad \forall X \in L. \qquad (6.12)$$

Proof. Let us write α instead of $1 \oplus \alpha$. Then, for $X, Y \in L$, $u, v \in V_1$, we have,

$$\alpha\left(\left[\begin{pmatrix} X \\ u \end{pmatrix}, \begin{pmatrix} Y \\ v \end{pmatrix}\right]\right) = \alpha\begin{pmatrix} [X, Y] \\ \sigma_1(X)(v) - \sigma_1(Y)(u) \end{pmatrix} = \begin{pmatrix} [X, Y] \\ \alpha\sigma_1(X)(v) - \alpha\sigma_1(Y)(u) \end{pmatrix},$$

$$\left[\alpha\begin{pmatrix} X \\ u \end{pmatrix}, \alpha\begin{pmatrix} Y \\ v \end{pmatrix}\right] = \left[\begin{pmatrix} X \\ \alpha(u) \end{pmatrix}, \begin{pmatrix} Y \\ \alpha(v) \end{pmatrix}\right] = \begin{pmatrix} [X, Y] \\ \sigma_2(X)\alpha(v) - \sigma_2(Y)\alpha(u) \end{pmatrix},$$

and the Lemma follows.   ∎

Proof of Theorem 6.6. By Lemma 6.11, $\Theta = 1 \oplus \overline{\Theta}$; and by Lemma 6.12, $\overline{\Theta}$ must intertwine representations of D_n on $l_\lambda(n)$ and $l_\mu(n)$. Let us take $u = (u_j) \in l_\lambda(n)$, with $u_j \in V_{\lambda_j}$. Let V_{μ_j}-component of $\overline{\Theta}(u)$ be given as

$$[\overline{\Theta}(u)]_i = \sum a_{ij}^v v_j^{(v)}, \quad a_{ij}^v \in K_n. \tag{6.13}$$

Using criterion (6.12) and definition (6.2), we obtain

$$[\sigma_2(X)\overline{\Theta}(u)]_i = \left(\sum X_k \partial_k + \mu_i \operatorname{div}(X)\right) \sum a_{ij}^v u_j^{(v)},$$

$$[\overline{\Theta}\sigma_1(X)(u)]_i = \sum a_{ij}^v \partial^v \left(\sum X_k \partial_k + \lambda_j \operatorname{div}(X)\right)(u_j)$$

so that

$$\sum_k (X_k \partial_k + \mu_i X_{k,k}) \sum_v a_{ij}^v \partial^v = \sum_v a_{ij}^v \partial^v \circ \sum_k (X_k \partial_k + \lambda_j X_{k,k}). \tag{6.14}$$

Comparing highest-order derivations acting on X_k in (6.13) we conclude first that $\Sigma_v a_{ij}^v \partial^v = a_{ij}$, so that (6.14) becomes

$$(X_k \partial_k + \mu_i X_{k,k})a_{ij} = a_{ij}(X_k \partial_k + \lambda_j X_{k,k}). \tag{6.15}$$

This equality, in turn, is equivalent to a pair of equations

$$a_{ij,k} = 0 \Rightarrow a_{ij} \in k, \tag{6.16}$$

$$a_{ij}(\mu_i - \lambda_j) = 0. \tag{6.17}$$

Thus, all a_{ij}'s are constant, and $a_{ij} = 0$ unless $\mu_i = \lambda_j$. This means that $\overline{\Theta}$ is the direct sum of linear invertible transformations $\overline{\Theta}_\lambda$ on $\bigoplus_{j:\lambda_j = \lambda} V_{\lambda_j}$.   ∎

7. CANONICAL ISOMORPHISMS BETWEEN HAMILTONIAN STRUCTURES $B_\lambda(n)$

Although Lie algebras $L_\lambda(n)$ are rarely isomorphic to each other (Theorem 6.6), there are no more than two isomorphism classes of the corresponding Hamiltonian structures $B_\lambda(n): \underline{\lambda} = (1, \ldots, 1)$ and $\underline{\lambda} = (0, \ldots, 0)$. Let us show that this is indeed the case.

THEOREM 7.1

If $\underline{\lambda} \neq (1, \ldots, 1)$ then $B_\lambda(n)$ is canonically equivalent to $B_0(n)$.

Proof. Let first $m = 1$, so that $\underline{\lambda} = \lambda \neq 1$. Set

$$\psi = \rho^{1/(1-\lambda)}. \tag{7.1}$$

Transforming in (6.10) from the variables $(\mathbf{M}, \rho)$ to the variables $(\mathbf{M}, \psi)$ and using formula

(5.6), we obtain

$$\dot{\psi} = \sum \frac{d\psi}{d\rho} (\partial_j\rho - \lambda\rho\partial_j) \left(\frac{\delta H}{\delta M_j}\right) = \sum \partial_j\psi\left(\frac{\delta H}{\delta M_j}\right),$$

since

$$\frac{d\psi}{d\rho} (\partial_j\rho - \lambda\rho\partial_j) = \frac{d\psi}{d\rho} \rho_{,j} + \frac{d\psi}{d\rho} (1 - \lambda)\rho\partial_j = \partial_j(\psi) + \psi\partial_j = \partial_j\psi.$$

Hence, we see from (6.10) that the variables $(\mathbf{M}, \psi)$ are indeed associated to $B_0(n)$.

Now let $m > 1$. Since $\underline{\lambda} \neq (1, \ldots, 1)$, there exists at least one $\lambda_j \neq 1$. Using Theorem 6.6, we can move this λ_j into the first place. So, let $\lambda_1 = \lambda \neq 1$. Set $\rho = \rho_1$, and define

$$\psi_j = \rho_j\rho^{\gamma_j}, \quad 0 \neq \gamma_j \in k, \quad j = 2, \ldots, m. \tag{7.2}$$

Transforming in (6.10) from coordinates $(\mathbf{M}, \rho, \rho_2, \ldots, \rho_m)$ to $(\mathbf{M}, \rho, \psi_2, \ldots, \psi_m)$ and using (5.6), we get, for $j = 2, \ldots, m$:

$$\dot{\psi}_j = \sum_s [\gamma_j\rho_j\rho^{\gamma_j-1}(\partial_s\rho - \lambda\rho\partial_s) + \rho^{\gamma_j}(\partial_s\rho_j - \lambda_j\rho_j\partial_s)]\left(\frac{\delta H}{\delta M_s}\right)$$

$$= \sum_s \{\partial_s\psi_j - [\lambda_j + \gamma_j(\lambda - 1)]\psi_j\partial_s\}\left(\frac{\delta H}{\delta M_s}\right). \tag{7.3}$$

Thus, we transformed the index $\underline{\lambda}$ in $B_{\underline{\lambda}}(n)$ from $\underline{\lambda} = (\lambda; \lambda_2, \ldots, \lambda_m)$ into

$$\underline{\lambda}' = (\lambda; \ldots \lambda_j + \gamma_j(\lambda - 1), \ldots). \tag{7.4}$$

In particular, choosing $\gamma_j = \lambda_j/(1 - \lambda)$, we obtain $\underline{\lambda}' = (\lambda; 0, \ldots, 0)$. Finally, transforming from the variables $(\mathbf{M}, \rho, \psi_2, \ldots, \psi_m)$ into $(\mathbf{M}, \psi, \psi_2, \ldots, \psi_m)$ by (7.1), we move $\underline{\lambda}'$ into $(0; 0, \ldots, 0) = \mathbf{0}$. ∎

Remark 7.2. Now we can appreciate the reason why entropy per unit volume $\sigma = \rho\eta$ is used in (5.2) instead of entropy per unit mass η. Indeed, setting $\psi_2 = \eta$, $\rho_2 = \sigma$, we see that $\psi_2 = \rho_2\rho^{-1}$; hence, $\gamma_2 = 1$ by (7.2). Thus, $\lambda_2 + \gamma_2(\lambda - 1) = 0 + (-1)(0 - 1) = 1$. Therefore, the variables (ρ, η) correspond to $\underline{\lambda} = (0, 1)$ (in arbitrary number of dimensions n). As we shall see later, discretization of $B_{\underline{\lambda}}$ is done for $\underline{\lambda} = \mathbf{0}$ and this is why we prefer σ to η.

Remark 7.3. For $n = 1$, there exist canonical isomorphisms different from those of the type (7.1), (7.2). They are of the form (for $m = 1$):

$$\psi = M^{1-\mu}\rho^{(1-\mu)/(\lambda-1)}, \quad 1 \neq \lambda \mapsto \mu \neq 1. \tag{7.5}$$

In particular,

$$\psi = M\rho^{1/(\lambda-1)}, \quad 1 \neq \lambda \to 0, \tag{7.6}$$

$$\psi = M^{1-\lambda}v^{-1}, \quad 1 \neq \lambda \to \lambda. \tag{7.7}$$

The map (7.5) does not generalize for $m > 1$. On the other hand, if $\underline{\lambda} = (1, \ldots, 1)$, the following canonical automorphism of $B_{\underline{\lambda}}$

$$\psi_j = \psi_j(\rho_j) \tag{7.8}$$

is good also for $B_{\underline{\lambda}}(n)$ with $n > 1$.

Remark 7.4. The converse to Theorem 7.1, i.e., that $B_{(1,...,1)}(n) \not\approx B_0(n)$, is undoubtedly true but not easy to prove, except in the cases $m = n = 1$ and $m \geq n$:

Proposition 7.5. $B_1 \not\approx B_0$.

Proof. dim ker $B_1 = 0$, while ker $B_0 = \{c_1 M/\rho + c_2 \rho | c_i \in k\}$. ∎

THEOREM 7.6

Let $\underline{\lambda} = (1, \ldots, 1)$ and $m \geq n$. Then $B_{\underline{\lambda}}(n) \not\approx B_0(n)$.

Proof. Since ker $B_0(n) \ni \{\Sigma c_i \rho_i | c_i \in k\}$, dim ker $B_0(n) \geq m$. On the other hand, dim ker $B_{\underline{\lambda}}(n) = 0$, which can be seen as follows. Let

$$b = \begin{pmatrix} 0 & -1 \\ 1 & 0 \end{pmatrix}$$

be the canonical Hamiltonian matrix in the space with variables $(p_1, \ldots, p_m, q_1, \ldots, q_m)$, and let a differential map (i.e. commuting with $\partial_1, \ldots, \partial_n$) Φ be given by the formulae

$$M_i = \sum p_k q_{k,i}, \quad \rho_j = q_j. \tag{7.9}$$

Using formula (5.6), it is easy to see that Φ is a canonical map between $(\mathbf{p}, \mathbf{q})$-space and $(\mathbf{M}, \rho)$-space. Since $m \geq n$, the map Φ is an epimorphism (considered as a map between infinite jet bundles); equivalently, the dual map Φ^* is injective (considered as a differential homomorphism of differential rings). In particular, since Φ is canonical, Φ^* sends ker $B_{\underline{\lambda}}(n)$ injectively into ker b. But ker $b = \{0\}$. Hence, ker $B_{\underline{\lambda}}(n) = \{0\}$ as well. ∎

Remark 7.7. For $m = n$, the map (7.9) is a (rational) change of variables. In particular, the matrix $B_{(1,...,1)}(n)$ is immediately discretizable for $m = n$.

Remark 7.8. The module V_λ for $\lambda = 1$ has a simple geometric interpretation: this is the module of volume forms, $K_n \, dx_1 \wedge \cdots \wedge dx_n$.

We now examine 2-cocycles on some of the Lie algebras $L_{\underline{\lambda}}(n)$. Recall that a bilinear differential operator $w(X, Y)$ on a Lie algebra L is called a (generalized) 2-cocycle if $w(X, Y) \sim -w(Y, X)$ and $w(X, [Y, Z]) + w(Y, [Z, X]) + w(Z, [X, Y]) \sim 0$, $\forall X, Y, Z \in L$, were $a \sim 0$ means that $a \in \Sigma_s \mathrm{Im} \, \partial_s$ (or $\Sigma_s \mathrm{Im}(\Delta_s - 1)$ in the discrete case).

LEMMA 7.9

Let $\underline{\lambda} = (1, \ldots, 1)$. Define a bilinear form w_j on $L_{\underline{\lambda}}(n)$ by the formula

$$w_j\left(\begin{pmatrix} X \\ u \end{pmatrix}, \begin{pmatrix} Y \\ v \end{pmatrix}\right) = \sum X_i v_{j,i} + u_j \, \mathrm{div}(Y) = X(v_j) - u_j \, \mathrm{div}(Y) \sim X(v_j) - Y(u_j). \tag{7.10}$$

Then, w_j is a 2-cocycle on $L_{\underline{\lambda}}(n)$.

Proof. First,

$$w_j\left(\begin{pmatrix} X \\ u \end{pmatrix}, \begin{pmatrix} Y \\ v \end{pmatrix}\right) + w_j\left(\begin{pmatrix} Y \\ v \end{pmatrix}, \begin{pmatrix} X \\ u \end{pmatrix}\right) = \sum X_i v_{j,i} + u_j \, \mathrm{div}(Y)$$

$$+ \sum Y_i u_{j,i} + v_j \, \mathrm{div}(X) = \sum (X_i v_j)_{,i} + \sum (u_j Y_i)_{,i} \sim 0,$$

so that w_j is skew-symmetric. Next,

$$w_j\left(\left[\begin{pmatrix} X \\ u \end{pmatrix}, \begin{pmatrix} Y \\ v \end{pmatrix}\right], \begin{pmatrix} Z \\ w \end{pmatrix}\right) + \mathrm{c.p.} = w_j\left(\begin{pmatrix} [X, Y] \\ \Sigma(X_i v)_{,i} - \Sigma(Y_i u)_{,i} \end{pmatrix}, \begin{pmatrix} Z \\ w \end{pmatrix}\right) + \mathrm{c.p.}$$

$$= \sum [X, Y]_i w_{j,i} + \sum (X_i v_j - Y_i u_j)_{,i} Z_{k,k} + \mathrm{c.p.}$$

$$\sim -w_j \, \mathrm{div}([X, Y]) + u_j Y(\mathrm{div}(Z)) - v_j X(\mathrm{div}(Z))$$

$$+ \mathrm{c.p.}[\text{by } (6.2)] = w_j X(\mathrm{div}(Y)) + w_j Y(\mathrm{div}(X))$$

$$+ u_j Y(\mathrm{div}(Z)) - v_j X(\mathrm{div}(Z)) + \mathrm{c.p.} = 0,$$

where c.p. stands for cyclic permutation.

Denote by b_j the Hamiltonian matrix associated to the 2-cocycle w_j. Recall that b_j is defined via the formula

$$\begin{pmatrix} X \\ u \end{pmatrix}' b_j \begin{pmatrix} Y \\ v \end{pmatrix} \sim w_j\left(\begin{pmatrix} Y \\ u \end{pmatrix}, \begin{pmatrix} Y \\ v \end{pmatrix} \right). \tag{7.11}$$

Since w_j is a 2-cocycle on $L_{\underline{\lambda}}(n)$ (for $\underline{\lambda} = (1, \dots, 1)$), the matrix

$$B_c = B_{\underline{\lambda}}(n) + \sum c_j b_j, \quad c_j \in k \tag{7.12}$$

is Hamiltonian for any vector $\mathbf{c} = (c_1, \dots, c_m) \in k^m$ (see Gel'fand and Dorfman[19], Kupershmidt[11,18]).

THEOREM 7.10

If $\mathbf{c} \neq \mathbf{0}$ then B_c is canonically equivalent to $B_{\underline{\lambda}}(n)$ with $\lambda_j = 1 - c_j$.

COROLLARY 7.11.

If $\mathbf{c}_1, \mathbf{c}_2 \neq \mathbf{0}$ then $B_{\mathbf{c}_1}$ is canonically equivalent to $B_{\mathbf{c}_2}$.

Proof of Corollary. By Theorem 7.10, any matrix $B_{\mathbf{c}}$ with $\mathbf{c} \neq 0$ is equivalent to $B_{\underline{\lambda}}(n)$, with $\underline{\lambda} \neq (1, \dots, 1)$ (since $\mathbf{c} \neq \mathbf{0}$). By Theorem 7.1, the latter matrix $B_{\underline{\lambda}}(n)$ is canonically equivalent to $B_0(n)$. ∎

Proof of Theorem. Changing the variables $(\mathbf{M}, \rho)$ into the variables $(\mathbf{M}, \underline{\Psi})$, with

$$\Psi_j = \ln \rho_j, \quad j = 1, \dots, m, \tag{7.13}$$

transforms an arbitrary matrix $B_{\underline{\lambda}}(n)$ into $B_{(1,\dots,1)}(n) + \Sigma(1 - \lambda_j)b_j$. ∎

8. 2-COCYCLES ON LIE ALGEBRAS L_C

In this section we compute all 2-cocycles on Lie algebras of the form L_C. Recall that the commutator in L_C is given according to (6.3), (6.5) as

$$\left[\begin{pmatrix} X \\ u \end{pmatrix}, \begin{pmatrix} Y \\ v \end{pmatrix} \right] = \begin{pmatrix} XY' - X'Y \\ Xv' + X'Cv - Yu' - Y'Cu \end{pmatrix}, \tag{8.1}$$

where $X, Y \in D_1$, $u, v \in K_1^m$, C is a constant matrix from k^m, and we write $(\cdot)'$ instead of $(\cdot)^{(1)}$. (The matrix C may be put into its Jordan canonical form, but this is not important at the moment.) We seek skew-symmetric constant-coefficient operators of the form

$$b = \sum r_s \partial^s, \quad r_s \in \mathrm{Mat}_{m+1}(k) \tag{8.2}$$

whose associated bilinear forms [see (7.11)] are 2-cocycles on L_C. Since the multiplication (8.1) in L_C is a bilinear homogeneous first order differential operator, it follows that each summand $r_s \partial^s$ in the sum (8.2) is a 2-cocycle on L_C, so we can and will look only for homogeneous 2-cocycles. Further, since D_1 can be realized as a Lie subalgebra in L_C consisting of elements $\{\binom{X}{0}\}$, the restriction onto D_1 of a 2-cocycle from L_C is a 2-cocycle on D_1. (This is a general property of semidirect products.) Let us first see what such a 2-cocycle looks like.

THEOREM 8.1

The space of 2-cocycles on D_1 is generated over k by ∂ and ∂^3.

Proof. We look for all $i \in \mathbf{Z}_+$ such that

$$[X_1, X_2]\partial^{2i+1}(X_3) + \text{c.p.} \sim 0, \quad \forall X_1, X_2, X_3 \in D_1.$$

In the long hand, we have

$$0 \sim (X_1X_2' - X_1'X_2)X_3^{(2i+1)} + (X_2X_3' - X_2'X_3)X_1^{(2i+1)} + (X_3X_1' - X_3'X_1)X_2^{(2i+1)} \sim$$
$$X_3 \times [-(X_1X_2' - X_1'X_2)^{(2i+1)} - (X_2X_1^{(2i+1)})' - X_2'X_1^{(2i+1)} + X_1'X_2^{(2i+1)} + (X_1X_2^{(2i+1)})'],$$

so that the expression in the square brackets must vanish. It obviously vanishes when $2i + 1$ equals 1 or 3. On the other hand, if $2i + 1 \geq 5$, then this expression contains the term

$$X_1^{(2)}X_2^{(2i-1)}\left[-\binom{2i+1}{2} + \binom{2i+1}{1}\right] \neq 0. \qquad \blacksquare$$

Having found all 2-cocycles on D_1, we can restrict ourselves in searching only for those 2-cocycles on L_C which vanish on D_1. So let

$$b = \begin{pmatrix} 0 & \beta' \\ (-1)^{s+1}\beta & \gamma \end{pmatrix}\partial^s, \quad \gamma' = (-1)^{s+1}\gamma, \tag{8.3}$$

be a skew-symmetric operator, with a column-vector $\beta \in k^m$, and with $\gamma \in \mathrm{Mat}_m(k)$. For b to be a 2-cocycle, we must have

$$0 \sim \left[\begin{pmatrix} X_1 \\ u_1 \end{pmatrix}, \begin{pmatrix} X_2 \\ u_2 \end{pmatrix}\right]' b\begin{pmatrix} X_3 \\ u_3 \end{pmatrix} + \text{c.p.} = (X_1X_2' - X_1'X_2)\beta \cdot u_3^{(s)}$$

$$+ [X_1u_2' + X_1'Cu_2 - X_2u_1' - X_2'Cu_1] \cdot [(-1)^{s+1}\beta X_3^{(s)} + \gamma u_3^{(s)}] + \text{c.p.}$$

$$\sim (-1)^s(X_3X_1' - X_3'X_1)^{(s)}\beta \cdot u_2 + (-1)^s u_2 \cdot \beta(X_1X_3^{(s)})'$$

$$- (-1)^s Cu_2 \cdot \beta X_1'X_3^{(s)} - (-1)^s u_2 \cdot \beta(X_3X_1^{(s)})'$$

$$+ (-1)^s Cu_2 \cdot \beta X_3'X_1^{(s)} + \text{c.p.} \tag{8.4a}$$

$$+ X_1\{u_2' \cdot \gamma u_3^{(s)} - [Cu_2 \cdot \gamma u_3^{(s)}]' - u_3' \cdot \gamma u_2^{(s)} + [Cu_3 \cdot \gamma u_2^{(s)}]'\} + \text{c.p.} \tag{8.4b}$$

Since (8.4a) is bilinear in X while (8.4b) is bilinear in u, we can consider (8.4a) and (8.4b) separately.

Starting with (8.4b), we see that since no derivatives of X_1 are present, the expression in the curly brackets must vanish. If $s \geq 1$, there is nothing to compensate for the $Cu_2 \cdot \gamma u_3^{(s+1)}$ term. Hence, $C'\gamma = 0$. If $s > 1$ then $u_2' \cdot \gamma u_3^{(s)}$ must vanish, so $\gamma = 0$:

$$s > 1, \quad \gamma = 0. \tag{8.5}$$

If, on the other hand, $s = 1$, the remaining terms $u_2' \cdot \gamma u_3' - u_3' \cdot \gamma u_2'$ cancel each other out since $\gamma' = \gamma$ by (8.3), so:

$$s = 1, \quad C'\gamma = 0. \tag{8.6}$$

Finally, if $s = 0$ then $\gamma' = -\gamma$, so the curly bracket expression becomes

$$\partial\{u_2 \cdot [\gamma - C'\gamma + \gamma'C]u_3\} = \partial\{u_2 \cdot [\gamma - C'\gamma - \gamma C]u_3\},$$

so

$$s = 0 : \gamma = C'\gamma + \gamma C. \tag{8.7}$$

In a similar fashion, working out the expression (8.4a) we get

$$0 = u_2 \cdot \beta\{(X_3X_1' - X_1'X_1)^{(s)} + (X_1X_3^{(s)})' - (X_3X_1^{(s)})'\} + Cu_2 \cdot \beta\{-X_1'X_3^{(s)} + X_3'X_1^{(s)}\}.$$

Analyzing the latter equality, we consider separately various values of s. For $s > 3$, looking at the $X_3^{(2)}X_1^{(s-1)}$ coefficient, $\binom{s}{2} - \binom{s}{1}$, we see that $\beta = 0$:

$$s > 3, \quad \beta = 0. \tag{8.8}$$

If $s = 3$, we obtain

$$0 = u_2 \cdot \beta\partial\{(X_3X_1'' - X_3''X_1)' + X_1X_3^{(3)} - X_3X_1^{(3)}\} + Cu_2 \cdot \beta\partial\{-X_1'X_3'' + X_3'X_1''\}$$
$$= (u_2 \cdot \beta\partial + Cu_2 \cdot \beta\partial)(X_3'X_1'' - X_3''X_1'),$$

so that $\beta + C'\beta = 0$:

$$s = 3, \quad (\mathbf{1} + C')\beta = 0. \tag{8.9}$$

If $s = 2$, we have

$$0 = u_2 \cdot \beta\partial\{X_3X_1'' - X_3''X_1 + X_1X_3'' - X_3X_1''\} + Cu_2 \cdot \beta\{-X_1'X_3'' + X_3'X_1''\}$$
$$= Cu_2 \cdot \beta\{-X_1'X_3'' + X_3'X_1''\},$$

so

$$s = 2, \quad C'\beta = 0. \tag{8.10}$$

If $s = 1$, there are no conditions on β,

$$s = 1, \quad \beta \text{ is arbitrary.} \tag{8.11}$$

Finally, for $s = 0$, we have

$$0 = (X_3X_1' - X_1'X_1)\{u_2 \cdot \beta - Cu_2 \cdot \beta\},$$

so

$$s = 0, \quad (\mathbf{1} - C')\beta = 0. \tag{8.12}$$

Formulae (8.5)–(8.12), together with Theorem 8.1, describe all 2-cocycles on the Lie algebra L_C.

Remark 8.2. Throughout the paper we treat K_n as $C^\infty(\mathbf{R}^n)$ only to avoid lengthy detours into differential algebra. All the results (and the proofs) have their natural algebraic counterparts when K_n is considered as an arbitrary commutative ring with n commuting derivations $\partial_1, \ldots, \partial_n$: $K_n \mapsto K_n$.

9. A DISCRETIZATION OF ONE-DIMENSIONAL FLUID DYNAMICS

In this section we derive a formula [(9.8) below] which discretizes one-dimensional hydrodynamics.

We start with the Lie algebra L_0. As we know, $B_{0,0}$ is the Hamiltonian matrix of adiabatic fluid dynamics (5.2). By formula (8.6) for $C = 0$, we see that (for $\gamma = \epsilon\mathbf{1}$)

$$b = \epsilon\begin{pmatrix} 0 & \mathbf{0}' \\ \mathbf{0} & \mathbf{1} \end{pmatrix}\partial, \quad \epsilon \in k, \tag{9.1}$$

is a 2-cocycle on L_0. Hence, the matrix

$$B = B_0 + b \tag{9.2}$$

is Hamiltonian.

THEOREM 9.1

The following change of variables,

$$Q = M - \rho^2/2\epsilon, \quad \underline{\rho} = \rho, \tag{9.3}$$

where $\rho^2 = \Sigma\rho_j\rho_j$, transforms the Hamiltonian matrix B (9.2) into the form

$$\begin{pmatrix} Q\partial + \partial Q & \mathbf{0}^t \\ \mathbf{0} & \epsilon\mathbf{1}\partial \end{pmatrix}. \tag{9.4}$$

Proof. Multiplying subsequently the Jacobian of the map (9.3),

$$J = \begin{pmatrix} 1 & -\underline{\rho}^t/\epsilon \\ 0 & 1 \end{pmatrix}. \tag{9.5}$$

by the matrix B, followed by the matrix

$$J^\dagger = \begin{pmatrix} 1 & \mathbf{0}^t \\ -\underline{\rho}/\epsilon & 1 \end{pmatrix}, \tag{9.6}$$

results in (9.4).   ∎

Now it is clear how to proceed. If we first perturb our matrix B_0 by adding to it a small 2-cocycle (9.1), the resulting matrix (9.2) is equivalent to the matrix (9.4) whose discretization is easy: Indeed, the matrix (9.4) is a direct sum of $B(D_1)$ whose discretization is given by the formula (3.22), and a constant coefficient matrix $\epsilon\mathbf{1}\partial$, whose discretization presents no problem (e.g. $\epsilon[(\Delta - \Delta^{-1})/2]\mathbf{1}$ will do). Reverting back to the variables $(M, \underline{\rho})$ by (9.3), we obtain a discretization of adiabatic fluid dynamics by the following matrix:

$$\begin{pmatrix} \sqrt{M - \rho^2/2\epsilon}(\Delta - \Delta^{-1})\sqrt{M - \underline{\rho}^2/2\epsilon} + \underline{\rho}^t(\Delta - \Delta^{-1})\underline{\rho}/2\epsilon & \tfrac{1}{2}\underline{\rho}^t(\Delta - \Delta^{-1}) \\ \tfrac{1}{2}(\Delta - \Delta^{-1})\underline{\rho} & \tfrac{1}{2}\epsilon\mathbf{1}(\Delta - \Delta^{-1}) \end{pmatrix} \tag{9.7}$$

The discrete motion equations generated by (9.7) are

$$\dot{M}_n = \sqrt{M_n - \rho_n^2/2\epsilon}\left[\sqrt{M_{n+1} - \rho_{n+1}^2/2\epsilon}(\partial H/\partial M_{n+1}) \right.$$
$$\left. - \sqrt{M_{n-1} - \frac{\rho_{n-1}^2}{2\epsilon}}\left(\frac{\partial H}{\partial M_{n-1}}\right) \right] + \sum_j \frac{1}{2}\rho_{jn}\left[\frac{\partial H}{\partial\rho_{jn+1}} - \frac{\partial H}{\partial\rho_{jn-1}} \right],$$
$$\dot{\rho}_{jn} = \tfrac{1}{2}[\rho_{jn+1}(\partial H/\partial M_{n+1}) - \rho_{jn-1}(\partial H/\partial M_{n-1})]$$
$$+ \tfrac{1}{2}\epsilon[\partial H/\partial\rho_{jn+1} - \partial H/\partial\rho_{jn-1}], \tag{9.8}$$

where

$$H = \sum_n H_n(M, \rho) \tag{9.9}$$

is a discretization of the total energy; for example, discretizing (5.3) one may take

$$H = \sum_n \left[\frac{M_n^2}{2\rho_{1n}} + \rho_{1n}e(\rho_{1n}, \rho_{2n}/\rho_{1n}) \right], \tag{9.10}$$

with $\rho_2 = \sigma$, $\rho_1 = \rho$.

Remark 9.2. The matrix (9.7) is not regular in ϵ, but its continuous limit (9.2) is.

Remark 9.3. For the barotropic case (5.4), we have a special discretization scheme (5.8) which is quite different from (9.7).

Remark 9.4. The equations (9.8) have a drawback in that the physical dimensions of all the ρ's are the same, namely, equal to the dimension of $\sqrt{\epsilon M}$. In practice, of course, this is not the case: $\rho_1 = \rho$ and $\rho_2 = \sigma = \rho\eta$ have different dimensions. This problem is easy to rectify by taking instead of (9.1) the following 2-cocycle

$$b' = \begin{pmatrix} 0 & \mathbf{0}^t \\ \mathbf{0} & E \end{pmatrix}\partial, \quad E = \mathrm{diag}(\epsilon_1, \epsilon_2, \ldots), \tag{9.11}$$

and generalizing (9.3) into

$$Q = M - \Sigma\rho_i^2/2\epsilon_i, \quad \underline{\rho} = \rho, \tag{9.12}$$

which results in the following analog of (9.4):

$$\begin{pmatrix} Q\partial + \partial Q & \mathbf{0}^t \\ \mathbf{0} & E\partial \end{pmatrix}. \tag{9.13}$$

Discretizing (9.13) and inverting the map (9.12), we obtain the desired discretization scheme of one-dimensional fluid dynamics:

$$\dot{M} = \left[\sqrt{M - \sum \frac{\rho_i^2}{2\epsilon_i}}\,(\Delta - \Delta^{-1})\sqrt{M - \sum \frac{\rho_i^2}{2\epsilon_i}} \right.$$
$$\left. + \sum \rho_i(\Delta - \Delta^{-1})\frac{\rho_i}{2\epsilon_i} \right]\left(\frac{\delta H}{\delta M}\right) + \sum \rho_i(\Delta - \Delta^{-1})\left(\frac{\delta H}{\delta \rho_i}\right) \Big/ 2,$$
$$\dot{\rho}_i = \frac{1}{2}(\Delta - \Delta^{-1})\rho_i\left(\frac{\delta H}{\delta M}\right) + \frac{1}{2}\epsilon_i(\Delta - \Delta^{-1})\left(\frac{\delta H}{\delta \rho_i}\right) \tag{9.14}$$

In the lattice notations, we get

$$\dot{M}_n = \left\{ \sqrt{M_n - \sum_i \frac{\rho_{in}^2}{2\epsilon_i}}\left[\sqrt{M_{n+1} - \sum_i \frac{\rho_{in+1}^2}{2\epsilon_i}}\left(\frac{\partial H}{\partial M_{n+1}}\right) \right.\right.$$
$$\left. - \sqrt{M_{n-1} - \sum_i \frac{\rho_{in-1}^2}{2\epsilon_i}}\left(\frac{\partial H}{\partial M_{n-1}}\right) \right]$$
$$\left. + \sum_i \frac{1}{2\epsilon_i}\rho_{in}\left[\rho_{in+1}\left(\frac{\partial H}{\partial M_{n+1}}\right) - \rho_{in-1}\left(\frac{\partial H}{\partial M_{n-1}}\right) \right] \right\}$$
$$+ \sum_i \frac{1}{2}\rho_{in}\left[\frac{\partial H}{\partial\rho_{in+1}} - \frac{\partial H}{\partial\rho_{in-1}} \right]; \tag{9.15}$$
$$\dot{\rho}_{kn} = \frac{1}{2}\left[\rho_{kn+1}\left(\frac{\partial H}{\partial M_{n+1}}\right) - \rho_{kn-1}\left(\frac{\partial H}{\partial M_{n-1}}\right) \right] + \frac{1}{2}\epsilon_k\left[\frac{\partial H}{\partial\rho_{kn+1}} - \frac{\partial H}{\partial\rho_{kn-1}} \right]$$

Acknowledgments—I am thankful to K. C. Reddy for keeping alive my interest in the subject and for volunteering to test numerically the behavior of proposed discretization schemes. I also thank Jyh-Chyang Wang for pointing out the problem with the dimension of ϵ in the system (9.8). This work was supported in part by the National Science Foundation and by the U.S. Department of Energy.

REFERENCES

1. D. D. Holm, B. A. Kupershmidt, and C. D. Levermore, Canonical maps between Poisson brackets in Eulerian and Lagrangian description of continuum mechanics. *Phys. Lett.* **98A**, 389–395 (1983).
2. D. D. Holm, B. A. Kupershmidt, and C. D. Levermore, Hamiltonian differencing of fluid dynamics. *Adv. Appl. Math.* **6**, 52–84 (1985).
3. I. E. Dzyaloshinskii and G. E. Volovick, Poisson brackets in condensed matter physics. *Ann. Phys.* (NY) **125**, 67–97 (1980).
4. D. D. Holm and B. A. Kupershmidt, Poisson structure of superfluids, *Phys. Lett.* **91A**, 425–430 (1982).
5. D. D. Holm and B. A. Kupershmidt, Poisson structure of superconductors, *Phys. Lett.* **93A**, 177–181 (1983).
6. J. Gibbons, D. D. Holm, and B. A. Kupershmidt, Gauge-invariant Poisson brackets for chromohydrodynamics. *Phys. Lett.* **90A**, 281–283 (1982).
7. D. D. Holm and B. A. Kupershmidt, Poisson brackets and Clebsch representations for magnetohydrodynamics, multifluid plasmas, and elasticity. *Physica* **6D**, 347–363 (1983).
8. J. Gibbons, D. D. Holm, and B. A. Kupershmidt, The Hamiltonian structure of classical chromohydrodynamics. *Physica* **6D**, 177–193 (1983).
9. D. D. Holm and B. A. Kupershmidt, Relativistic fluid dynamics as a Hamiltonian system. *Phys. Lett.* **101A**, 23–26 (1984).
10. D. D. Holm and B. A. Kupershmidt, Planar incompressible Yang–Mills magnetohydrodynamics. *Lett. Nuovo Cim.* **40**, 70–72 (1984).
11. B. A. Kupershmidt, *Discrete Lax Equations and Differential-Difference Calculus*, Asterisque **123**, Paris (1985).
12. J. F. Ritt, Differential groups of order two. *Ann. Math.* **53**, 491–519 (1951).
13. P. J. Cassidy, Differential algebraic Lie groups. *Trans. Am. Math. Soc.* **247**, 247–273 (1979).
14. B. A. Kupershmidt, Integrable and Superintegrable Systems, and Differential and Difference Lie Algebras and Superalgebras. In *Open Problems in the Structure Theory of Nonlinear Integrable Differential and Difference Systems*. Nagoya University, Nagoya (1984).
15. B. A. Kupershmidt, On dual spaces of differential Lie algebras. *Physica* **7D**, 334–337 (1983).
16. B. A. Kupershmidt, On algebraic models of dynamical systems. *Lett. Math. Phys.* **6**, 85–89 (1982).
17. B. A. Kupershmidt and Yu. I. Manin, Equations of long waves with a free surface. II. Hamiltonian structure and higher equations. *Funct. Anal. Appl.* **12**(1), 20–29 (1978).
18. B. A. Kupershmidt, A review of superintegrable systems. In *Lect. Appl. Math.* **23** Part 1, 83–120 (1985).
19. I. M. Gel'Fand and I. Ya. Dorfman, Hamiltonian operators and infinite-dimensional Lie algebras. *Funct. Anal. Appl.* **15**, 23–40 (1981) (Russian); 173–187 (English).

Comp. & Maths. with Appls. Vol. 12A, Nos. 4/5, pp. 413–432, 1986
Printed in Great Britain.

0886-9553/86 $3.00 + .00
© 1986 Pergamon Press Ltd.

LINEARIZED FORM OF IMPLICIT TVD SCHEMES FOR THE MULTIDIMENSIONAL EULER AND NAVIER–STOKES EQUATIONS

H. C. YEE†
Computational Fluid Dynamics Branch, MS 202A-1, NASA Ames Research Center,
Moffett Field, CA 94035, U.S.A.

Abstract—Linearized alternating direction implicit (ADI) forms of a class of total variation diminishing (TVD) schemes for the Euler and Navier–Stokes equations have been developed. These schemes are based on the second-order-accurate TVD schemes for hyperbolic conservation laws developed by Harten[1,2]. They have the property of not generating spurious oscillations across shocks and contact discontinuities. In general, shocks can be captured within 1–2 grid points. These schemes are relatively simple to understand and easy to implement into a new or existing computer code. One can modify a standard three-point central-difference code by simply changing the conventional numerical dissipation term into the one designed for the TVD scheme. For steady-state applications, the only difference in computation is that the current schemes require a more elaborate dissipation term for the explicit operator; no extra computation is required for the implicit operator. Numerical experiments with the proposed algorithms on a variety of steady-state airfoil problems illustrate the versatility of the schemes.

1. INTRODUCTION

Harten's method of constructing high-resolution TVD schemes involves starting with a first-order TVD scheme and applying it to a modified flux. The modified flux is chosen so that the scheme is second order at regions of smoothness and first order at points of extrema. This technique is sometimes referred to as the modified flux approach. Although the scheme is an upwind scheme, it is written in a symmetric form; i.e. central difference plus a "smart" numerical dissipation term. This symmetric form is especially advantageous for systems of higher than one dimension. It results in less storage and a smaller operation count than its upwind form[3]. The modified flux approach is relatively simple to understand and easy to implement into a new or existing computer code. One can modify a standard three-point central-difference code by simply changing the conventional numerical dissipation term into the one designed for the TVD scheme. However, for non-Cartesian grids, care must be taken to preserve freestream. A formulation closer to finite volume would be more desirable.

In [4], a preliminary study was done on an implicit TVD scheme for a two-dimensional gas dynamics problem in a Cartesian grid. It was found that further improvement in computational efficiency and convergence rate is required for practical application.

The objective of this paper is to formulate linearized forms and to develop the various solution strategies for the implicit TVD schemes for two-dimensional Euler and Navier–Stokes equations. In particular, the study of more efficient formulations for two-dimensional steady-state applications is emphasized. Numerical experiments with some of the 1984 AGARD Fluid Dynamics Panel Working Group 07 airfoil test cases[5] are included.

2. EXPLICIT TVD SCHEMES

First-order explicit TVD scheme

Consider the scalar hyperbolic conservation law

$$\frac{\partial u}{\partial t} + \frac{\partial f(u)}{\partial x} = 0, \tag{2.1}$$

where f is the flux and $a(u) = \partial f/\partial u$ is the characteristic speed. Let u_j^n be the numerical solution of (2.1) at $x = j\Delta x$ and $t = n\Delta t$, with Δx the spatial mesh size and Δt the time step.

---

†Research Scientist.

A general three-point explicit-difference scheme in conservation form can be written as

$$u_j^{n+1} = u_j^n - \lambda(h_{j+1/2}^n - h_{j-1/2}^n), \tag{2.2}$$

where $h_{j+1/2}^n = h(u_j^n, u_{j+1}^n)$, and $\lambda = \Delta t/\Delta x$. Here, h, commonly called a numerical flux function, is required to be consistent with the conservation law in the following sense:

$$h(u_j, u_j) = f(u_j). \tag{2.3}$$

Consider a numerical scheme with a numerical flux function of the following form:

$$h_{j+1/2} = \frac{1}{2}[f_j + f_{j+1} - \psi(a_{j+1/2})\Delta_{j+1/2}u], \tag{2.4a}$$

where $f_j = f(u_j)$, $\Delta_{j+1/2}u = u_{j+1} - u_j$ and

$$a_{j+1/2} = \begin{cases} (f_{j+1} - f_j)/\Delta_{j+1/2}u & \Delta_{j+1/2}u \neq 0, \\ a(u_j) & \Delta_{j+1/2}u = 0. \end{cases} \tag{2.4b}$$

Here ψ is a function of $a_{j+1/2}$ and λ. The function ψ is sometimes referred to as the coefficient of numerical viscosity. A scheme with a numerical flux of the form (2.4) is the first-order accurate upwind scheme[6,7]

$$u_j^{n+1} = u_j^n - \frac{\lambda}{2}[1 - \operatorname{sgn}(a_{j+1/2}^n)](f_{j+1}^n - f_j^n) - \frac{\lambda}{2}[1 + \operatorname{sgn}(a_{j-1/2}^n)](f_j^n - f_{j-1}^n). \tag{2.5}$$

If we define $a_{j+1/2}$ as (2.4b), then (2.5) can be written as

$$u_j^{n+1} = u_j^n - \frac{\lambda}{2}[f_{j+1}^n - f_{j-1}^n - |a_{j+1/2}^n|\Delta_{j+1/2}u^n + |a_{j-1/2}^n|\Delta_{j-1/2}u^n]. \tag{2.6a}$$

Here, the numerical flux function is

$$h_{j+1/2} = \frac{1}{2}[f_j + f_{j+1} - |a_{j+1/2}|\Delta_{j+1/2}u] \tag{2.6b}$$

with

$$\psi(a_{j+1/2}) = |a_{j+1/2}|. \tag{2.6c}$$

This scheme is sometimes known as the Huang Scheme[6], or the Roe Scheme[7]. It is well known that (2.5) and (2.6) are not consistent with an entropy inequality, and the scheme might converge to a nonphysical solution. A slight modification of the numerical viscosity term

$$\psi(z) = \begin{cases} |z| & |z| \geq \epsilon, \\ (z^2 + \epsilon^2)/2\epsilon & |z| < \epsilon \end{cases} \tag{2.7}$$

can remedy the entropy violating problem[1], where $\epsilon > 0$ is a parameter. A formula for ϵ can be found in Ref. [8]. The $\psi(z)$ in (2.7) is a continuously differentiable positive approximation to $|z|$ in (2.6c). The notation of the numerical flux function (2.4a) will be used heavily for the rest of the paper.

If one defines

$$C^{\pm}(z) = \frac{1}{2}[\psi(z) \pm z], \tag{2.8}$$

then Eq. (2.2) together with (2.4) can be written as

$$u_j^{n+1} = u_j^n + \lambda C^-(a_{j+1/2}^n)\Delta_{j+1/2}u^n - \lambda C^+(a_{j-1/2}^n)\Delta_{j-1/2}u^n. \tag{2.9}$$

The total variation of a mesh function u^n is defined to be

$$TV(u^n) = \sum_{j=-\infty}^{\infty} |u_{j+1}^n - u_j^n| = \sum_{j=-\infty}^{\infty} |\Delta_{j+1/2}u^n|. \tag{2.10}$$

The numerical scheme (2.2) for an initial-value problem of (2.1) is said to be TVD if

$$TV(u^{n+1}) \leq TV(u^n). \tag{2.11}$$

It can be shown that sufficient conditions[1] for (2.2), together with (2.4), to be a TVD scheme
are

$$\lambda C^-(a_{j+1/2}) \geq 0, \tag{2.12a}$$

$$\lambda C^+(a_{j+1/2}) \geq 0, \tag{2.12b}$$

$$\lambda[C^-(a_{j+1/2}) + C^+(a_{j+1/2})] \leq 1. \tag{2.12c}$$

Applying the above conditions to Eq. (2.6), it can be easily shown that (2.6) is a TVD scheme.
Therefore, for the scheme (2.2) with a general ψ in (2.4a) [other than (2.6c)] to be TVD, we
have to pick ψ such that (2.12) is satisfied.

Second-order explicit TVD scheme

In [1], Harten converted the first-order scheme (2.2) into a second-order TVD scheme by
applying it to a modified flux $\tilde{f}(u) = [f(u) + g(u)]$. The new numerical flux function $\tilde{h}_{j+1/2}$
depends on $(f + g)$ instead of f alone, the coefficient of the numerical viscosity term ψ is a
function of a modified characteristic speed $(a + \gamma)$, and $\tilde{h}_{j+1/2}$ can be written as

$$\tilde{h}_{j+1/2} = \frac{1}{2}[\tilde{f}_j + \tilde{f}_{j+1} - \psi(a_{j+1/2} + \gamma_{j+1/2})\Delta_{j+1/2}u], \tag{2.13a}$$

where $\tilde{f}_j = f_j + g_j$ and $\gamma_{j+1/2}$ is defined almost the same way as $a_{j+1/2}$ except that it is a
function of the g_j's instead of the f_j's. It can be expressed as

$$\gamma_{j+1/2} = \begin{cases} (g_{j+1} - g_j)/\Delta_{j+1/2}u & \Delta_{j+1/2}u \neq 0, \\ 0 & \Delta_{j+1/2}u = 0. \end{cases} \tag{2.13b}$$

The requirements on g are (i) the function g should have a bounded γ in (2.13b) so that the
scheme (2.2) together with (2.13) is TVD with respect to the modified flux $(f + g)$, and (ii)
the modified scheme should be second-order accurate (except at points of extrema). In [1,2],
Harten devised a recipe for g that satisfies the above two requirements. We will use this particular
form of g for the discussion here. It can be written as

$$g_j = S \cdot \max[0, \min(\sigma_{j+1/2}|\Delta_{j+1/2}u|, S \cdot \sigma_{j-1/2}\Delta_{j-1/2}u)], \tag{2.13c}$$
$$S = \text{sgn}(\Delta_{j+1/2}u),$$

with $\sigma_{j+1/2} = \sigma(a_{j+1/2})$, and we choose

$$\sigma(z) = \frac{1}{2}[\psi(z) - \lambda z^2] \geq 0 \tag{2.13d}$$

for time-accurate calculations. With this choice of $\sigma(z)$, the scheme (2.2) together with (2.13) is second-order accurate in both time and space (except at points of extrema); see [1,2]. Also, with this choice of the g function, the second-order TVD scheme will automatically switch itself to first-order at points of extrema. This is one of the vehicles to prevent spurious oscillation near a shock. Other more general forms for g can be obtained following the line of argument of Sweby[9] and Roe[10]. The function g_j is sometimes referred to as the "minmod" function of the argument indicated. It is a form of the so-called "limiter" for the control of unwanted oscillations in numerical schemes. A more uniform second-order nonoscillatory scheme is under development by Harten and Osher[11]. Their new scheme requires a higher operations count than the current one.

The second-order TVD scheme can be written as

$$u_j^{n+1} = u_j^n + \lambda C^-(\tilde{a}_{j+1/2}^n)\Delta_{j+1/2}u^n - \lambda C^+(\tilde{a}_{j-1/2}^n)\Delta_{j-1/2}u^n, \tag{2.14}$$

with $\tilde{a}_{j+1/2}^n = a_{j+1/2}^n + \gamma_{j+1/2}^n$. As a side remark, with the choice of g_j above, $\mathrm{sgn}(\tilde{a}_{j+1/2}^n) = \mathrm{sgn}(a_{j+1/2}^n)$.

Another interesting observation is that Eq. (2.13) together with (2.6c) can be rewritten as

$$u_j^{n+1} = u_j^n - \frac{\lambda}{2}[1 - \mathrm{sgn}(\tilde{a}_{j+1/2}^n)](\tilde{f}_{j+1}^n - \tilde{f}_j^n) - \frac{\lambda}{2}[1 + \mathrm{sgn}(\tilde{a}_{j-1/2}^n)](\tilde{f}_j^n - \tilde{f}_{j-1}^n). \tag{2.15}$$

This is a straightforward extension of Huang or Roe's entropy-violating first-order upwind scheme to second-order accuracy. The scheme looks identical to their first-order scheme (2.5) except the arguments a's and f's are different. Here

$$\tilde{h}_{j+1/2} = \frac{1}{2}[1 - \mathrm{sgn}(\tilde{a}_{j+1/2})](\tilde{f}_{j+1} - \tilde{f}_j) + \tilde{f}_j. \tag{2.16}$$

Equation (2.16) is identical to (2.13a) if one defines $\psi(z) = |z|$ in (2.13). Since $\mathrm{sgn}(\tilde{a}_{j+1/2}^n) = \mathrm{sgn}(a_{j+1/2}^n)$, Eq. (2.15) and (2.16) are also equal to

$$u_j^{n+1} = u_j^n - \frac{\lambda}{2}[1 - \mathrm{sgn}(a_{j+1/2}^n)](\tilde{f}_{j+1}^n - \tilde{f}_j^n) - \frac{\lambda}{2}[1 + \mathrm{sgn}(a_{j-1/2}^n)](\tilde{f}_j^n - \tilde{f}_{j-1}^n), \tag{2.17}$$

with

$$\tilde{h}_{j+1/2} = \frac{1}{2}[1 - \mathrm{sgn}(a_{j+1/2})](\tilde{f}_{j+1} - \tilde{f}_j) + \tilde{f}_j. \tag{2.18}$$

With this formulation, one does not have to calculate $\gamma_{j+1/2}$ at all.

For the numerical flux (2.16), the g function of (2.13c) can be defined in a slightly different form:

$$g_j = S \cdot \max[0, \min(\overline{\sigma}_{j+1/2}|\Delta_{j+1/2}f|, S \cdot \overline{\sigma}_{j-1/2}\Delta_{j-1/2}f)], \tag{2.19a}$$

$$S = \mathrm{sgn}(\Delta_{j+1/2}f),$$

with $\overline{\sigma}_{j+1/2} = \overline{\sigma}(a_{j+1/2})$ and

$$\overline{\sigma}(z) = \frac{1}{2}[\mathrm{sgn}(z) - \lambda z]. \tag{2.19b}$$

Here the identities $\mathrm{sgn}(\Delta_{j+1/2}f)\,\mathrm{sgn}(a_{j+1/2}) = \mathrm{sgn}(u_{j+1} - u_j)$ and $|\Delta_{j+1/2}f| = |a_{j+1/2}||u_{j+1} - u_j|$ are used. The above limiter g_j of (2.19) can be considered as a flux limiter since the flux f is limited. Equations (2.13c,d) are preferred over (2.19) for its straightforward extension to system cases because u appears rather than f.

3. IMPLICIT TVD SCHEMES

First-order implicit TVD scheme

Now consider a one-parameter family of three-point conservative schemes of the form

$$u_j^{n+1} + \lambda\theta(h_{j+1/2}^{n+1} - h_{j-1/2}^{n+1}) = u_j^n - \lambda(1 - \theta)(h_{j+1/2}^n - h_{j-1/2}^n), \qquad (3.1)$$

where $0 \leq \theta \leq 1$ is a parameter, $h_{j+1/2}^n = h(u_j^n, u_{j+1}^n)$, $h_{j+1/2}^{n+1} = h(u_j^{n+1}, u_{j+1}^{n+1})$, and $h(u_j, u_{j+1})$ is the numerical flux (2.4). This one-parameter family of schemes contains implicit as well as explicit schemes. When $\theta = 0$, (3.1) reduces to (2.2), the explicit method. When $\theta \neq 0$, (3.1) is an implicit scheme. For example, if $\theta = 1/2$, the time differencing is the trapezoidal formula, and if $\theta = 1$, the time differencing is the backward Euler method. To simplify the notation, rewrite (3.1) as

$$L \cdot u^{n+1} = R \cdot u^n, \qquad (3.2)$$

where L and R are the following finite-difference operators:

$$(L \cdot u)_j = u_j + \lambda\theta(h_{j+1/2} - h_{j-1/2}), \qquad (3.3a)$$

$$(R \cdot u)_j = u_j - \lambda(1 - \theta)(h_{j+1/2} - h_{j-1/2}). \qquad (3.3b)$$

Sufficient conditions for (3.1) to be a TVD scheme are that

$$TV(R \cdot u^n) \leq TV(u^n) \qquad (3.4a)$$

and

$$TV(L \cdot u^{n+1}) \geq TV(u^{n+1}). \qquad (3.4b)$$

A sufficient condition for (3.4) is the CFL-like restriction

$$|\lambda a_{j+1/2}| \leq \lambda\psi(a_{j+1/2}) \leq \frac{1}{1 - \theta}, \qquad (3.5)$$

where $a_{j+1/2}$ is defined in Eq. (2.5b). Therefore, for the scheme to be TVD, one has to pick $\psi(a_{j+1/2})$ such that (3.5) is satisfied. For a detailed proof of Eqs. (3.4) and (3.5), see [2]. Observe that the backward Euler implicit scheme, $\theta = 1$ in (3.1), is unconditionally TVD, while the trapezoidal formula, $\theta = 1/2$ is TVD under the CFL-like restriction of 2. The forward Euler explicit scheme, $\theta = 0$ or Eq. (2.2), is TVD under the CFL restriction of 1.

Second-order implicit TVD scheme

One can obtain a second-order-accurate implicit TVD scheme by replacing the numerical flux function h of (3.1) with $\tilde{h}$ of Eqs. (2.13) or (2.18); i.e.

$$u_j^{n+1} + \lambda\theta(\tilde{h}_{j+1/2}^{n+1} - \tilde{h}_{j-1/2}^{n+1}) = u_j^n - \lambda(1 - \theta)(\tilde{h}_{j+1/2}^n - \tilde{h}_{j-1/2}^n). \qquad (3.6)$$

However, $\sigma_{j+1/2}$ is different from (2.13d). Instead, choose

$$\sigma(z) = \frac{1}{2}\psi(z) + \lambda\left(\theta - \frac{1}{2}\right)z^2 \qquad (3.7a)$$

for time-dependent calculations, or

$$\sigma(z) = \frac{1}{2}\psi(z) \qquad (3.7b)$$

for either steady-state or time-accurate calculations.

For the first choice (3.7a), the scheme is second-order-accurate in space and time regardless of θ. However, the steady-state solution depends on the time step. The second choice in (3.7b) makes the scheme second-order accurate in space but first-order accurate in time if $\theta = 1$. This choice of $\sigma(z)$ ensures that the steady-state solution does not depend on the time step Δt. Second-order in space and time can be achieved for (3.7b) if $\theta = 1/2$ is chosen. For example, an unconditionally TVD backward Euler scheme is of the form

$$u_j^{n+1} + \lambda(\tilde{h}_{j+1/2}^{n+1} - \tilde{h}_{j-1/2}^{n+1}) = u_j^n. \tag{3.8}$$

This is a highly nonlinear implicit scheme. An efficient procedure to solve this set of nonlinear equations is needed. The following focuses on linearized forms of the implicit scheme (3.6).

Linearized nonconservative implicit (LNI) form

For steady-state calculations, the following linearized version of (3.6) in delta formulation[4,12] can be used:

$$E_1 d_{j-1} + E_2 d_j + E_3 d_{j+1} = -\lambda[\tilde{h}_{j+1/2}^n - \tilde{h}_{j-1/2}^n] \tag{3.9a}$$

with

$$E_1 = -\lambda\theta(C_{j-1/2}^+)^n, \tag{3.9b}$$

$$E_2 = 1 + \lambda\theta[(C_{j+1/2}^-)^n + (C_{j-1/2}^+)^n], \tag{3.9c}$$

$$E_3 = -\lambda\theta(C_{j+1/2}^-)^n, \tag{3.9d}$$

where $d_j = u_j^{n+1} - u_j^n$, $\tilde{h}_{j+1/2}$ from (2.13) or (2.18), and

$$(C^{\pm})^n = \frac{1}{2}[\psi(a + \gamma) \pm (a + \gamma)]^n. \tag{3.9e}$$

One can obtain Eq. (3.9) by simply rewriting (3.6) in an upwind form so that the resulting equation is a function of $C^{\pm}(a + \gamma)_{j+1/2}^{n+1}$, $C^{\pm}(a + \gamma)_{j+1/2}^n$, $\Delta_{j+1/2}u^{n+1}$ and $\Delta_{j+1/2}u^n$, rearranging terms, and dropping the time index of the $C^{\pm}$ from $(n + 1)$ to n. Equation (3.9) will be referred to as the linearized nonconservative implicit (LNI) form.

Although (3.9) is formally a five-point scheme, the coefficient matrix associated with it is tridiagonal with a dominant diagonal. Another TVD linearized form can also be obtained by setting $\gamma = 0$ in (3.9e); i.e., redefining (3.9e) by

$$(C^{\pm})^n = \frac{1}{2}[\psi(a) \pm a]^n. \tag{3.10}$$

Scheme (3.9a) together with (3.10) is spatially first-order accurate for the implicit operator and spatially second-order accurate for the explicit operator. It can be shown that (3.9a) together with (3.10) is still TVD. The LNI forms (3.9) and (3.10) are mainly useful for steady-state calculations, since the scheme is only conservative after the solution reaches steady state, i.e. if a steady state is attained the solution is the same as would have been obtained with a conservative scheme.

Linearized conservative implicit (LCI) form

A linearized conservative implicit (LCI) form can be obtained by rewriting (2.13a) as

$$\tilde{h}_{j+1/2}^{n+1} = \frac{1}{2}[f_j^{n+1} + f_{j+1}^{n+1}]$$

$$+ \frac{1}{2}\left[\left(\frac{g_j + g_{j+1}}{\Delta_{j+1/2}u}\right)\Delta_{j+1/2}u - \psi(a_{j+1/2} + \gamma_{j+1/2})\Delta_{j+1/2}u\right]^{n+1}, \tag{3.11}$$

and using a local Taylor expansion about u^n:

$$f^{n+1} - f^n = a^n(u^{n+1} - u^n) + O(\Delta t^2), \tag{3.12}$$

where $a = \partial f/\partial u$. Applying the first-order approximation of (3.12) and locally linearizing the coefficients of $\Delta_{j+1/2}u$ in the second and third terms on the right-hand-side by dropping the time index from $(n + 1)$ to n, one gets the LCI form in delta formulation:

$$\overline{E}_1 d_{j-1} + \overline{E}_2 d_j + \overline{E}_3 d_{j+1} = -\lambda[\tilde{h}^n_{j+1/2} - \tilde{h}^n_{j-1/2}], \tag{3.13a}$$

where

$$\overline{E}_1 = \frac{\lambda\theta}{2}\,[-a^n_{j-1} + \beta^n_{j-1/2} - \psi((a + \gamma)^n_{j-1/2})], \tag{3.13b}$$

$$\overline{E}_2 = 1 + \frac{\lambda\theta}{2}\,[-\beta^n_{j+1/2} + \psi((a + \gamma)^n_{j+1/2}) - \beta^n_{j-1/2} + \psi((a + \gamma)^n_{j-1/2})], \tag{3.13c}$$

$$\overline{E}_3 = \frac{\lambda\theta}{2}\,[a^n_{j+1} + \beta^n_{j+1/2} - \psi((a + \gamma)^n_{j+1/2})], \tag{3.13d}$$

and

$$\beta^n_{j+1/2} = \left(\frac{g_j + g_{j+1}}{\Delta_{j+1/2}u}\right)^n. \tag{3.13e}$$

Again, this is a five-point scheme but with a tridiagonal coefficient matrix. An LCI form which is spatially first-order accurate can also be obtained for the implicit operator by simply setting $\gamma_{j+1/2} = \beta_{j+1/2} = 0$ in (3.13b)–(3.13d). These LCI forms preserve the conservative form of the differencing scheme; thus they are applicable to unsteady as well as steady-state calculations. The main drawback is that the resulting scheme may no longer be unconditionally TVD for $\theta = 1$. There is no proof to show that the LCI form is still unconditionally TVD. Indeed, numerical experiments with the Burgers' equation show that the scheme (3.13) is not unconditionally TVD. This shortcoming of the LCI form appears to be disturbing at first. In practice, however, unconditional stability arising from linear stability analysis rarely carries over to nonlinear system cases with discontinuous solutions. Our main interest is a less restrictive time-step bound compared with explicit methods. Therefore, if the LCI form can offer a larger time-step bound with high accuracy, it should be sufficient for practical applications. As can be seen later, application of a variant of (3.1) for the Euler equation of gas dynamics in two dimensions shows that the LCI is a fairly useful tool for practical calculations. The method remains stable for fairly large CFL.

4. FORMAL EXTENSION OF ALGORITHM FOR TWO-DIMENSIONAL SYSTEM OF HYPERBOLIC CONSERVATION LAWS

Before going on to the next section, it should be emphasized that all the second-order TVD schemes are constructed so that no spurious oscillations are generated for one-dimensional nonlinear scalar hyperbolic conservation laws and constant-coefficient hyperbolic systems. None of the theory says anything about nonlinear systems or two-dimensional scalar hyperbolic conservation laws. Moreover, Goodman and Leveque[13] have obtained the following result: for a specific norm, a two-dimensional scalar approximation cannot be TVD and still be more than first-order accurate. Maybe what one needs is a different norm or a more relaxed definition than the TVD property for one dimension. But in practice, it is straightforward to formally extend the scheme to one- or two-dimensional nonlinear hyperbolic systems. Surprisingly enough, numerical experiments with these schemes which are based on a one-dimensional concept applied to two-dimensional problems via local one-dimensional splitting show that these types of schemes do perform well for fairly complex shock structures[12,14].

Therefore, for the rest of the paper it is understood that the properties of all the schemes under discussion are for one-dimensional nonlinear scalar hyperbolic conservation laws and one-dimensional constant-coefficient hyperbolic systems. The schemes are then formally extended to one- or two-dimensional systems of conservation laws and are evaluated by numerical experiments.

Extension of the scalar TVD scheme to systems of conservation laws can be accomplished by defining at each point a "local" system of characteristic fields, and then applying the scheme to each of the m scalar characteristic equations. Here m is the dimension of the hyperbolic system. The formulation described here is valid for both two- and three-dimensional systems of conservation laws. Only the two-dimensional case will be described. For three-dimensional formulations, one only has to add an extra dimension and the corresponding numerical flux. Also, for simplicity of presentation, only conservation laws in Cartesian coordinates with equal spatial step size will be discussed. The variable step size or generalized coordinate formulations with application to airfoil calculations are reported in separate papers[15,16].

Consider a two-dimensional system of hyperbolic conservation laws

$$\frac{\partial U}{\partial t} + \frac{\partial F(U)}{\partial x} + \frac{\partial G(U)}{\partial y} = 0. \tag{4.1}$$

Here U, $F(U)$ and $G(U)$ are column vectors of m components. Let $A = \partial F/\partial U$ and $B = \partial G/\partial U$. Let the eigenvalues of A be $(a_x^1, a_x^2, \ldots, a_x^m)$ and the eigenvalues of B be $(a_y^1, a_y^2, \ldots, a_y^m)$. Denote R_x and R_y as the matrices whose columns are eigenvectors of A and B, and denote R_x^{-1} and R_y^{-1} as the inverses of R_x and R_y. In the case of the compressible Euler equation of gas dynamics

$$U = \begin{bmatrix} \rho \\ \rho u \\ \rho v \\ e \end{bmatrix}; \quad F = \begin{bmatrix} \rho u \\ \rho u^2 + p \\ \rho u v \\ u(e + p) \end{bmatrix}; \quad G = \begin{bmatrix} \rho v \\ \rho u v \\ \rho v^2 + p \\ v(e + p) \end{bmatrix}. \tag{4.2a}$$

The variables are the density ρ, the velocity components u and v, and the pressure p. The total energy per unit volume, e, is related to p by the equation of state for a perfect gas:

$$p = (\gamma - 1)\left[e - \frac{(\rho u)^2 + (\rho v)^2}{2\rho} \right], \tag{4.2b}$$

where γ is the ratio of specific heats and should not be confused with the $\gamma_{j+1/2}$ in (2.13b). The eigenvalues of A are

$$(a_x^1, a_x^2, a_x^3, a_x^4) = (u - c, u, u + c, u), \tag{4.2c}$$

where c is the local speed of sound. The right eigenvectors of A are given by

$$R_x = \begin{bmatrix} 1 & 1 & 1 & 0 \\ u - c & u & u + c & 0 \\ v & v & v & 1 \\ H - uc & (u^2 + v^2)/2 & H + uc & v \end{bmatrix}, \tag{4.2d}$$

where

$$H = \frac{c^2}{\gamma - 1} + \frac{u^2 + v^2}{2}. \tag{4.2e}$$

Similarly, the eigenvalues of B are

$$(a_y^1, a_y^2, a_y^3, a_y^4) = (v - c, v, v + c, v). \tag{4.2f}$$

The right eigenvectors of B are given by

$$R_y = \begin{bmatrix} 1 & 1 & 1 & 0 \\ u & u & u & 1 \\ v - c & v & v + c & 0 \\ H - vc & (u^2 + v^2)/2 & H + vc & u \end{bmatrix}. \tag{4.2g}$$

Let the grid spacing be denoted by Δx and Δy such that $x = j\Delta x$ and $y = k\Delta y$. Denote $U_{j+1/2,k}$ as some symmetric average of $U_{j,k}$ and $U_{j+1,k}$ (for example, the arithmetic mean average or the Roe's average[7] for gas dynamics). Let $a_{j+1/2}^l$, $R_{j+1/2}$, $R_{j+1/2}^{-1}$ denote the quantities a_x^l, R_x, R_x^{-1} evaluated at $U_{j+1/2,k}$. Similarly, let $a_{k+1/2}^l$, $R_{k+1/2}$, $R_{k+1/2}^{-1}$ denote the quantities a_y^l, R_y, R_y^{-1} evaluated at $U_{j,k+1/2}$.

Define

$$\alpha_{j+1/2} = R_{j+1/2}^{-1}(U_{j+1,k} - U_{j,k}) \tag{4.3a}$$

as the difference of the characteristic variables in the locally x direction, and define

$$\alpha_{k+1/2} = R_{k+1/2}^{-1}(U_{j,k+1} - U_{j,k}) \tag{4.3b}$$

as the difference of the characteristic variables in the locally y direction.

With the above notation, a one-parameter family of TVD schemes (3.6) together with (2.13a–c) and (3.7b) in two dimensions can be written as

$$U_{j,k}^{n+1} + \lambda^x\theta(\tilde{F}_{j+1/2,k}^{n+1} - \tilde{F}_{j-1/2,k}^{n+1}) + \lambda^y\theta(\tilde{G}_{j,k+1/2}^{n+1} - \tilde{G}_{j,k-1/2}^{n+1})$$
$$= U_{j,k}^n - \lambda^x(1 - \theta)(\tilde{F}_{j+1/2,k}^n - \tilde{F}_{j-1/2,k}^n) - \lambda^y(1 - \theta)(\tilde{G}_{j,k+1/2}^n - \tilde{G}_{j,k-1/2}^n), \tag{4.4a}$$

with $\lambda^x = \Delta t/\Delta x$, $\lambda^y = \Delta t/\Delta y$, and the numerical flux function $\tilde{F}_{j+1/2,k}$ defined as

$$\tilde{F}_{j+1/2,k} = \frac{1}{2}[F_{j,k} + F_{j+1,k} + R_{j+1/2}\Phi_{j+1/2}], \tag{4.4b}$$

where the elements of the $\Phi_{j+1/2}$ denoted by $\phi_{j+1/2}^l$, $l = 1, \ldots, m$ are

$$\phi_{j+1/2}^l = g_j^l + g_{j+1}^l - \psi(a_{j+1/2}^l + \gamma_{j+1/2}^l)\alpha_{j+1/2}^l, \tag{4.4c}$$

$$g_j^l = S \cdot \max[0, \min(\sigma_{j+1/2}^l|\alpha_{j+1/2}^l|, S \cdot \sigma_{j-1/2}^l\alpha_{j-1/2}^l)], \tag{4.4d}$$

$$S = \operatorname{sgn}(\alpha_{j+1/2}^l),$$

with $\psi(z)$ defined in (2.7); $\sigma_{j+1/2}^l$ is (3.7a) or (3.7b) with $z = a_{j+1/2}^l$, and

$$\gamma_{j+1/2}^l = \begin{cases} (g_{j+1}^l - g_j^l)/\alpha_{j+1/2}^l & \alpha_{j+1/2}^l \neq 0, \\ 0 & \alpha_{j+1/2}^l = 0, \end{cases} \tag{4.4e}$$

where $\alpha_{j+1/2}^l$ are the elements of (4.3a). The numerical flux $\tilde{G}_{j,k+1/2}$ can be defined in a similar manner. Here, the formula (4.4) is only valid for $\sigma(z) = \psi(z)/2$. If one uses $\sigma(z) = \psi/2 - \lambda(\theta - 1/2)z^2$, then the effect of the cross derivative U_{xt} is neglected in formulation (4.4).

Alternate form for the numerical fluxes

Extension of the scalar TVD scheme to nonlinear system cases is not unique. Take, for example, the case where the numerical flux $\bar{h}_{j+1/2}$ in (2.13) is identical to $\tilde{h}_{j+1/2}$ in (2.16) if one sets $\psi(z) = |z|$ in (2.13). The corresponding numerical fluxes for the system case has a different form depending on which of the scalars $\tilde{h}_{j+1/2}$ one started with. If one started with (2.16), $\tilde{F}_{j\pm1/2,k}$ can be of the form

$$\tilde{F}_{j+1/2,k} = \frac{1}{2}[I - \mathcal{A}_{j+1/2}](\mathcal{F}_{j+1,k} - \mathcal{F}_{j,k}) + \mathcal{F}_{j,k}, \tag{4.5a}$$

$$\tilde{F}_{j-1/2,k} = \mathcal{F}_{j,k} - \frac{1}{2}[I + \mathcal{A}_{j-1/2}](\mathcal{F}_{j,k} - \mathcal{F}_{j-1,k}), \tag{4.5b}$$

where

$$\mathcal{A}_{j+1/2} = R_{j+1/2}^{-1}\Lambda_{j+1/2}R_{j+1/2}, \tag{4.5c}$$

$$\Lambda = \mathrm{diag}[\mathrm{sgn}(a_{j+1/2})] \tag{4.5d}$$

and

$$\mathcal{F}_{j,k} = F_{j,k} + R_j\Omega_j, \tag{4.5e}$$

with $\Omega = (g_j^1, g_j^2, \ldots, g_j^m)^T$. Here $\mathrm{diag}(z^l)$, denotes a diagonal matrix with diagonal elements z^l. Numerical experiments with the use of the two forms of the numerical fluxes (4.4b) and (4.5a,b) together with $\psi(z) = |z|$ on a quasi-one-dimensional divergent nozzle problem show that there is no visible improvement in accuracy of one over the other. By inspection, the operations count of the numerical flux (4.5a,b) is higher than (4.4b). Therefore, (4.4b) is favored over (4.5a,b).

5. EXTENSION OF THE IMPLICIT SCHEME BY THE ALTERNATING DIRECTION IMPLICIT (ADI) METHOD

In order to solve for U^{n+1} in (4.4a), one needs to solve a set of nonlinear algebraic equations. For computational efficiency, consider the following solution strategies. First, linearize the implicit operator in several ways as in the scalar case. Then construct an alternating direction implicit (ADI) form for the linearized implicit algorithms. The final step is to either use the ADI form as the solution algorithm, or use the ADI form as the predictor step and the original nonlinear algorithm (4.4) as the corrector step (with the numerical flux values evaluated at the predicted solution). The predictor–corrector method is proposed mainly because of its use (a) as a vehicle to compensate for the simplification of the various linearized implicit operators, and (b) for the possible improvement in convergence rate for steady-state applications.

Linearized nonconservative implicit (LNI) form
The corresponding LNI form (3.9) for the two-dimensional scheme (4.4) is

$$[I - \lambda^x\theta J_{j+1/2,k}^-\Delta_{j+1/2} + \lambda^x\theta J_{j-1/2,k}^+\Delta_{j-1/2} - \lambda^y\theta K_{j,k+1/2}^-\Delta_{k+1/2}$$
$$+ \lambda^y\theta K_{j,k-1/2}^+\Delta_{k-1/2}](U^{n+1} - U^n)$$
$$= -\lambda^x[\tilde{F}_{j+1/2,k}^n - \tilde{F}_{j-1/2,k}^n] - \lambda^y[\tilde{G}_{j,k+1/2}^n - \tilde{G}_{j,k-1/2}^n], \tag{5.1a}$$

where

$$J_{j+1/2,k}^\pm = (R_x \, \mathrm{diag}(C_x^\pm)R_x^{-1})_{j+1/2,k}^n, \tag{5.1b}$$

$$K_{j,k+1/2}^\pm = (R_y \, \mathrm{diag}(C_y^\pm)R_y^{-1})_{j,k+1/2}^n \tag{5.1c}$$

and

$$(C_x^{\pm})^n_{j+1/2,k} = \frac{1}{2}\,[\psi(a_x^l + \gamma_x^l) \pm (a_x^l + \gamma_x^l)]^n_{j+1/2,k}, \tag{5.2a}$$

$$(C_y^{\pm})^n_{j,k\pm1/2} = \frac{1}{2}\,[\psi(a_y^l + \gamma_y^l) \pm (a_y^l + \gamma_y^l)]^n_{j,k+1/2}, \quad l = 1,\ldots,m. \tag{5.2b}$$

Here the operator $\Delta_{j+1/2}$, operating on U, means $\Delta_{j+1/2}U = U_{j+1,k} - U_{j,k}$.

Notice that in each coordinate direction (5.1) is a spatially second-order-accurate, five-point scheme, yet the iteration matrix associated with (5.1) in that direction is block tridiagonal. Normally, the matrix associated with a five-point stencil scheme would have been a block pentadiagonal matrix.

To calculate (5.1b,c) at every time step is quite costly. For steady-state applications, (5.1) can be simplified even more by setting $\gamma = 0$ in (5.2) since it is only necessary for the scheme to be second order after it reaches steady state. The time integration and the entire implicit operator can be viewed as a relaxation procedure for the steady-state solution.

A numerical experiment for a one-dimensional gas dynamics problem shows that the LNI form with $\gamma \neq 0$ or $\gamma = 0$ on the left-hand side has a fairly rapid convergence rate and gives good shock resolution. However, numerical experiments with an ADI form of (5.1)[4] show that the LNI form does not have a good convergence rate. One possible way of solving (5.1) is by a different type of relaxation method. This will be the subject of a future investigation. For the rest of the paper, the LNI approach will be abandoned and in favor of the linearized conservative implicit form (3.13).

Linearized conservative implicit (LCI) form

The LCI form corresponding to (3.13) for the two-dimensional scheme (4.4) is

$$[I + \lambda^x\theta H^x_{j+1/2,k} - \lambda^x\theta H^x_{j-1/2,k} + \lambda^y\theta H^y_{j,k+1/2} - \lambda^y\theta H^y_{j,k-1/2}](U^{n+1} - U^n)$$
$$= -\lambda^x[\tilde{F}^n_{j+1/2,k} - \tilde{F}^n_{j-1/2,k}] - \lambda^y[\tilde{G}^n_{j,k+1/2} - \tilde{G}^n_{j,k-1/2}], \tag{5.3a}$$

where

$$H^x_{j+1/2,k} = \frac{1}{2}\,[A_{j+1,k} + \Omega^x_{j+1/2,k}]^n \tag{5.3b}$$

$$H^y_{j,k+1/2} = \frac{1}{2}\,[B_{j,k+1} + \Omega^y_{j,k+1/2}]^n, \tag{5.3c}$$

with A and B equal to the Jacobian of the fluxes F and G, and

$$\Omega^x_{j+1/2,k} = (R_x\,\text{diag}[\beta^l - \psi(a^l + \gamma^l)]R_x^{-1})_{j+1/2}\Delta_{j+1/2}, \tag{5.4a}$$

$$\Omega^y_{j,k+1/2} = (R_y\,\text{diag}[\beta^l - \psi(a^l + \gamma^l)]R_y^{-1})_{k+1/2}\Delta_{k+1/2}, \tag{5.4b}$$

where

$$\beta^l_{j+1/2} = \frac{(g^l_j + g^l_{j+1})}{\alpha^l_{j+1/2}}; \quad \beta^l_{k+1/2} = \frac{(g^l_k + g^l_{k+1})}{\alpha^l_{k+1/2}}. \tag{5.4c}$$

The nonstandard notation

$$H^x_{j+1/2,k}(U^{n+1} - U^n) = \frac{1}{2}\,[A^n_{j+1,k}(U^{n+1} - U^n)_{j+1,k} + \Omega^x_{j+1/2,k}(U^{n+1} - U^n)] \tag{5.4d}$$

is used and the difference operator $\Omega^x_{j+1/2,k}$ is defined in (5.4a).

For steady-state application, one way to simplify (5.4) is to use a spatially first-order implicit operator; i.e. by redefining (5.4a) and (5.4b) as

$$\Omega^x_{j+1/2,k} = (R_x \, \mathrm{diag}[-\psi(a^l)]R_x^{-1})_{j+1/2}\Delta_{j+1/2}, \qquad (5.5a)$$

$$\Omega^y_{j,k+1/2} = (R_y \, \mathrm{diag}[-\psi(a^l)]R_y^{-1})_{k+1/2}\Delta_{k+1/2}. \qquad (5.5b)$$

The computation can be reduced even more if $\Omega^x_{j+1/2,k}$ and $\Omega^y_{j,k+1/2}$ are simplified to diagonal matrices. For example, redefine (5.5) as

$$\Omega^x_{j+1/2,k} = \left(\mathrm{diag}[-\max_l \psi(a^l)]\right)_{j+1/2}\Delta_{j+1/2}, \qquad (5.6a)$$

$$\Omega^y_{j,k+1/2} = \left(\mathrm{diag}[-\max_l \psi(a^l)]\right)_{k+1/2}\Delta_{k+1/2}. \qquad (5.6b)$$

From here on, algorithm (5.3) together with (5.6) is referred to as the linearized conservative diagonal form. It turns out that this linearized conservative diagonal form is quite an attractive method for steady-state applications.

ADI form

Even with the above simplifications, it is still very costly to solve the two-dimensional difference equation (5.3). An ADI form of (5.3) will be adopted as

$$[I + \lambda^x\theta H^x_{j+1/2,k} - \lambda^x\theta H^x_{j-1/2,k}]D^*$$
$$= -\lambda^x[\tilde{F}^n_{j+1/2,k} - \tilde{F}^n_{j-1/2,k}] - \lambda^y[\tilde{G}^n_{j,k+1/2} - \tilde{G}^n_{j,k-1/2}], \qquad (5.7a)$$

$$[I + \lambda^y\theta H^y_{j,k+1/2} - \lambda^y\theta H^y_{j,k-1/2}]D = D^*, \qquad (5.7b)$$

$$U^{n+1} = U^n + D. \qquad (5.7c)$$

Observe that (5.7) is the original Beam and Warming[17] algorithm if $\Omega^x_{j+1/2,k} = \Omega^y_{j,k+1/2} = 0$ in (5.3b,c) and $\Phi_{j+1/2}R_{j+1/2}$ in (4.4b) is replaced by the conventional fourth-order dissipation term. The implementation of this ADI scheme into an existing central difference code (such as the code based on the Beam and Warming algorithm) is relatively simple. All one has to do is add the extra matrices $\Omega^x_{j+1/2,k}$ and $\Omega^y_{j,k+1/2}$ for the implicit operator and a more sophisticated dissipation term $\Phi_{j+1/2}R_{j+1/2}$ (4.4b) for the explicit operator. For the case of the linearized conservative diagonal form (5.3) together with (5.6), no extra work is involved on the implicit operator, since Ω^x and Ω^y in (5.6) are diagonal matrices with equal elements, and can be saved while computing for the right-hand side.

From numerical experiments with the NACA0012 airfoil steady-state calculations, the linearized conservative diagonal ADI form is the most efficient scheme among the various proposed linearized methods for the case of $\theta = 1$. No comparison has been made for time-accurate calculations or for any other values of θ. A study on the predictor–corrector step method (as mentioned at the beginning of the section) does not show a drastic change in convergence rate for the same airfoil calculations. In the next section, some numerical results for the linearized conservative ADI algorithm with $\theta = 1$ (the backward Euler time differencing) will be shown.

6. NUMERICAL RESULTS FOR THE EULER EQUATIONS

The results presented here utilized a coordinate transformation transforming a general curvilinear physical space into a rectangle with uniform spacing of unit length. For the airfoil calculations, the actual geometry is mapped onto the computational rectangle such that all the boundary surfaces are edges of the rectangle. Figure 1 shows the transformation for a "C" mesh topology where a branch cut (wake cut) is used at the trailing edge of the airfoil. In "O" mesh topologies the wake cut boundary is periodic and slightly more computation is involved for the block tridiagonal inversion.

The generalized coordinates formulation of the algorithm, treatment of boundary conditions,

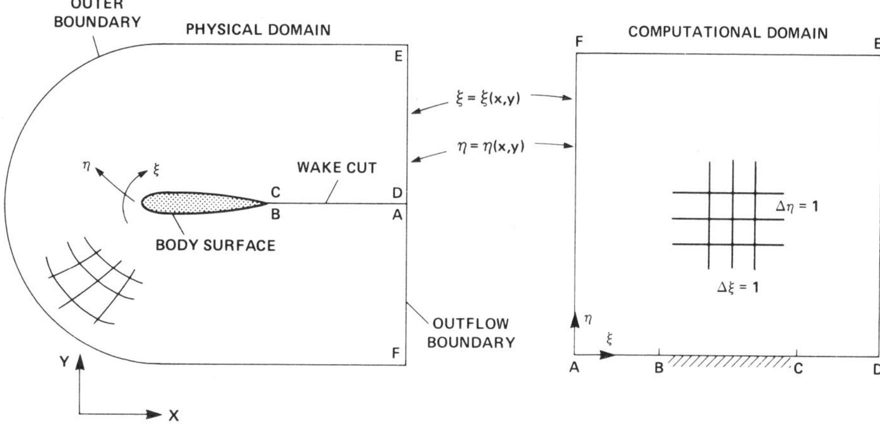

Fig. 1. Generalized coordinate transformation with "C" grid topology for airfoil calculations.

and more detailed description of the applications to the Euler equations of gas dynamics are described in [15]. Here some of the numerical results are shown to illustrate the accuracy and efficiency of the method for $\theta = 1$.

Since the actual grids have widely varying cell sizes, a space-varying Δt similar to the one used by Pulliam and Steger[18] is used as a vehicle to improve the convergence rate. The ϵ in (2.7) is set to 0.125 for all cases. Numerical experiments show that the solutions are insensitive to the value of ϵ between 0.06 to 0.25. A more sophisticated ϵ has been proposed in [8]. Computational experiments again show no visible improvement in efficiency or accuracy in using the more sophisticated formula while more computations are required. No artificial compression term as discussed in [4] is needed in all of the airfoil calculations.

The inviscid cases considered here are the NACA0012 airfoil with (a) $M_\infty = 0.8$, $\alpha = 1.25$, (b) $M_\infty = 0.85$, $\alpha = 1.00$, (c) $M_\infty = 0.95$, $\alpha = 0.0$, and (d) $M_\infty = 1.2$, $\alpha = 7.0$. Here M_∞ is the freestream Mach number and α is the angle of attack. These are four of the cases which have been considered in the AGARD Fluid Dynamics Panel Working Group 07[5].

To show the accuracy of the scheme, a 249 × 41 C grid with no special clusterings on the upper or lower surface near the vicinity of the shocks is used. Figure 2 shows the grid distribution around the airfoil. The outer boundary is 24 chord lengths away from the body. Each case was initialized with a uniform freestream flow at the prescribed Mach number and angle of attack and used the same grid as shown in Fig. 2. Figures 3–6 show the pressure coefficients and Mach contours for all four cases. The symbol "+" on the pressure coefficient plots is used to indicate the computed values. The solid (upper surface) and the dashed (lower surface) lines are just connectors between grid points. The value C_p^* in all of the pressure coefficient figures indicates the critical pressure coefficient. One can see in all cases that shocks can be captured within 1–2 grid points. When the same cases are run with the FLO52R code of Jameson[19] and the improved ARC2D code (version 150) of Pulliam and Steger[18], 3–4 points in the shock transition are generally observed. Away from the shocks, the three methods produce almost identical results. As a side remark, the accuracy and efficiency of FLO52R and the improved version of ARC2D are comparable; see Ref. [18] for more details. Both codes use central difference in space with similar numerical dissipation terms but they use different time-stepping methods for steady-state applications. These two codes are widely circulated. Figures 7 and 8 show the comparison of the current scheme with ARC2D (version 150) for cases (a) and (b) using the same mesh as in Fig. 2. One can see that the current method captures the shock better than ARC2D, especially on the lower surface.

Here, as a guideline, the results of Pulliam and Barton[5] (using ARC2D, version 150) and the not-yet-published results of the AGARD Fluid Dynamics Panel Working Group 07 are used as the "exact" solutions. Pulliam and Barton computed all the cases with very fine grids of 561 × 65 and with very dense clustering near the shocks. Shock strengths and shock locations of the current calculations coincide very well with the fine-grid results of Pulliam and Barton. The present numerical experiments indicate that, in order to produce the same accuracy as the TVD scheme, special clustering of grid points near the shocks and/or denser grid has to be

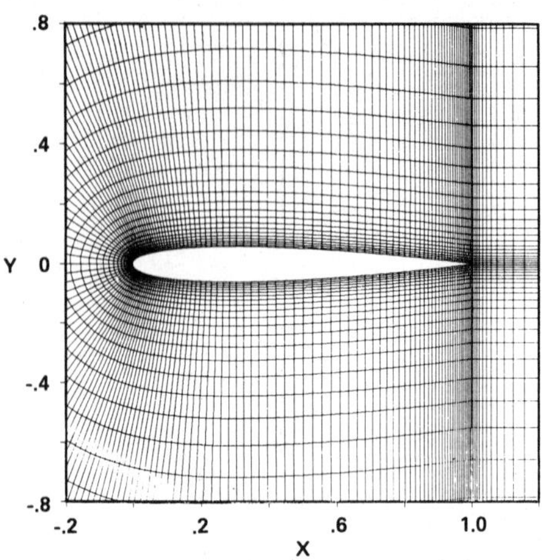

Fig. 2. The 249 × 41 C grid for the NACA0012 airfoil

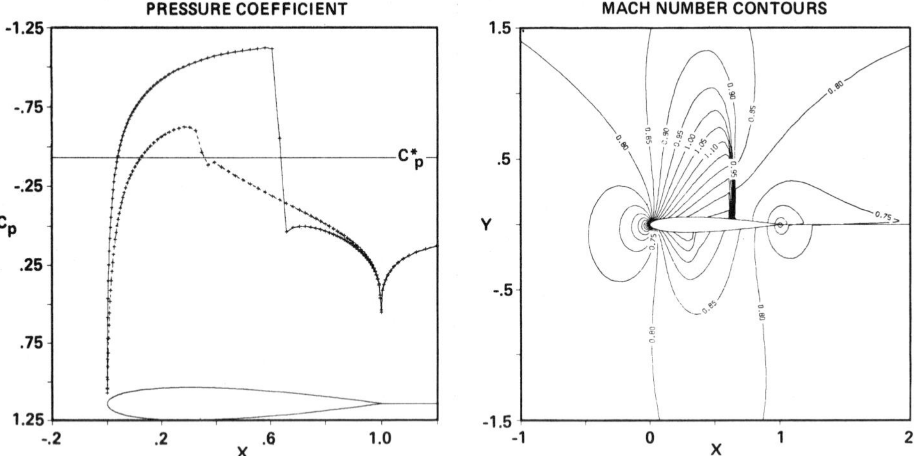

Fig. 3. Pressure coefficient and Mach contours for the NACA0012 airfoil with $M_\infty = 0.8$, $\alpha = 1.25$.

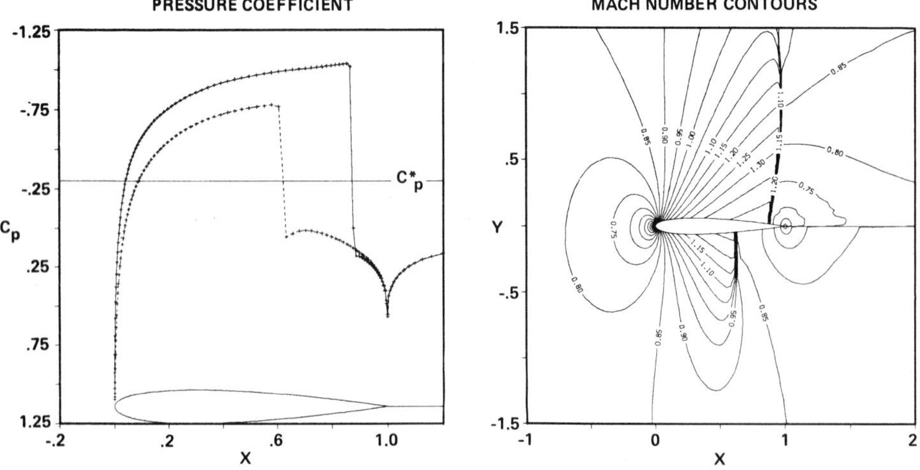

Fig. 4. Pressure coefficient and Mach contours for the NACA0012 airfoil with $M_\infty = 0.85$, $\alpha = 1.0$.

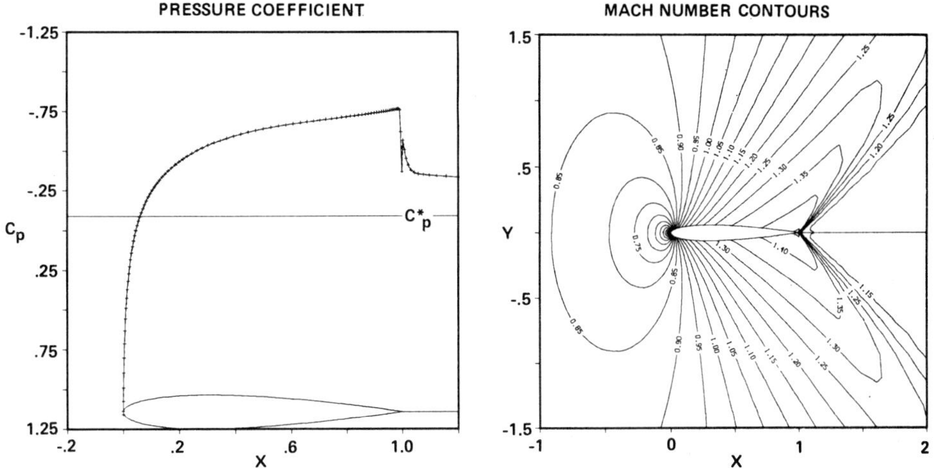

Fig. 5. Pressure coefficient and Mach contours for the NACA0012 airfoil with $M_\infty = /0.95$, $\alpha = 0.0$

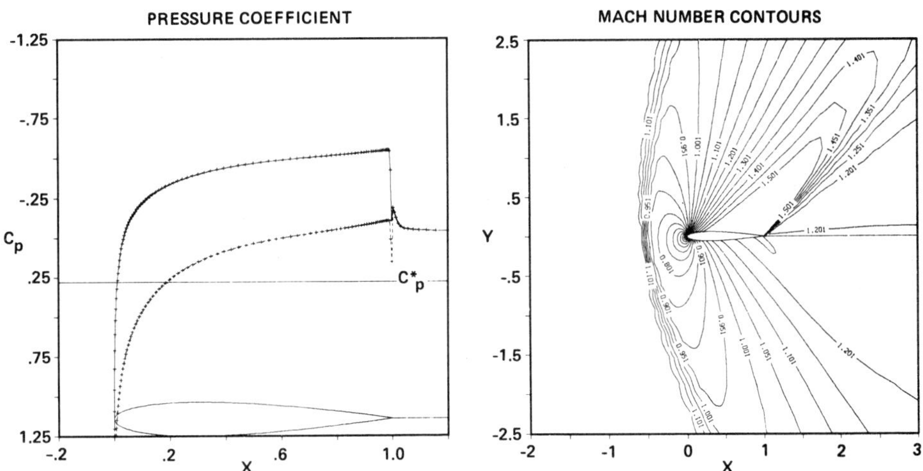

Fig. 6. Pressure coefficient and Mach contours for the NACA0012 airfoil with $M_\infty = 1.2$, $\alpha = 7.0$.

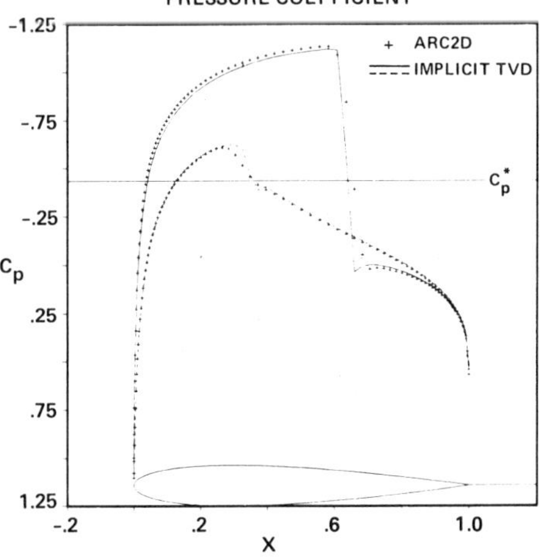

Fig. 7. Comparison of the current scheme with ARC2D for the NACA0012 airfoil with $M_\infty = 0.8$, $\alpha = 1.25$.

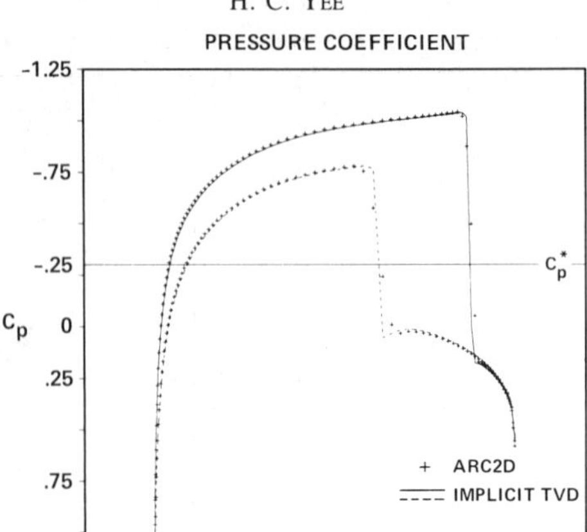

Fig. 8. Comparison of the current scheme with ARC2D for the NACA0012 airfoil with M_x = 0.85, α = 1.0.

employed by FLO52R and ARC2D. Thus, the advantage of the current scheme over FLO52R and ARC2D is that the unfavorable condition in computations which arises from the high aspect ratio of grid spacing can be avoided. Moreover, the present method requires less storage and lower operation count than other TVD schemes[3].

For cases (a) and (d), the L_2-norm residual (of the explicit operator) reaches machine zero at around 3000 steps. A residual of 10^{-7} can be reached in around 800 steps. Cases (b) and (c) are slightly more difficult. The shocks appear to be oblique and not aligned with the C grid coordinate system. The convergence rates are slower. Since the computer code is not fully vectorized and is not coded in an optimized fashion, it requires 0.53 sec per time step on the Cray XMP at the NASA Ames Research Center [based on the (249 × 41) C grid]. A careful recoding could possibly increase the speed by a factor of 2 or more. When the vectorized option is turned off, the current code requires 1.69 sec per time step, while ARC2D with the diagonal form[18] requires 1.01 sec per time step. This indicates that the present method requires 68% more computation time than ARC2D.

The form of the modified flux function

In all of the inviscid calculations, the linearized conservative diagonal form (5.3) together with (5.6) is used, with a slightly simplified form of the modified flux with

$$g_j^l + g_{j+1}^l = \sigma_{j+1/2}^l(\tilde{g}_j^l + \tilde{g}_{j+1}^l) \tag{6.1a}$$

and

$$\tilde{g}_j^l = S \cdot \max[0, \min(|\alpha_{j+1/2}^l|, S \cdot \alpha_{j-1/2}^l)], \tag{6.1b}$$

where $\sigma_{j+1/2}^l$, S, and $\alpha_{j+1/2}^l$ are defined as before.

Figure 9 shows the pressure coefficient with $(g_j^l + g_{j+1}^l)$ defined in (6.1a) compared with the ones defined in Eqs. (4.4c,d). Solid and dashed lines are numerical results using (4.4c,d), and +'s are numerical results using Eq. (6.1). It is found that a definite improvement in accuracy can be obtained by using (6.1). The definition of g_j^l's in (6.1) is identical to (4.4c,d) in the constant-coefficient cases. As a side remark, there is no visible difference in accuracy between (4.4c,d) and (6.1) for one-dimensional applications.

Evaluation of the symmetric averages $U_{j+1/2,k}$ *and* $U_{j,k+1/2}$

For a perfect gas, numerical experiments have been performed with two types of averaging for $U_{j+1/2,k}$ and $U_{j,k+1/2}$. The simplest form of $U_{j+1/2,k}$ is the arithmetic average $U_{j+1/2,k} = 0.5(U_{j+1,k} + U_{j,k})$. The other, Roe's averaging[7], is only applicable to a perfect gas. It has the computational advantage of perfectly resolving stationary discontinuities. For detailed implementation, see Ref. [4]. However, under certain conditions, such as highly irregular grids

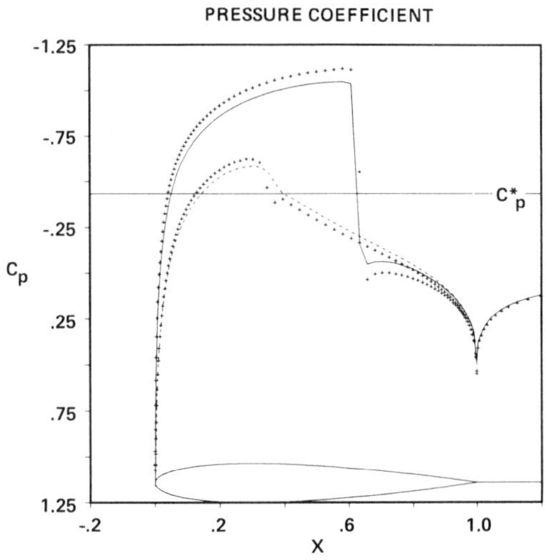

Fig. 9. Comparison of pressure coefficients between Egs. (6.1) (+) and 4.4c,d) (— and - -) for the NACA0012 airfoil with $M_x = 0.8$ and $\alpha = 1.25$

or special flow conditions, the characteristic speeds $a^l_{j+1/2}$ can lie outside the interval (a_j, a_{j+1}). Consequently, under this situation the direction of upwinding, which is determined solely by the sign of the $a^l_{j+1/2}$'s on Roe's schemes, might be the opposite of what is desired. This special property of the Roe's averaging was first observed by M. Vinokur of NASA Ames Research Center. Roe's averaging has been tested on a variety of one- and two-dimensional gas dynamics problems[12,14–16], and no sign of ill condition or instability was encountered. Numerical experiments with these two averages show no visible difference in numerical solutions for the above airfoil test cases. However, Roe's average requires slightly more computation.

7. APPLICATION TO THE NAVIER–STOKES EQUATIONS

Consider the two-dimensional mixed hyperbolic–parabolic system of conservation laws

$$\frac{\partial U}{\partial t} + \frac{\partial F(U)}{\partial x} + \frac{\partial G(U)}{\partial y} = \frac{\partial F_v(U, U_x, U_y)}{\partial x} + \frac{\partial G_v(U, U_x, U_y)}{\partial y}. \tag{7.1}$$

Here U, $F(U)$ and $G(U)$ are the same as in (4.1). The additional vectors F_v and G_v are vector functions of not only the components of U, but also of U_x and U_y, where $U_x = \partial U/\partial x$ and $U_y = \partial U/\partial y$. The compressible Navier–Stokes equations have the form (7.1). For a nondimensional form of the Naiver–Stokes equations,

$$F_v = \frac{1}{\text{Re}} \begin{bmatrix} 0 \\ \tau_{xx} \\ \tau_{xy} \\ e_x \end{bmatrix}, \quad G_v = \frac{1}{\text{Re}} \begin{bmatrix} 0 \\ \tau_{xy} \\ \tau_{yy} \\ e_y \end{bmatrix}, \tag{7.2}$$

with

$$\tau_{xx} = \mu(4u_x - 2v_y)/3,$$

$$\tau_{xy} = \mu(u_y + v_x),$$

$$\tau_{yy} = \mu(-2u_x + 4v_y)/3, \tag{7.3}$$

$$e_x = u\tau_{xx} + v\tau_{xy} + \mu Pr^{-1}(\gamma - 1)^{-1} \frac{\partial c^2}{\partial x},$$

$$e_y = u\tau_{xy} + v\tau_{yy} + \mu Pr^{-1}(\gamma - 1)^{-1} \frac{\partial c^2}{\partial y},$$

where γ is the ratio of specific heats. The dynamic viscosity is μ and typically consists of a constant plus a computed turbulent eddy viscosity. Re and Pr are the Reynolds number and Prandtl number.

A thin-layer approximation of the Navier–Stokes equation is made by resolving the viscous terms in a thin layer near the body[20]. Viscous terms in x, which is the direction along the solid body, are neglected, and terms in y are retained. Equation (7.1) thus simplifies to

$$\frac{\partial U}{\partial t} + \frac{\partial F(U)}{\partial x} + \frac{\partial G(U)}{\partial y} = \frac{\partial G_v(U, U_x, U_y)}{\partial y}. \tag{7.4}$$

In general, for complex configurations, one does not have sufficient computer power to resolve the full Navier–Stokes equation. For sufficiently high Reynolds number, the thin-layer Navier–Stokes equations prove to be a useful approximation in a variety of applications.

For steady-state application, a simple algorithm utilizing the TVD scheme for the Navier–Stokes equations is to difference the hyperbolic terms the same way as before, and then central difference the viscous term. The final algorithm is the same as Eqs. (5.7) except that the spatial central differencing of the viscous term is added to the right-hand side of (5.7). It will be shown later that this simple-minded algorithm produces a fairly good solution for the case of a RAE2822 airfoil calculation.

For time-accurate calculations, time accuracy is just as important as spatial accuracy for both the implicit and the explicit operator, and it is not apparent how to extend the current implicit method efficiently with the viscous term included. One way is to use a different time differencing, such as the Runge–Kutta method. A different approach is proposed here.

The unsteady compressible Navier–Stokes equation with turbulence models generally lead to an extremely "stiff" nonlinear system. In the numerical solution of such problems, it is sometimes more advantageous to treat the terms responsible for the severe time-step restriction implicitly and handle the remaining terms explicitly. The approach proposed here is similar to that of MacCormack[21].

The proposed method is to time split the equation into a hyperbolic part and a parabolic part. The hyperbolic part is solved with an explicit second-order TVD scheme via the locally-one-dimensional (LOD) time-splitting method of Strang type (a fractional-step approach as described in Ref. [4]). The parabolic part is solved with an ADI implicit method with central difference in space. For the thin-layer Navier–Stokes equations, there is only one viscous term to worry about. Therefore, the parabolic part is especially easy to solve with one simple matrix inversion.

Let L_x^h, L_y^h be the LOD split hyperbolic finite-difference operator in the x and y directions. Furthermore, let L_{yv} be the split parabolic operator in the y direction. Then a fractional step method for the thin-layer Navier–Stokes Eqs. (7.4) will look like

$$U_{j,k}^{n+2} = L_{yv}^h L_y^h L_x^{2h} L_y^h L_{yv}^h U_{j,k}^n, \tag{7.5}$$

where $h = \Delta t$, and $L_x^h U_{j,k}^n$, for example, is

$$L_x^h U_{j,k}^n = U_{j,k}^n - \lambda^x [\tilde{F}_{j+1/2,k} - \tilde{F}_{j-1/2,k}], \tag{7.6}$$

with $\tilde{F}_{j+1/2,k}$ defined in (4.4b) and $\sigma(z)$ in (3.7a). The operator L_{yv}, which solves for the viscous (parabolic) term G_v, can be any one of the stable implicit schemes designed specially for parabolic equations (see, for example, MacCormack[21] and Beam and Warming[17]). Details and numerical experiments with this approach for transient calculations will appear in a separate article. Here we only illustrate some results for the first approach for steady-state calculations.

8. NUMERICAL RESULTS FOR THE NAVIER–STOKES EQUATIONS

The viscous case considered here is the RAE2822 airfoil with $M_\infty = 0.73$, $\alpha = 2.79$ and the Reynolds number Re $= 6.5 \times 10^6$. The grid used is a 249×51 O grid [see Fig. 10(a)].

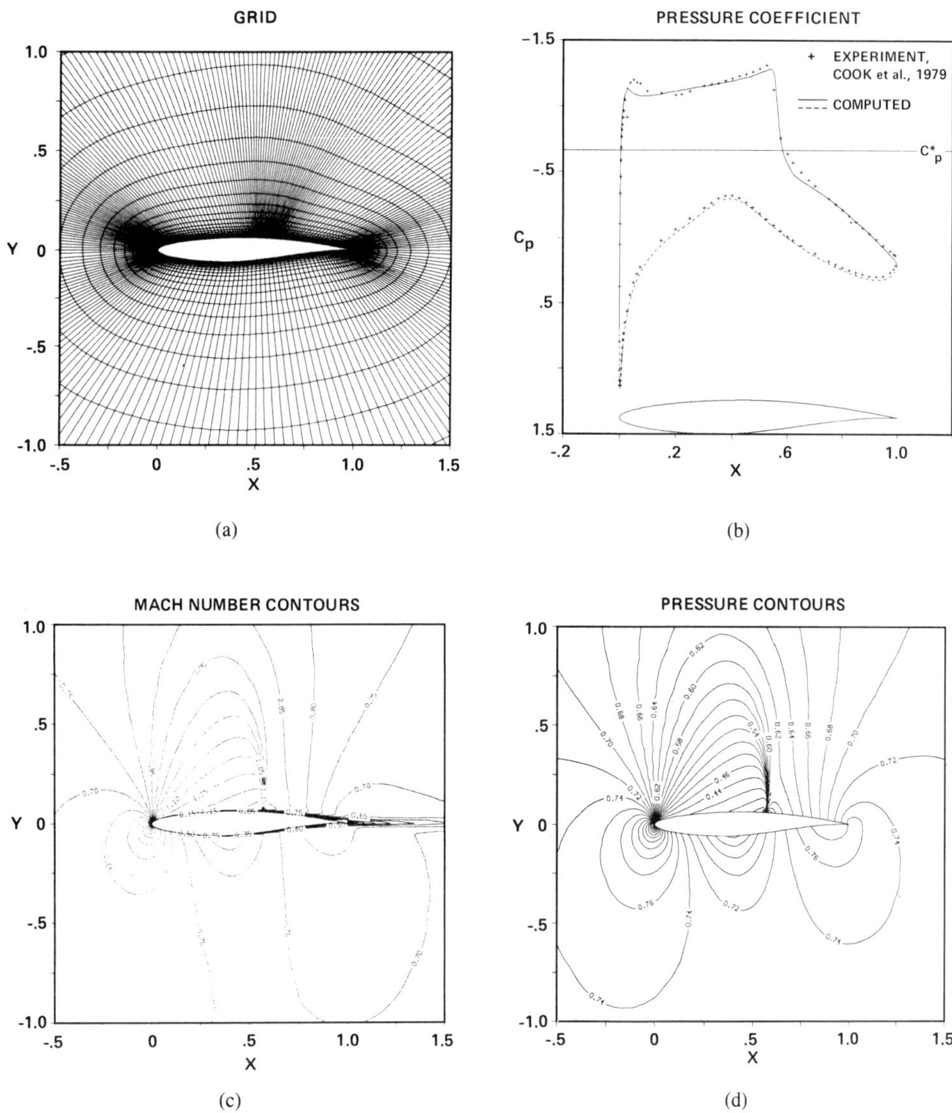

Fig. 10. The 249 × 51 "O" grid, pressure coefficient, Mach contours, and pressure contours for the RAE2822 airfoil with $M_\alpha = 0.73$, $\alpha = 2.79$, Re $= 6.5 \times 10^6$.

The thin-layer Navier–Stokes equations with the algebraic turbulence model of Baldwin and Lomax[20] are used and the transition is fixed at 3% of chord. Experimental data due to Cook *et al.*[22] are used for comparison. Figure 10(b) shows the pressure coefficients compared with experiment. The solid (upper-surface) and dashed (lower-surface) lines are the numerical solution for the present method. The $+$'s are the experimental data. The overall comparison with experiment is quite good. Figures 10(c,d) show the Mach contours and pressure contours. The L_2-norm residual of 10^{-7} can be reached in around 900 steps. For this case, it requires approximately 1 sec per time step. The extra CPU time required for the viscous case is mainly because of the additional computation due to the viscous term, the periodic boundary condition for the implicit operator, and the slightly bigger grid than in the inviscid case.

Again a space-varying Δt similar to that of the inviscid case is used. The calculations shown in Fig. 10 use the expression (6.1) for g. The same difference method for the convection terms as the inviscid airfoil cases is used. The effect of the various turbulence models and the use of smoother transition operators as studied by Mehta *et al.*[23] have not been tried. They have found, in certain problems, that their proposed modification can improve convergence rate and accuracy. This will be a subject of a future investigation.

CONCLUSION

Numerical experiments for the airfoil calculations show that the application of the proposed linearized second-order implicit TVD scheme generates good shock resolution for steady-state computations. The method is also applicable to viscous calculations. Very good agreement with experimental data was obtained for a case of the RAE2822 airfoil. The numerical dissipation is built in and has an automatic feedback mechanism. The current algorithm requires less storage and operation count than other TVD schemes[3]. The method is quite robust, fairly efficient, and can easily be implemented into an existing central difference code. In addition, the same formulation is also applicable to the three-dimensional Euler and Navier–Stokes equations.

Acknowledgment—The author wishes to thank Thomas Pulliam for his generous assistance on the airfoil geometries. The permission to use his well-documented computer code ''ARC2D'' as a base code for the development of a research code for the present study is gratefully acknowledged.

REFERENCES

1. A. Harten, A high resolution scheme for the computation of weak solutions of hyperbolic conservation laws. NYU Report, Oct., 1981; *J. Comp. Phys.* **49**, 357–393 (1983).
2. A. Harten, On a class of high resolution total-variation-stable finite-difference schemes. NYU Report, Oct., 1982; *SIAM J. Numer. Anal.* **21**, 1–23 (1984).
3. H. C. Yee, On the implementation of a class of upwind schemes for system of hyperbolic conservation law. NASA Technical Memorandum 86839 (Sept. 1985).
4. H. C. Yee, R. F. Warming and A. Harten, Implicit total variation diminishing (TVD) schemes for steady-state calculations. AIAA Paper No. 83-1902, Proc. of the AIAA 6th Computational Fluid Dynamics Conference, Danvers, Mass., July, 1983; also in *J. Comp. Phys.* **57**, 327–360 (1985).
5. T. H. Pulliam and J. T. Barton, Euler computations of AGARD working group 07 airfoil test cases. AIAA Paper No. 85-0018 (1985).
6. L. C. Huang, Psuedo-unsteady difference schemes for discontintuous solutions of steady-state, one-dimensional fluid dynamics problems. *J. Comp. Phys.* **42**, 195–211 (1981).
7. P. L. Roe, Approximate Riemann solvers, parameter vectors, and difference schemes. *J. Comp. Phys.* **43**, 357–372 (1981).
8. A. Harten and J. M. Hyman, A self-adjusting grid for the computation of weak solutions of hyperbolic conservation laws. *J. Comp. Phys.* **50**, 235–269 (1983).
9. P. K. Sweby, High resolution schemes using flux limiters for hyperbolic conservation laws. *SIAM J. Num. Analy.*, **21**, 995–1011 (1984).
10. P. L. Roe, Some contributions to the modelling of discontinuous flows, in *Proceedings of the AMS–SIAM Summer Seminar on Large-Scale Computation in Fluid Mechanics, June 27–July 8, 1983, Lectures in Applied Mathematics* Vol. 22. AMS (1985).
11. A. Harten and S. Osher, Uniformly high order nonoscillatory schemes. *UCLA Math. Report* (1985).
12. H. C. Yee, R. F. Warming and A. Harten, Application of TVD schemes for the Euler equations of gas dynamics, in *Proc. of the AMS—SIAM Summer Seminar on Large-Scale Computation in Fluid Mechanics, June 27–July 8, 1983, Lectures in Applied Mathematics* Vol. 22. AMS (1985).
13. J. B. Goodman and R. J. LeVeque, On the accuracy of stable schemes for 2D scalar conservation laws. NYU Report, New York (May 1983).
14. H. C. Yee and P. Kutler, Application of second-order-accurate total variation diminishing (TVD) schemes to the Euler equations in general geometries. NASA TM-85845 (August 1983).
15. H. C. Yee and A. Harten, Implicit TVD schemes for hyperbolic conservation laws in curvilinear coordinates, AIAA Paper No. 85-1513-CP, *Proc. of the AIAA 7th Computational Fluid Dynamics Conference, Cinn., Ohio, July* 15–17, 1985. AIAA (1985).
16. H. C. Yee, On Symmetric and Upwind TVD Schemes, *Proceedings of the 6th GAMM Conference on Numerical Methods in Fluid Mechanics*, (Sept. 25–27, 1985); also NASA Technical Memorandum 86842 (Sept. 1985).
17. R. Beam and R. F. Warming, An implicit finite-difference algorithm for hyperbolic systems in conservation law form. *J. Comp. Phys.* **22**, 87–110 (1976).
18. T. H. Pulliam and J. Steger, Recent improvements in efficiency, accuracy and convergence for implicit approximate factorization algorithms. AIAA Paper No. 85-0360 (1985).
19. A. Jameson *et al.*, Numerical solutions of the Euler equations by finite volume methods using Runge–Kutta time-stepping schemes. AIAA Paper No. 81-1259 (1981).
20. B. Baldwin and H. Lomax, Thin layer approximation and algebraic model for separated turbulent flows. AIAA Paper No. 78-257 (1978).
21. R. W. MacCormack, A rapid solver for hypervolic systems of equations, in *Proceedings of the Fifth International Conference on Numerical Methods in Fluid Dynamics, Lecture Notes in Physics* Vol. 59. Springer-Verlag (1976).
22. P. H. Cook *et al.*, Aerofoil RAE2822—Pressure distributions and boundary layer and wake measurements. AGARD-AR-138 (1979).
23. Mehta *et al.*, A Comparison of Interactive boundary layer and thin-layer Navier–Stokes procedures, in *Proceedings of the 3rd Symposium on Numerical and Physical Aspects of Aerodynamic Flows,''* Long Beach, Jan. 21–24, 1985.

Comp. & Maths. with Appls. Vol. 12A, Nos. 4/5, pp. 433–455, 1986
Printed in Great Britain.

0886–9553/86 $3.00 + .00
© 1986 Pergamon Press Ltd.

A RIEMANN PROBLEM IN GAS DYNAMICS
WITH BIFURCATION

D. Marchesin and P. J. Paes-Leme

Departamento de Matemática, Pontifícia Universidade Católica do Rio de Janeiro,
22.453 Rio de Janeiro, RJ, Brasil

Abstract—We construct the solution of the Riemann problem for 2×2 isothermal gas dynamics in a
duct with discontinuous diameter. Besides shocks and rarefactions, there are standing waves. The solution
exists globally. It is obtained as an asymptotic solution for an appropriate Cauchy problem with continuous
data. In certain cases bifurcation occurs and there are three solutions, one of which is unstable. This is
an example of a Riemann problem whose solution depends discontinuously on the initial data.

1. INTRODUCTION

In this paper, we construct the solution of the Riemann problem for the one-dimensional
isothermal equations of gas dynamics in a duct with discontinuous cross section. The flow is
modeled by a system of quasilinear hyperbolic partial differential equations of the form

$$\frac{\partial U}{\partial t} + \frac{\partial F(x, U)}{\partial x} = 0, \tag{1.1}$$

where U is the vector of conserved quantities and F is the flux vector. The novel feature in
this problem is the presence of standing waves in addition to shock and rarefaction waves.
These waves are defined by solving the ordinary differential equations

$$\frac{\partial F(x, U)}{\partial x} = 0. \tag{1.2}$$

When combining these waves to construct the solution of the Riemann problem bifurcation may
occur. In this case there are three solutions for the Riemann problem with the same initial data.
One of these solutions is unstable under perturbations. Among the other two, the correct solution
is chosen utilizing other physical considerations. Thus this is an example of a Riemann problem
whose solution depends discontinuously on the initial data.

One-dimensional nonhomogeneous systems of the form (1.1) arise from homogeneous
problems in higher dimensions after symmetries or simplifying hypotheses are taken into account.
Such reductions to one-dimensional problems are advantageous since solutions of the hyperbolic
system can then be constructed from its characteristics. These nonhomogeneous systems also
appear in the contexts of nonequilibrium flow, multiphase flow and chemically reacting flow.
Some results on global existence and uniqueness for the associated Cauchy problem may be
found in [1–3].

Of particular interest is the situation in which the flux function F has a sharp discontinuity
at $x = 0$ and is independent of x for $x < 0$ and for $x > 0$. This occurs when the diameter $a(x)$
of the duct jumps at $x = 0$ from a value a_L to another value a_R. In this case the stationary
waves become standing discontinuities[4–7]. Their interaction with the already known classical
waves, composed of shocks and rarefactions, leads to cases of nonexistence[8] and of bifurcations[3]. The solution of the Riemann problem for the initial data $U = U_L, a = a_L$ for $x < 0$
and $U = U_R, a = a_R$ for $x > 0$ is defined in the following way. First, taking into account the
appropriate entropy condition, we solve the Cauchy problem for the data $U = U_L, a = a_L$ for
$x < -\epsilon, U = U_R, a = a_R$ for $x > \epsilon$ with $U(x, t = 0)$ and $a(x)$ interpolated monotonously in
$-\epsilon \leq x \leq \epsilon$. Then we calculate the weak limit of this solution as ϵ goes to zero.

Problems with nonunique solutions such as the one considered in this work are also studied
in [1,6,9–13].

The solution of this Riemann problem may serve as the main building block of a stable numerical scheme that resolves sharp discontinuities well. For this purpose extentions of Glimm–Chorin's method[14,15] or Godunov's method[16] are necessary. A method closely related to such an extension is described in [17,18].

In Sec. 2 we present a construction of the solution of the Riemann problem for 2×2 isothermal gas dynamics in a duct with constant cross section that will serve as a basis for the general construction presented later. In Sec. 3 we study the stationary waves that arise when a duct with discontinuous cross section is considered. The full Riemann solution is then constructed in Secs. 4, 5, and 6. In Sec. 4 we show how to find the solution using the ideas of Sec. 2. We reduce the problem to the geometrical problem of finding the intersection of two curves in state space. In Sec. 5 we discuss the construction of all such curves. Finally, Sec. 6 is devoted to the construction of the solution in all cases. The main idea of the construction is to attach the standing wave to the waves originating to the left or right according to whether the particle velocity is positive or negative. The interpretation of this construction is obtained by following a particle path.

2. FLOW IN A DUCT WITH CONSTANT AREA

The equations for isothermal gas dynamics are

$$\rho_t + (\rho v)_x = 0,$$
$$(\rho v)_t + (\rho v^2)_x + c^2 \rho_x = 0, \tag{2.1}$$

where ρ is the density, v is the velocity and c is the (constant) speed of sound. These equations model the flow of gas at constant temperature, as well as flow of fluids for pressure ranges in which the speed of sound may be taken as constant.

We review briefly the solution of the initial-value problem consisting of (2.1) together with the discontinuous data:

$$(\rho(x, t = 0), v(x, t = 0)) = \begin{cases} (\rho_L, v_L) & \text{for } x < 0, \\ (\rho_R, v_R) & \text{for } x > 0. \end{cases} \tag{2.2}$$

Here ρ_L, v_L, ρ_R and v_R are constants (see [16], Sec. 2 or [20,21]).

The solution of this Riemann problem consists of a left wave (or 1-wave) and a right wave (or 2-wave). These waves are either rarefaction or shock waves. Because (2.1)–(2.2) are invariant under the scale transformation $x' = ax$, $t' = at$ $(a > 0)$, the Riemann solution is constant along the lines $x/t = \text{const}$.

2.1 Rarefaction waves

Rarefaction waves correspond to smooth sections of certain solutions of (2.1)–(2.2) for $t > 0$ that we now describe.

For smooth ρ and v system (2.1) is equivalent to

$$\rho_t + v\rho_x + \rho v_x = 0,$$
$$v_t + (c^2/\rho)\rho_x + vv_x = 0. \tag{2.3}$$

Multiplying the first equation in (2.3) by $\pm c/\rho$ and adding the second one, we get the system in characteristic form

$$s_t + (v + c)s_x = 0, \tag{2.4a}$$

$$r_t + (v - c)r_x = 0, \tag{2.4b}$$

where

$$s = (v + c \ln \rho)/2, \tag{2.5a}$$

$$r = (v - c \ln \rho)/2. \tag{2.5b}$$

The functions s and r are called the Riemann invariants. By definition a 1-rarefaction wave is a smooth scale-invariant solution on which s is constant. Indeed, ρ and v are constant along the lines $x/t = $ const and so is r from (2.5b). If we choose v on each line satisfying $x/t = v - c$ and ρ according to (2.5a), we obtain a 1-rarefaction wave: (2.4a) is satisfied trivially and (2.4b) because

$$0 = \frac{\mathrm{d}}{\mathrm{d}t} r(x = (v - c)t, t) = r_x(v - c) + r_t.$$

Similarly, a 2-rarefaction wave is a smooth scale-invariant solution on which r is constant.

A state $u = (c \ln \rho, v)$ is represented by a point in phase space. In this space we define the 1- and 2-rarefaction curves as follows:

$$R_1(u_L) = \{u_M / r(u_M) \geq r(u_L), s(u_M) = s(u_L)\}$$
$$= \{u_M / v_L - v_M = -z \quad \text{for } z = c \ln \rho_L - c \ln \rho_M \geq 0\}, \tag{2.6}$$

$$R_2(u_R) = \{u_M / r(u_M) = r(u_R), s(u_M) \leq s(u_R)\}$$
$$= \{u_M / v_M - v_R = -z \quad \text{for } z = c \ln \rho_R - c \ln \rho_M \geq 0\}. \tag{2.7}$$

Thus $R_1(u_L)$ is the set of states u_M which can be connected to the left state u_L by a 1-rarefaction and $R_2(u_R)$ is the set of states u_M which can be connected to the right state u_R by a 2-rarefaction. The inequality restrictions in the definitions above result from a geometric consideration which we explain for a 1-curve. On a left state, $x/t = v_L - c$ is less than $x/t = v_M - c$ at the state u_M to the right. This fact, together with the fact that s is constant, yields the inequality.

2.2 Shock waves

Typically nonlinear hyperbolic systems like (2.3) admit solutions with discontinuities (shocks). Therefore (2.3) has to be considered in the integral form[22]. This leads to a relationship between the speed of the shock σ and the states u_- and u_+ to the left and right of the shock. In our case this relationship, called the Rankine–Hugoniot jump condition, is

$$\sigma(u_-, u_+) = \frac{\rho_+ v_+ - \rho_- v_-}{\rho_+ - \rho_-} = \frac{\rho_+ v_+^2 + c^2 \rho_+ - \rho_- v_-^2 - c^2 \rho_-}{\rho_+ v_+ - \rho_- v_-}. \tag{2.8}$$

Given ρ_-, v_- and v_+ it is easy to verify that the second equality in (2.8) is satisfied by two values of ρ_+. We now define the 1- and 2-shock curves as follows:

$$S_1(u_L) = \{u_M / v_L - v_M = c(e^{-z/2c} - e^{z/2c})/2$$
$$\text{for } z = c \ln \rho_L - c \ln \rho_M \leq 0\}, \tag{2.9}$$

$$S_2(u_R) = \{u_M / v_M - v_R = c(e^{-z/2c} - e^{z/2c})/2$$
$$\text{for } z = c \ln \rho_R - c \ln \rho_M \leq 0\}. \tag{2.10}$$

Thus $S_1(u_L)$ is the set of states u_M which can be connected to the left state u_L by a 1-shock and $S_2(u_R)$ is the set of states u_M which can be connected to the right state u_R by a 2-shock. A straightforward computation shows the states u_L and u_M in (2.9) as well as u_M and u_R in (2.10) satisfy the Rankine–Hugoniot condition. The inequality restrictions in the definitions above result from entropy considerations, which in this case ensure that the characteristics enter the shock from both sides.

2.3 Construction of the Riemann solution

In order to solve the Riemann problem (2.1)–(2.2) we define the 1-M curve as the set of states u_M which can be connected to the left state u_L by a 1-wave. Similarly the 2-M curve is the set of states u_M which can be connected to the right state u_R by a 2-wave. In this case they

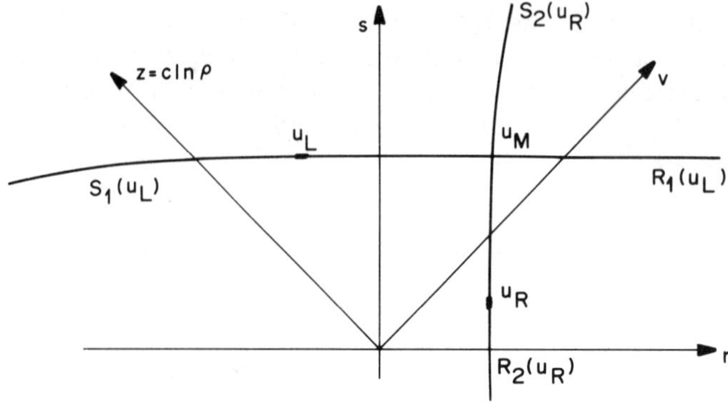

Fig. 1.

coincide with

$$T_1(u_L) = R_1(u_L) \cup S_1(u_L),$$

$$T_2(u_R) = R_2(u_R) \cup S_2(u_R).$$

(2.11)

The following result is true: all $T_1(T_2)$ curves are translates of one another. The T_2 curves are reflections of the T_1 curves with respect to the axis $v = 0$. Moreover, there exists a parametrization of $u = S_1(u_L)$ such that

$$0 < \frac{\mathrm{d}s(u(z))}{\mathrm{d}z} < \frac{\mathrm{d}r(u(z))}{\mathrm{d}z} \quad \text{and} \quad \frac{\mathrm{d}s(u(z))}{\mathrm{d}z} \bigg/ \frac{\mathrm{d}r(u(z))}{\mathrm{d}z}$$

is monotone decreasing from 1^- at $z = -\infty$ to 0^+ at $z = 0$. Furthermore, any T_1 curve always intercepts a T_2 curve precisely once.

To solve the Riemann problem for given u_L and u_R we find the intermediate state $u_M = T_1(u_L) \cap T_2(u_R)$ (see Fig. 1). The solution consists of the left state u_L, a 1-wave (shock or rarefaction), the intermediate state u_M, a 2-wave (shock or rarefaction) and the right state u_R. Examples are given in Figs. 2–5. The curves $\mathrm{d}x/\mathrm{d}t = v - c$ are the 1-characteristics and $\mathrm{d}x/\mathrm{d}t = v + c$ the 2-characteristics.

Note that the characteristics of one family are deflected when they pass through a region where the other family has a shock or a rarefaction.

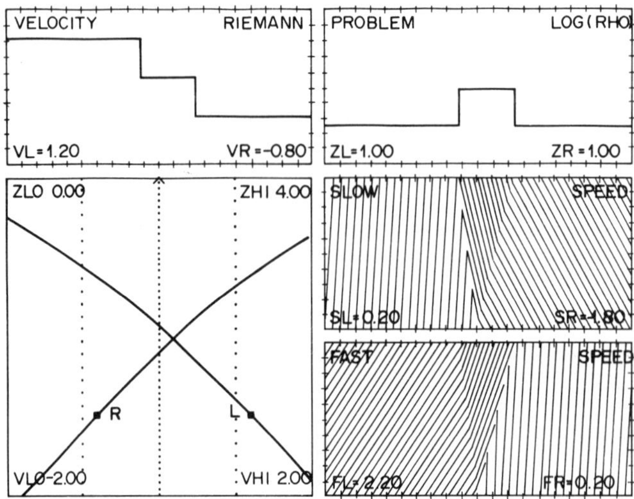

Fig. 2.

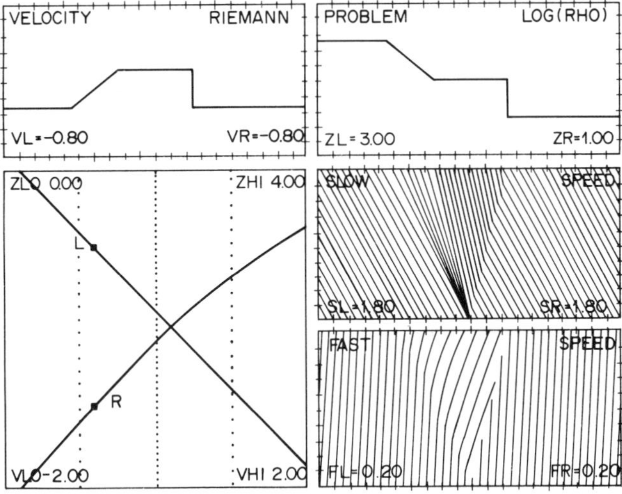

Fig. 3.

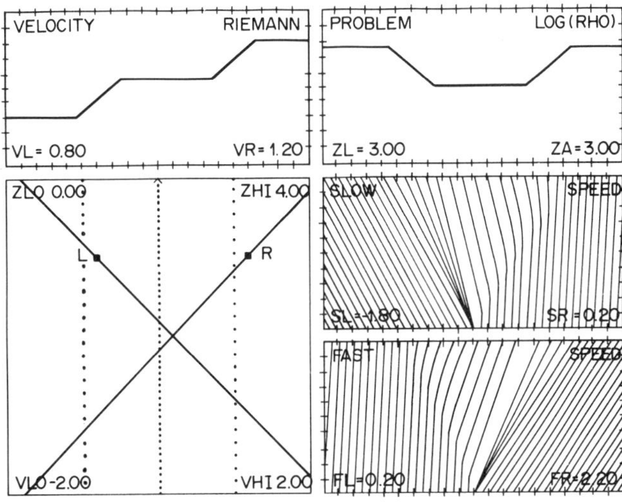

Fig. 4.

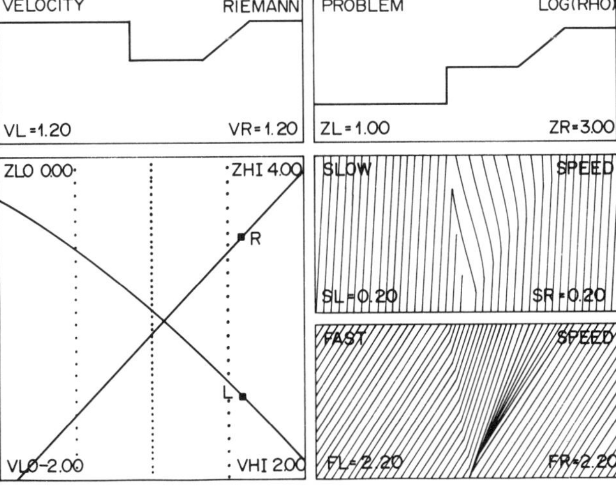

Fig. 5.

3. STATIONARY FLOW IN A DUCT WITH VARIABLE AREA: STATIONARY WAVES

The equations for isothermal gas dynamics in a duct with smooth area $a(x) > 0$ are

$$(a\rho)_t + (a\rho v)_x = 0,$$
$$(a\rho v)_t + (a\rho v^2)_x + ac^2\rho_x = 0. \tag{3.1}$$

For smooth flow (3.1) is equivalent to

$$(a\rho)_t + (a\rho v)_x = 0,$$
$$v_t + (v^2/2 + c^2 \ln \rho)_x = 0. \tag{3.2}$$

For stationary flow (3.2) becomes

$$a\rho v = \text{const}$$
$$v^2/2 + c^2 \ln \rho = \text{const}'. \tag{3.3}$$

Since $a > 0$ and $\rho > 0$, the sign of v is constant in (3.3). We introduce the change of variables

$$z = c \ln \rho,$$
$$b = \ln a. \tag{3.4}$$

Thus (3.3) may be rewritten as

$$z/c + \ln|v| + b = \text{const}'',$$
$$z + v^2/2c = \text{const}'''. \tag{3.5}$$

In this work we consider either expanding or contracting ducts. The area and the state at x_- are $a_- = a(x_-)$ and $u_- = (z_-, v_-)$; at x_+ they are a_+ and u_+. The area varies monotonically between x_- and x_+: we assume that $a'(x) \neq 0$ in this range.

A straightforward computation shows that for these ducts smooth stationary flow cannot become sonic, i.e.: $|v_-| < c$ if and only if $|v_+| < c$. [Differentiating (3.5) we obtain $db/dx = 0$ at the sonic point.]

The above considerations show that two states u_- and u_+ connected by smooth stationary flow in a monotone duct satisfy (i) $v_- = 0$ if and only if $v_+ = 0$; (ii) v_-, v_+ have the same sign; and (iii) either $|v_-|$, $|v_+|$ are both greater or both smaller than c.

We are ready to introduce stationary curves. Fix u_-, $b_- = \ln a_-$. We define $J(u_-, b_-)$ as the set of u_+ states that can be connected to u_- by a smooth stationary flow, with b monotone between b_- and b_+. From (3.5) we have

$$(z_+ - z_-)/c + \ln(v_+/v_-) + (b_+ - b_-) = 0, \tag{3.6a}$$

$$(z_+ - z_-) + (v_+^2 - v_-^2)/2c = 0. \tag{3.6b}$$

An easy computation using (3.6) shows that J is a curve which can be parametrized by b_+ in the range (b_-°, ∞) where

$$b_-^\circ = b_- + \ln(|v_-|/c) + (1 - v_-^2/c^2)/2. \tag{3.7}$$

A point in this curve is denoted as $u_+ = J(u_-, b_-; b_+)$. As b_+ tends to b_-°, $|v_+|$ tends to c. As b_+ tends to infinity, $|v_+|$ tends to zero if $|v_-| < c$ and it tends to infinity if $|v_-| > c$.

The construction of J is facilitated by observing Figs. 6 and 7, which display curves satisfying (3.6a) and (3.6b) respectively.

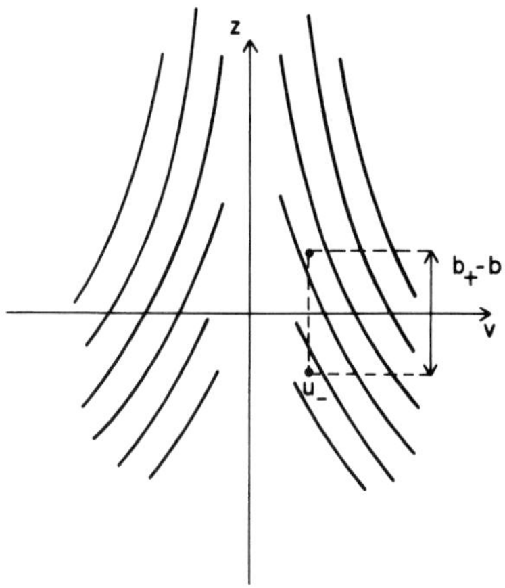

Fig. 6.

The members of each family are obtained by vertical translations of one representative. The second family consists of parabolas. The two families are tangent to each other at $v = \pm c$. In fact, from (3.5), there the slopes are $dz/dv = \mp 1$.

The stationary curves J are displayed in Fig. 8. We observe that given u_-, a connecting state u_+ can always be found for any $b_+ > b_-$, while this is not always true for $b_+ < b_-$; that is, smooth stationary flow is always possible for expanding ducts. This is true for contracting ducts provided the change in area is not too large.

Our discussion up to this point is valid irrespective of how close x_- and x_+ are. It is useful in many applications to consider the case where the area of the duct jumps from a_- to a_+. In this case the flow is obtained as a limit as $x_+ - x_-$ tends to zero of the solution corresponding to a smooth monotone interpolation of $a(x)$. Our considerations show that a *standing wave* is generated in the flow, as indicated in Fig. 9.

4. FLOW IN A DUCT WITH DISCONTINUOUS AREA

In this section we construct the solution for the initial-value problem consisting of Eq. (3.1) with initial data:

$$(\rho(x, t = 0), v(x, t = 0)) = \begin{cases} (\rho_L, v_L) & \text{for } x < 0, \\ (\rho_R, v_R) & \text{for } x > 0. \end{cases} \tag{4.1a}$$

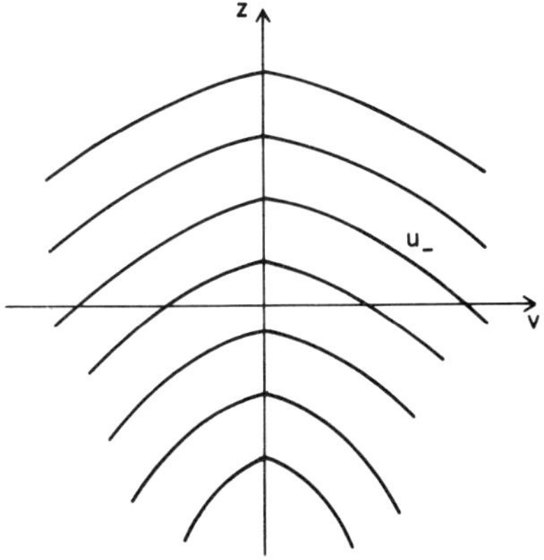

Fig. 7.

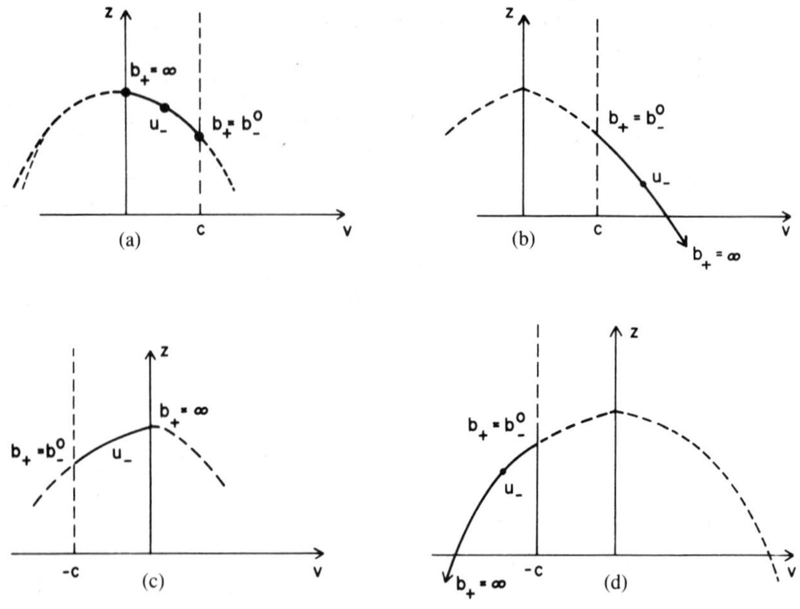

Fig. 8.

We assume that the duct has area

$$a(x) = \begin{cases} a_L & \text{for } x < 0, \\ a_R & \text{for } x > 0. \end{cases} \tag{4.1b}$$

As we will see, the solution consists of a sequence of 1- and 2-shocks and rarefactions, as well as one standing wave. As before, 1-waves are localized to the left of 2-waves in physical space. They are separated by a state $u_M = (z_M, v_M)$. We introduce a few definitions.

A 1-wave curve $W_1(u_L, u_M)$ is a curve in phase space $[(z, v)$-plane] starting at u_L and ending at u_M. It consists of a continuous succession of components with increasing speeds. The components, not necessarily all present, are 1-shock curves S_1, 1-rarefaction curves R_1 and a standing wave J.

A 2-wave curve $W_2(u_R, u_M)$ is defined analogously; it starts at u_R and ends at u_M. Its components are arranged with decreasing speeds. In physical space the u_L state is located at the left of u_M which is in turn at the left of u_R.

A (full) wave curve is a 1-wave curve from u_L to u_M followed by a 2-wave curve from u_M to u_R. The wave curve joining u_L to u_R is a precise representation of the Riemann problem (3.1), (4.1). Our algorithm for finding the Riemann solution has two steps. The first, which contains most of the difficulty, consists in finding u_M for given u_L and u_R. The second consists in constructing the 1- and 2- corresponding wave curves.

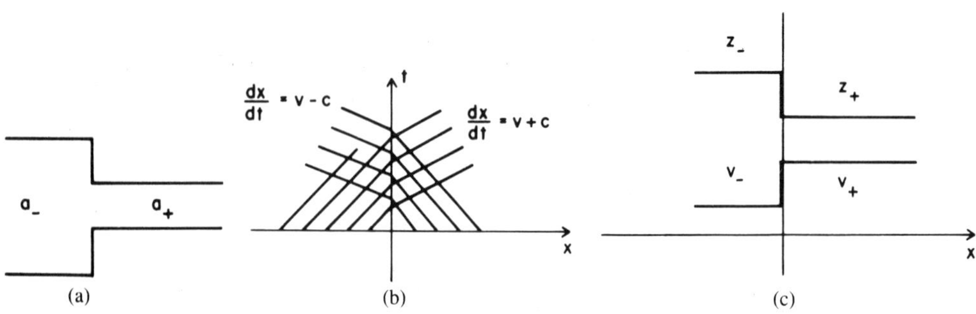

Fig. 9. (a) Duct; (b) characteristics; (c) solution.

To facilitate our construction, we introduce the notion of 1- and 2-M curves:

$$C_1(u_L) = \{u_M / u_M \text{ is connected to } u_L \text{ by a 1-wave curve}),$$

$$C_2(u_R) = \{u_M / u_M \text{ is connected to } u_R \text{ by a 2-wave curve}\}. \qquad (4.2)$$

Thus, to determine u_M for given u_L and u_R, we construct $C_1(u_L)$ and $C_2(u_R)$ and find their intersection $\{u_M\} = C_1(u_L) \cap C_2(u_R)$. As we will prove, this intersection always exists.

The M curves (4.2) and (2.11) are different, because of the presence of a standing wave. There is precisely one standing wave in a full-wave curve from u_L to u_R; it is located either in the 1-M curve or the 2-M curve. To locate the standing wave we use the following rule, which is also appropriate for nonisentropic gas dynamics: if $v_M > 0$ the standing wave is in the 1-wave curve, and if $v_M < 0$ it is in the 2-wave curve. (As we will see the standing wave vanishes for $v_M = 0$.)

There are six types of 1-M curves $C_1(u_L)$:

I. $v_L \leq c;\ a_L \geq a_R$
II. $v_L \leq c;\ a_L < a_R$
III. $v_L > c;\ a_L \leq a_R$
IV. $v_L > c;\ a_L \gtrsim a_R$, with bifurcation
V. $v_L > c;\ a_L \gg a_R$
VI. $v_L > c;\ a_L \gg a_R$, with bifurcation.

Similarly, there are six corresponding types of 2-M curves obtained from the 1-M curves through reflection with respect to the axis $v = 0$:

I. $v_R \geq -c;\ a_R \geq a_L$
II. $v_R \geq -c;\ a_R < a_L$
III. $v_R < -c;\ a_R \leq a_L$
IV. $v_R < -c;\ a_R \gtrsim a_L$, with bifurcation
V. $v_R < -c;\ a_R \gg a_L$
VI. $v_R < -c;\ a_R \gg a_L$, with bifurcation.

The next section is devoted to the construction of the 1-M curves, where the following notation will be used.

Given a state u_- with $v_- \geq c$, we define the zero speed shock map S_1^o by $u_+ = S_1^o(u_-)$, where $u_+ \in S_1$ and $\sigma(u_-, u_+) = 0$. Using (2.8) we see that

$$v_- v_+ = c^2. \qquad (4.3)$$

We also define the sets

$$S_1^-(u_-) = \{u_+ | u_+ \in S_1(u_-) \text{ and } \sigma(u_-, u_+) < 0\},$$

$$S_1^+(u_-) = \{u_+ | u_+ \in S_1(u_-) \text{ and } \sigma(u_-, u_+) > 0\}.$$

These sets are indicated in Fig. 10. For $v_- \leq -c$ we define in a similar way $S_2^o,\ S_2^+,\ S_2^-$.

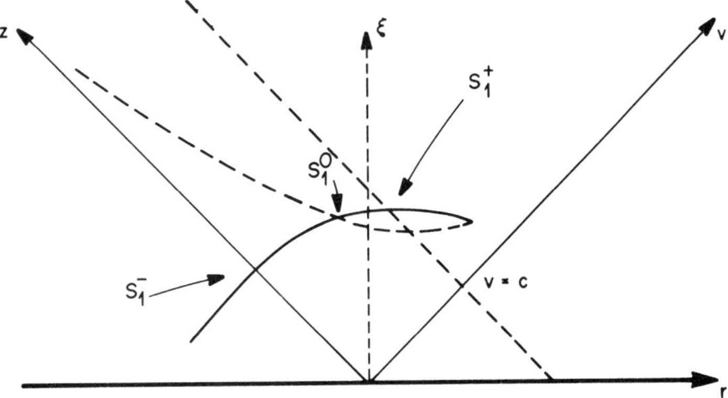

Fig. 10.

For given b_-, b_+, the map $u \to J(u, b_-; b_+)$ is called the standing-wave map. One should keep in mind that for $b_- > b_+$, a state is mapped closer to the sonic line with the same sign of v; the opposite is true for $b_- < b_+$.

5. THE BASIC CASES

In this section we construct the six types of 1-M curves. We note they are invariant under translations parallel to the z axis.

5.1 Case I: $v_L \leq c$; $b_L \geq b_R$

In this case the curve $C_1(u_L)$ consists of three parts: Q_1, Q_2 and Q_3. The first is the part of the curve $T_1(u_L)$ [see Eq. (2.11)] which corresponds to negative v, i.e.

$$Q_1(u_L) = \{u = (v, z) | u \in T_1(u_L) \text{ with } v \leq 0\}.$$

The second is part of the image of $T_1(u_L)$ under the standing-wave map. The part in question has v between 0 and c. To construct this part, we need the points $u_c \equiv (c, z_c)$ and $\hat{u}_c \in T_1(u_L)$ satisfying $u_c = J(\hat{u}_c, b_L; b_R)$ (see Fig. 11). Thus

$$Q_2(u_L) = \{u | u = J(u_-, b_L; b_R) \text{ with } u_- \in T_1(u_L),\, 0 < v_- \leq \hat{v}_c\}.$$

The third part is

$$Q_3(u_L) = \{u | u \in T_1(u_c) \text{ with } v > c\}.$$

A simple computation using (2.6), (2.9) and (3.6) shows that at u_c, the curves Q_2 and Q_3 have the same slope ($dz/dv = -1$). Similarly, the slope of Q_2 at the axis $v = 0$ lies between -1 and 0. It is crucial for our construction of the Riemann solution that the curve $C_1(u_L)$ defines z as a monotone decreasing function of v.

When solving the Riemann problem one finds $\{u_M\} = C_1(u_L) \cap C_2(u_R)$. If the point u_M belongs to $Q_1(u_L)$, the 1-wave curve is a shock in Cases I(A) and I(B), and either a 1-shock or a 1-rarefaction in Case I(C), according to whether $v_M < v_L$ or $v_M > v_L$ [see Figs. 12(a), 13(a)].

If u_M belongs to $Q_2(u_L)$, the 1-wave curve is either a 1-shock or a 1-rarefaction from u_L to $\hat{u}_M$ followed by a standing wave from $\hat{u}_M$ to u_M. [Here $\hat{u}_M = J(u_M, b_R; b_L)$ satisfies $u_M = J(\hat{u}_M, b_L; b_R)$.] In Case I(A) we have a 1-shock and in I(C) a 1-rarefaction. In Case I(B) it is a 1-shock for $v_M < \bar{v}_L$ and a 1-rarefaction for $v_M > \bar{v}_L$ [see Figs. 12(b), 13(b)].

If u_M belongs to $Q_3(u_L)$ the 1-wave curve consists of a 1-shock or a 1-rarefaction from u_L to $\hat{u}_c$, followed by a standing wave from $\hat{u}_c$ to u_c, and finally a 1-rarefaction wave from u_c to u_M. We have a 1-shock for $v_L > \hat{v}_c$ and a 1-rarefaction for $v_L < \hat{v}_c$ [Figs. 12(c), 13(c)].

In Figs. 12 and 13 we show the 1-characteristics for typical solutions in the absence of the 2-wave curve, i.e. when $u_R = u_M$.

5.2 Case II: $v_L \leq c$; $b_L < b_R$

In this case the curve $C_1(u_L)$ consists of four parts. The first two are analogous to Case I:

$$Q_1(u_L) = \{u | u \in T_1(u_L) \text{ with } v \leq 0\},$$

$$Q_2(u_L) = \{u | u = J(u_-, b_L; b_R) \text{ with } u_- \in T_1(u_L),\, 0 < v_- \leq c,\, v < c\}.$$

The shape of Q_2 in this case is different from the one in Case I because b_L and b_R have opposite relationships in these two cases. This curve ends at $\hat{u}_c = J(u_c, b_L; b_R)$ with $\hat{v}_c < c$. Here $u_c = T_1(u_L) \cap \{u | v = c\}$ (see Fig. 15).

The third part $Q_3(u_L)$ consists of states which are obtained from a succession of three stationary waves. This construction is obtained by introducing between a_L and a_R, in the duct, an intermediate diameter a where a zero speed shock is present [Fig. 14(a)].

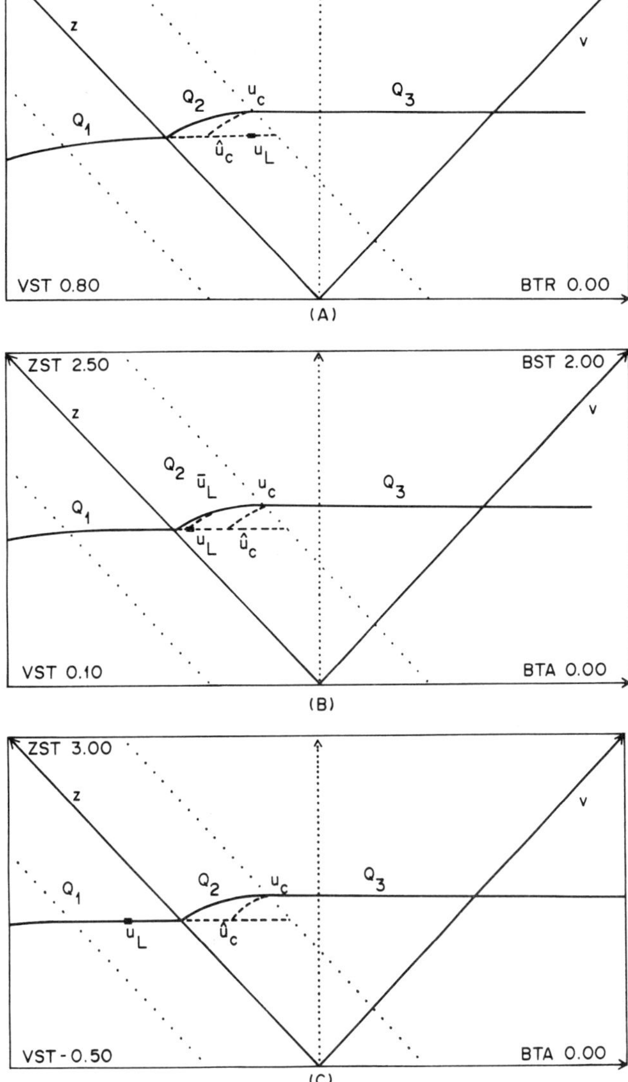

Fig. 11. 1-M curves for Case I. (A): $v_L \geq \hat{v}_c$; (B) $0 \leq v_L \leq \hat{v}_c$, $\bar{u}_L = J(u_L, b_L; b_R)$; (C): $v_L < 0$.

Thus $Q_3(u_L)$ is parametrized by b with $b_L \leq b \leq b_R$:

$$Q_3(u_L) = \{u | u = J(u_+, b; b_R) \text{ where } u_+ = S_1^0(u_-),$$
$$u_- = J(u_c, b_L; b) \text{ (with } v_- > c), b_L < b < b_R\}.$$

If $u_M \in Q_3(u_L)$, the 1-wave curve consists of a piece of $T_1(u_L)$, a standing wave, a zero speed shock and another standing wave, as in Fig. 14(b). Of course, in a Riemann problem, region 2 in Fig. 14 has zero width, so that the two standing waves and the zero speed shock coalesce. As shown in Fig. 15 the curve $Q_3(u_L)$ starts at $\hat{u}_c$ and ends at $\bar{\bar{u}}_c$. Here $\bar{\bar{u}}_c = S_1^0(\bar{u}_c)$, where $\bar{u}_c = J(u_c, b_L; b_R)$ with $\bar{v}_c > c$.

The last part of $C_1(u_L)$ is

$$Q_4(u_L) = \{u | u \in T_1(\bar{u}_c) \text{ with } v > \bar{\bar{v}}_c\}.$$

It consists of a 1-shock curve between $\bar{\bar{u}}_c$ and $\bar{u}_c$ with positive speed, and a 1-rarefaction curve from $\bar{u}_c$.

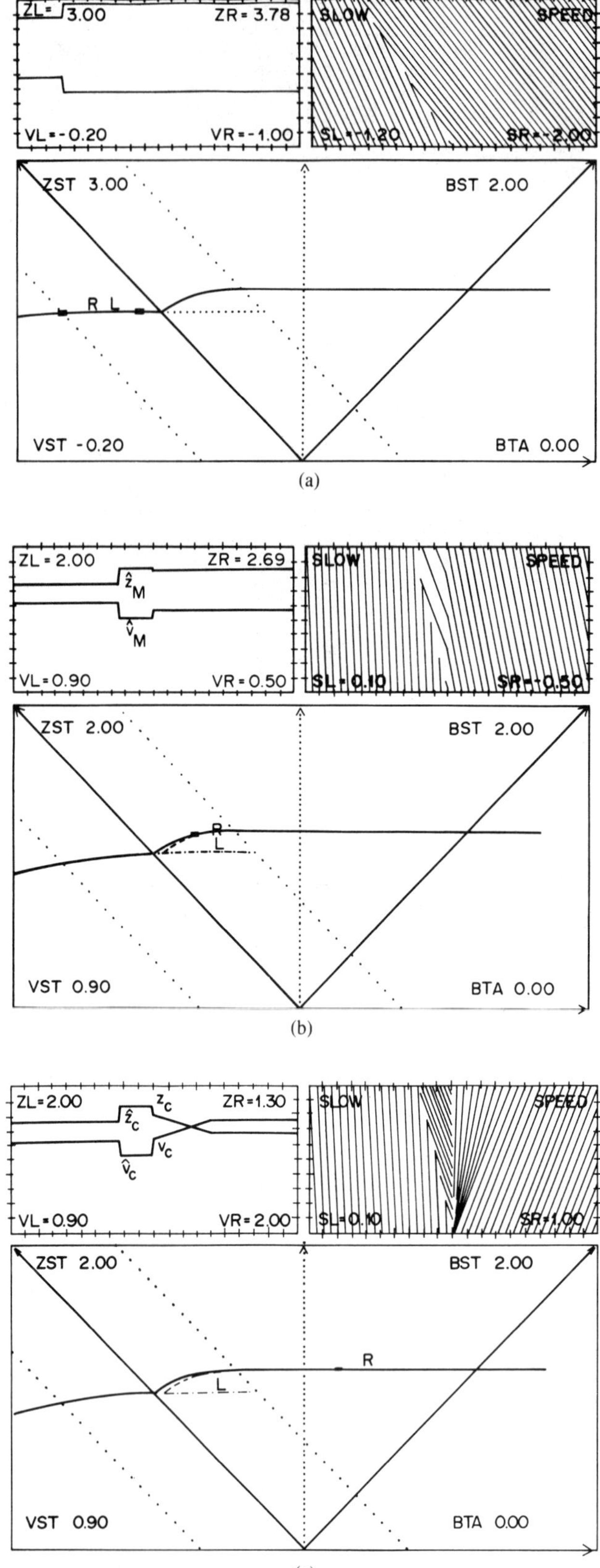

Fig. 12. (a) $u_M \in Q_1(u_L)$: Cases I(A), I(B) and I(C) for $v_M < v_L$. (b) $u_M \in Q_2(u_L)$: Cases I(A) and I(B) for $v_M < \overline{v}_L$. (c) $u_M \in Q_3(u_L)$: Case I(A).

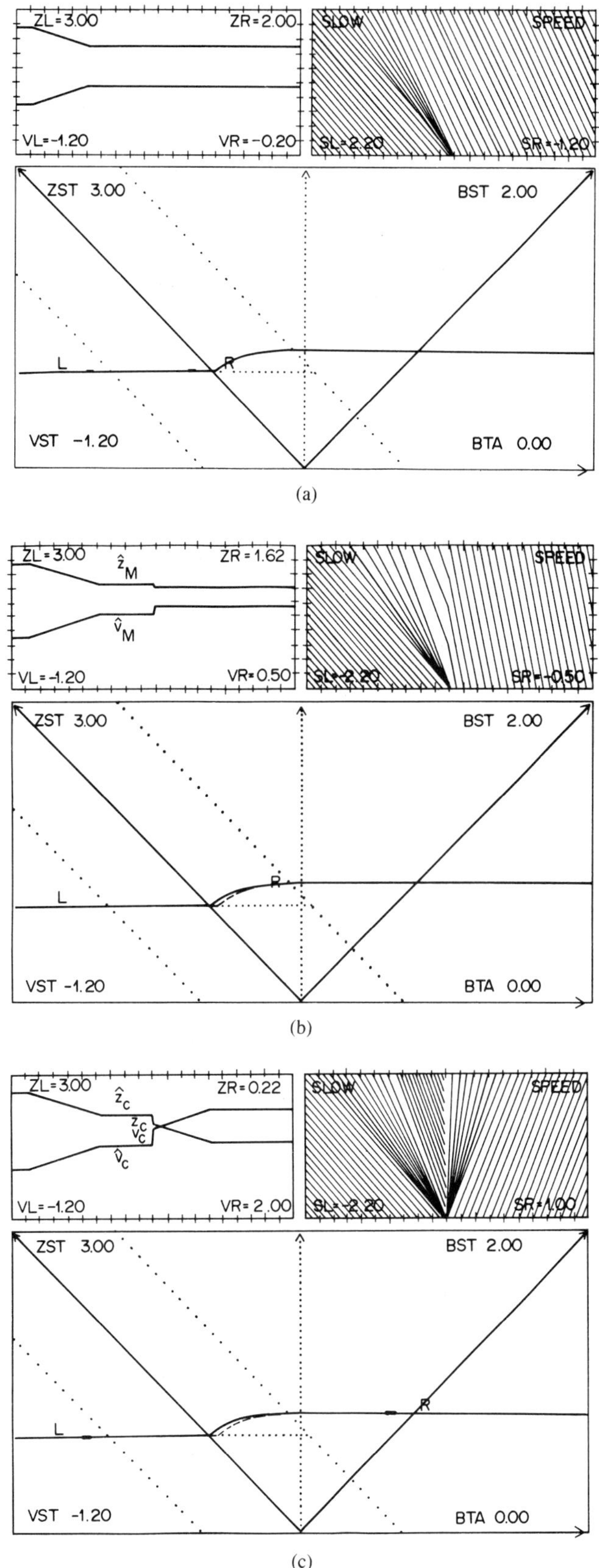

Fig. 13. (a) $u_M \in Q_1(u_L)$: Case I(C) for $v_L < v_M$. (b) $u_M \in Q_2(u_L)$: Case I(B) for $\bar{v}_L < v_M$ and Case I(C). (c) $u_M \in Q_3(u_L)$: Cases I(B) and I(C).

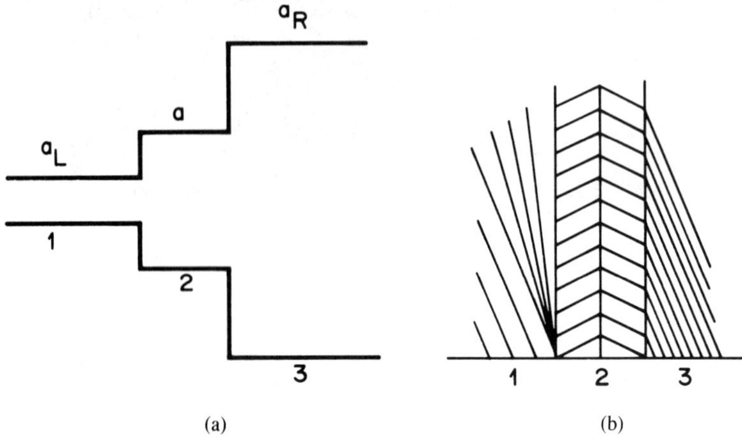

Fig. 14.

As seen in Fig. 15 the curve $C_1(u_L)$ defines z as a monotone-decreasing function of v.

The 1-wave curves when u_M belongs to $Q_1(u_L)$ or $Q_2(u_L)$ are similar to Case I. If u_M belongs to $Q_3(u_L)$ the 1-wave curve is a 1-rarefaction from u_L to u_c, followed by a standing wave from u_c to u_-, a zero speed shock from u_- to u_+ and a standing wave from u_+ to u_M [Fig. 16(a)].

If u_M belongs to $Q_4(u_L)$ the 1-wave curve is a 1-rarefaction from u_L to u_c, a standing wave from u_c to $\bar{u}_c$ followed by either a 1-shock or 1-rarefaction according to whether $v_M < \bar{v}_c$ or $v_M > \bar{v}_c$ [Figs. 16(b,c)].

5.3 *Case III*: $v_L > c$; $b_L \leq b_R$

In this case the curve $C_1(u_L)$ consists of four parts very similar to Case II (see Fig. 17):

$$Q_1(u_L) = \{u \,|\, u \in T_1(u_L) \text{ with } v < 0\},$$

$$Q_2(u_L) = \{u \,|\, u = J(u_-, b_L; b_R) \text{ with } u_- \in S_1^-(u_L), 0 < v_-\}.$$

The curve $Q_2(u_L)$ ends at $\hat{\bar{u}}_L = J(\bar{u}_L, b_L; b_R)$ where $\bar{u}_L = S_1^\circ(u_L)$:

$$Q_3(u_L) = \{u/u = J(u_+, b; b_R) \text{ with } u_+ = S_1^\circ(u_-)\text{where } u_- = J(u_L, b_L, b) \text{ and } b_L \leq b \leq b_R\}.$$

The curve $Q_4(u_L)$ ends at $\tilde{\bar{u}}_L = S_1^\circ(\bar{u}_L)$, where $\bar{u}_L = J(u_L, b_L; b_R)$. Finally

$$Q_4(u_L) = \{u \,|\, u \in T_1(\bar{u}_L) \text{ with } v > \tilde{\bar{v}}_L\}.$$

Figure 12(a) displays the 1-characteristics for $u_M \in Q_1(u_L)$. The other cases are displayed in Fig. 18.

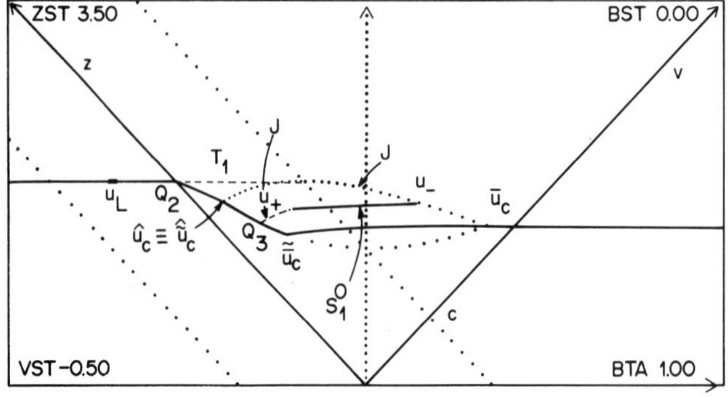

Fig. 15.

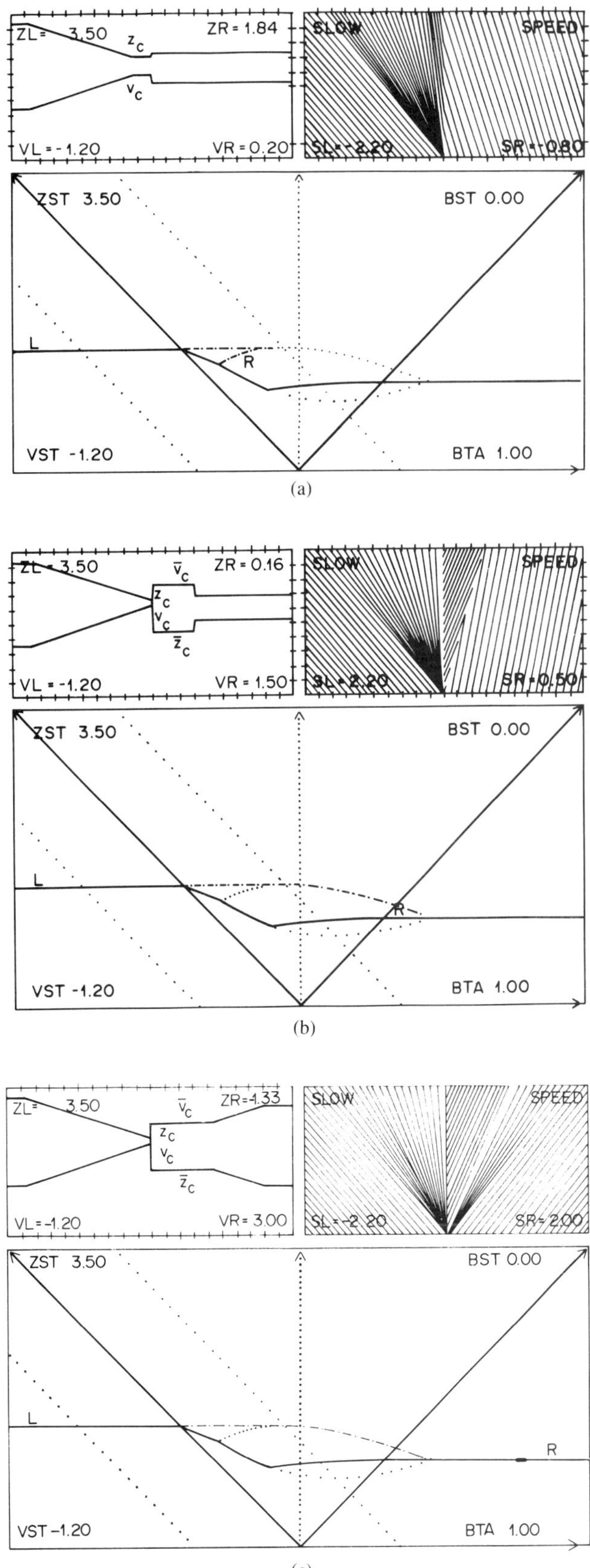

Fig. 16. (a) $u_M \in Q_3(u_L)$: Case II. (b) $u_M \in Q_4(u_L)$: Case II for $v_M < \bar{v}_c$. (c) $u_M \in Q_4(u_L)$: Case II for $v_M > \bar{v}_c$.

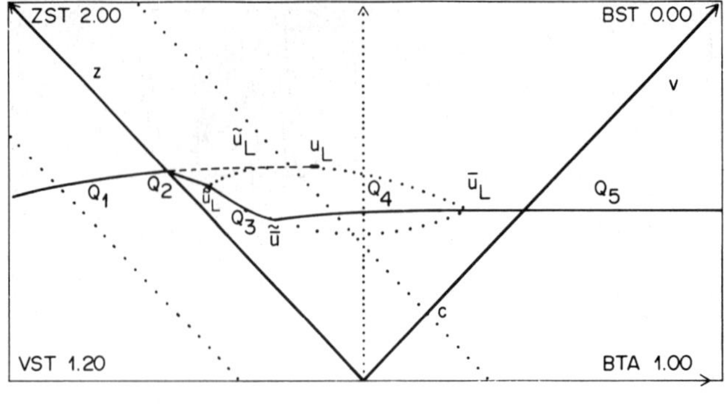

Fig. 17.

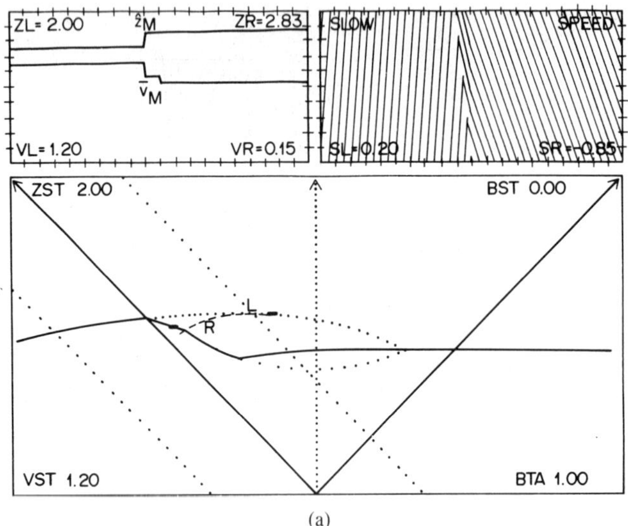

(a)

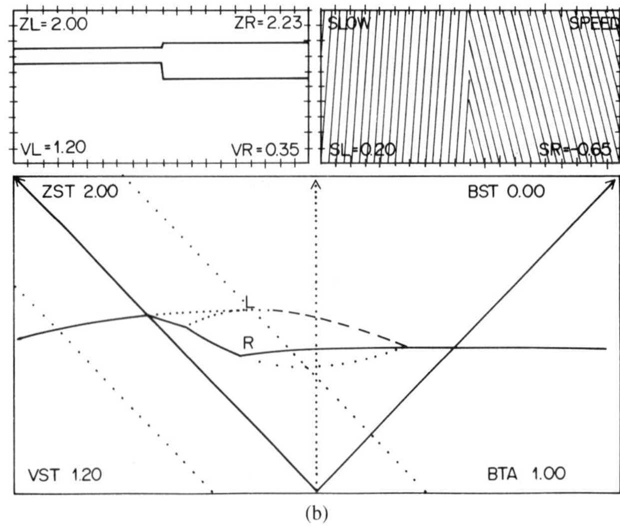

(b)

Fig. 18. (a) $u_M \in Q_2(u_L)$: Case III. (b) $u_M \in Q_3(u_L)$: Case III. (c) $u_M \in Q_4(u_L)$: Case III for $v_M < \bar{v}_L$. (d) $u_M \in$
$Q_4(u_L)$: Case III for $v_M > \bar{v}_L$.

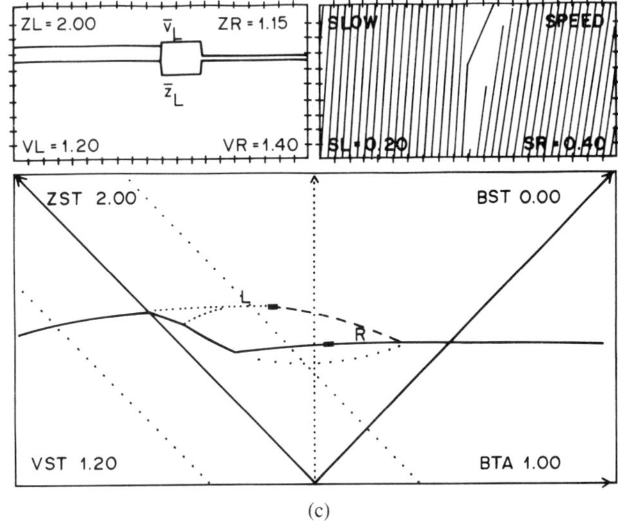

(c)

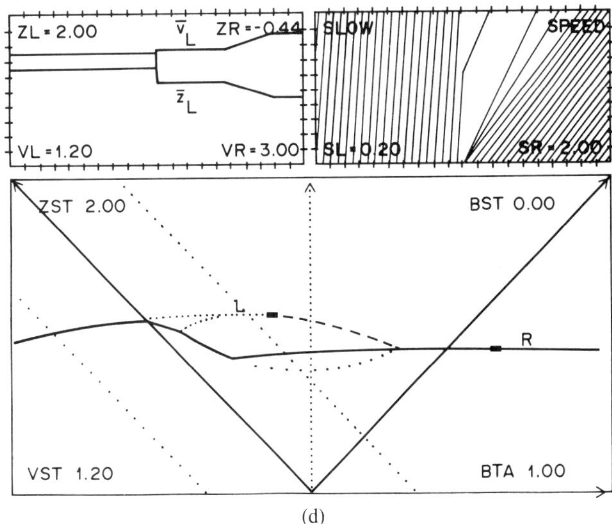

(d)

Fig. 18. *(Continued)*

5.4 *Case IV:* $v_L > c$; $b_L \geq b_R$, *with bifurcation*

This case occurs when both $J(u_L, b_L; b_R)$ and $J(S_1^\circ(u_L), b_L; b_R)$ exist. The curve $C_1(u_L)$ consists of four parts with the same definitions, as in Case III. Because $b_L \geq b_R$, the shape of this curve is different (see Fig. 19). We consider $C_1(u_L)$ as consisting of three branches. The first, $Q_1(u_L) \cup Q_2(u_L)$, ends at $\hat{\bar{u}}_L$. The second, $Q_3(u_L)$, begins at $\tilde{\bar{u}}_L$ and ends at $\hat{\bar{u}}_L$. The third, $Q_4(u_L)$, starts at $\tilde{\bar{u}}_L$. A simple computation using the definiton of $Q_3(u_L)$ and Eqs. (3.6), (4.3) and (2.9) shows that

$$Q_3(u_L) = \{(z, v) | z = z_L + \ln(v_L/v) - (b_R - b_L)\}.$$

Each of the branches define z as a monotone decreasing function of v.

The 1-wave curves and characteristics are similar to Case III.

5.5 *Case V:* $v_L > c$; $b_L > b_R$ *with no bifurcation*

This case occurs when $J(u_L, b_L; b_R)$ fails to exist. Using formulae (3.6) and (4.3) one can show that this implies that $J(S_1^\circ(u_L), b_L; b_R)$ also fails to exist. The curve $C_1(u_L)$ consists of

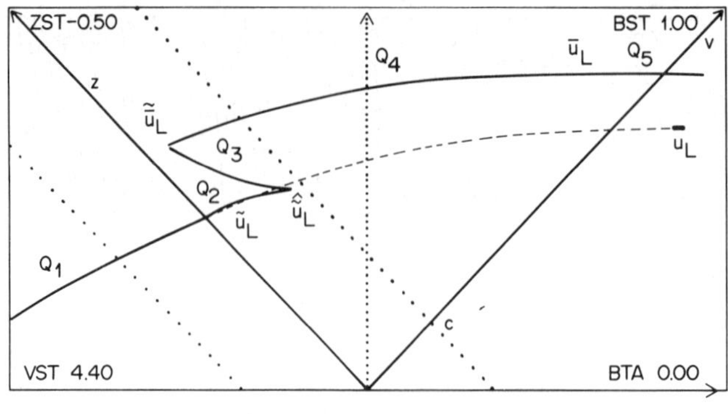

Fig. 19.

three parts:

$$Q_1(u_L) = \{u \mid u \in T_1(u_L) \text{ with } v \leq 0\},$$

$$Q_2(u_L) = \{u \mid u = J(u_-, b_L; b_R) \text{ with } u_- = S_1^-(u_L),\ 0 < v_-\},$$

$$Q_3(u_L) = \{u \mid u \in T_1(u_L) \text{ with } v > c\}$$

(see Fig. 20).

The 1-wave curves and characteristics are similar to Case I(A).

5.6 *Case VI*: $v_L > c$; $b_L > b_R$ *with bifurcation*

This case occurs when $J(u_L, b_L; b_R)$ exists but $J(S_1^o(u_L), b_L; b_R)$ does not. The curve $C_1(u_L)$ consists of two disconnected components. The first component is precisely the curve $C_1(u_L)$ of Case V. The second component consists of three parts, Q_4, Q_5 and Q_6:

$$Q_6(u_L) = \{u \mid u \in T_1(\overline{u}_L) \text{ with } v \geq \tilde{v}_L\}$$

[here $\overline{u}_L = J(u_L, b_L; b_R)$ and $\tilde{\overline{u}}_L = S_1^o(\overline{u}_L)$];

$$Q_5(u_L) = \{u \mid u = J(u_+, b; b_R) \text{ with } u_+ = S_1^o(u_-),\ u_- = J(u_L, b_L; b),\ b_R < b < b_L,\ v \leq c\}.$$

Let us denote by $\hat{\tilde{u}}_l$ the sonic point in Q_5. Then

$$Q_4(u_L) = \{u \mid u = T_1(\hat{\tilde{u}}_l) \text{ with } v > c\}$$

(see Fig. 21).

The curve $C_1(u_L)$ consists of three branches, $Q_1(u_L) \cup Q_2(u_L) \cup Q_3(u_L)$, $Q_5(u_L) \cup Q_4(u_L)$ and $Q_6(u_L)$, each of which defines z as a monotone decreasing function of v.

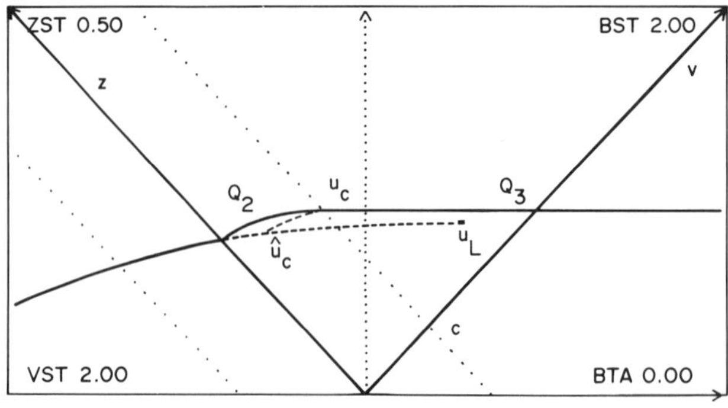

Fig. 20.

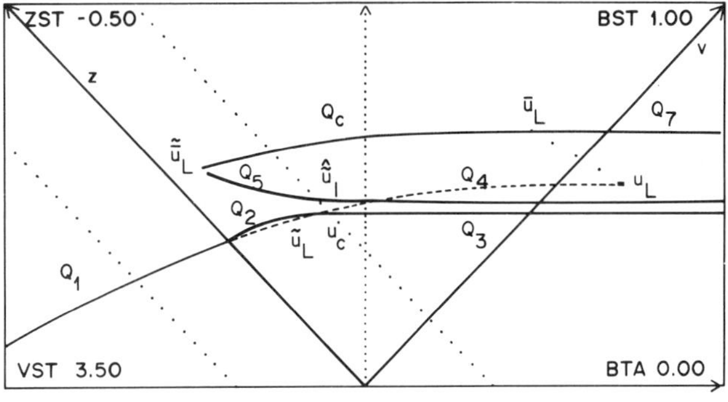

Fig. 21.

6. THE SOLUTION OF THE RIEMANN PROBLEM

Given two states u_L and u_R we construct $C_1(u_L)$ and $C_2(u_R)$. Their intersection always exists because each has a connected component whose asymptotic behavior at infinity is the same as if $b_L = b_R$ (Figs. 2–5). In the absence of bifurcations the monotonicity of the curves as a function of v guarantees that the intersection is unique. In the presence of bifurcations our construction insures that there are one, two or three solutions. This is so because the bifurcation is present in only one of the two curves.

Some typical solutions are in Figs. 22–29. In each figure, the top pictures represent the solution at time 1 for $-1 \le x \le 1$. The left picture contains the profile of v while the right contains the profile of z. For convenience we take $c = 1$. The large square picture shows the curves $C_1(u_L)$, $C_2(u_R)$ and their intersection in phase space. The dotted axes are r horizontally and s vertically [see Eqs. (2.5).] The z axis goes from bottom right to top left. Parallel to it are the two sonic lines (dotted). The v axis goes from bottom left to top right. The scale in s is indicated by the values at the left corners. The two families of characteristics are displayed in the remaining pictures, for $-1 \le x \le 1$ and $0 \le t \le 1$. The top one shows the 1-characteristics $dx/dt = v - c$ ("slow" speed) and the bottom one shows the 2-characteristics $dx/dt = v + c$ ("fast" speed). The slopes of the characteristics at the corners of the pictures are indicated. In Figs. 22–26, $b_L - b_R = 1.0$; in Figs. 27–29, $b_L - b_R = 0.5$.

In Fig. 22, $C_1(u_L)$ is a Case I(A) curve, while $C_2(u_R)$ is a Case III curve. The intersection u_M lies in $Q_1(u_L)$ and $Q_4(u_R)$. From left to right we have a 1-shock, a 2-shock and a standing wave. The state $\bar{u}_R$ is shown in Fig. 17. In Figs. 23 and 24 the M curves are the same as in Fig. 22. In Fig. 23 u_M belongs to $Q_1(u_L)$ and $Q_3(u_R)$. We have a 1-shock and a stationary wave.

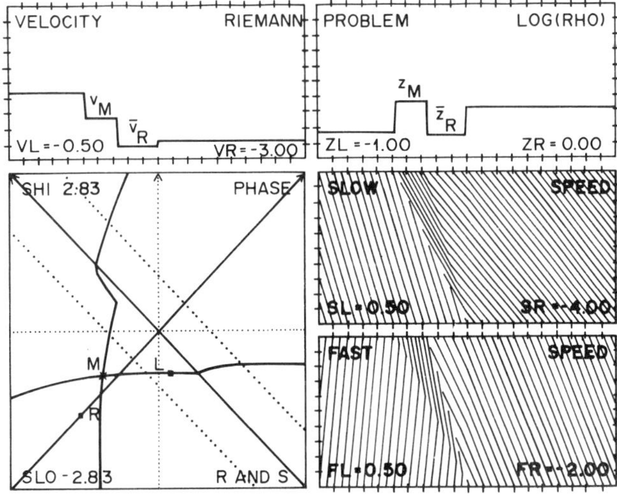

Fig. 22.

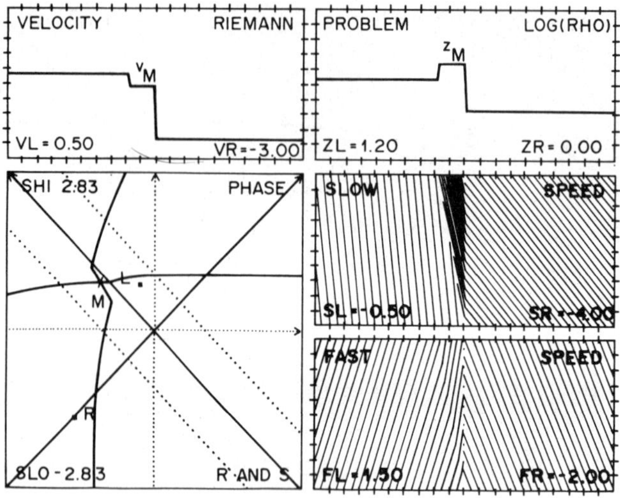

Fig. 23.

This stationary wave consists of a standing wave, a 2-shock with zero speed, and another standing wave which coalesce at $x = 0$. In Fig. 24 u_M belongs to $Q_3(u_L)$ and $Q_1(u_R)$. The state $\hat{u}_c$ is shown in Fig. 11(a). We have a 1-shock, a standing wave, a 1-rarefaction wave and finally a 2-shock.

In Figs. 25 and 26, $C_1(u_L)$ is a Case I(A) curve, while $C_2(u_R)$ is a Case II curve. In Fig. 25 the intersection u_M lies in $Q_1(u_L)$ and $Q_3(u_R)$. We have a 1-shock, a standing wave and a 2-rarefaction. The stationary wave is composed as before of three coalescing waves. In Fig. 26 u_M belongs to $Q_3(u_L)$ and $Q_1(u_R)$. The state $\hat{u}_c$ is shown in Fig. 11(a). We have a 1-shock, a standing wave, a 1-rarefaction and a 2-rarefaction wave.

In Figs. 27–29 $C_1(u_L)$ is a Case VI curve, while $C_2(u_R)$ is a Case II curve; u_M always lies in $Q_1(u_R)$. The difference between these three figures is due to the choice of the branch of $C_1(u_L)$ where u_M lies: $Q_3(u_L)$ in Fig. 27, $Q_4(u_L)$ in Fig. 28 and $Q_6(u_L)$ in Fig. 29. This is an example where the solution of the Riemann problem is nonunique. The configuration in Fig. 28 corresponding to the intermediate branch is unstable[4], so we do not consider it to be a genuine solution. (The states $\bar{u}_L$, u_c, $\hat{u}_l$ and $\bar{u}_L$ are indicated in Fig. 21.)

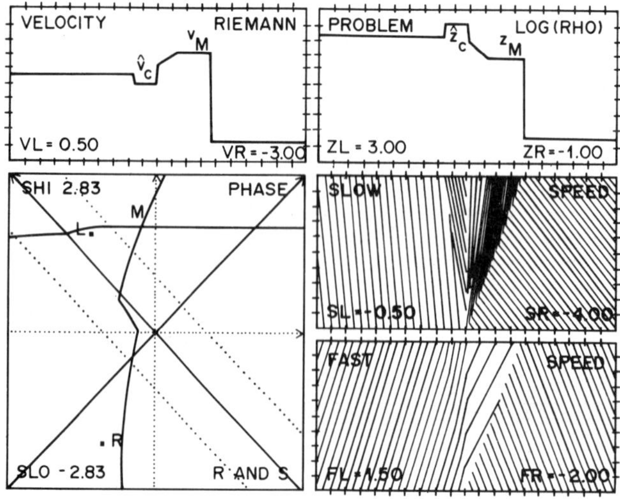

Fig. 24.

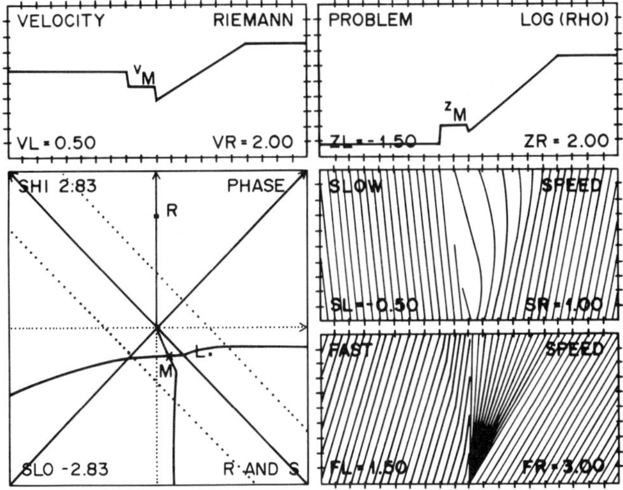

Fig. 25.

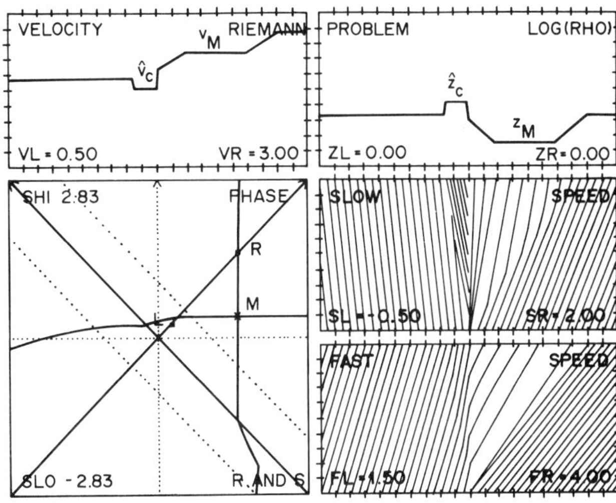

Fig. 26.

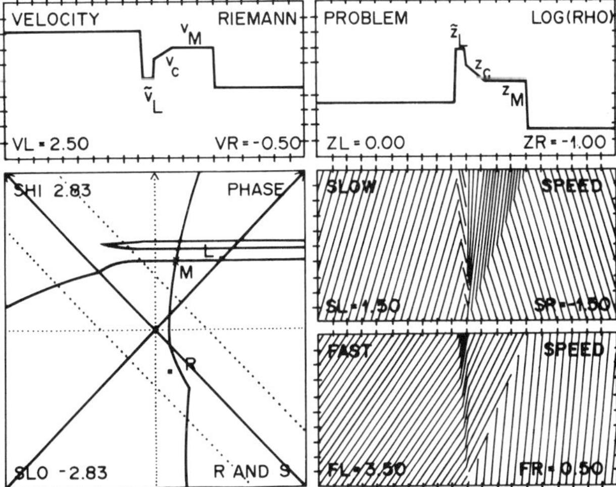

Fig. 27.

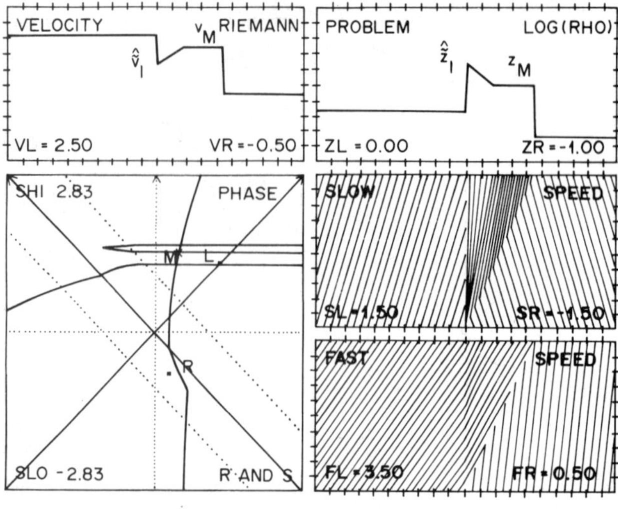

Fig. 28.

We observe that the solution of the Riemann problem is continuous (in the L^1-sense) as a function of the Cauchy data u_L, u_R and the coefficients b_L, b_R, as long as the intersection u_M remains in the same connected component of each of the two M curves. Because of the possibility of bifurcation we do not have global continuity.

We intend to use this Riemann problem to solve the general Cauchy problem for isothermal gas dynamics with variable coefficient $a(x)$ [Eq. (3.1)]. In the case of bifurcation, the intermediate branch is not used because it gives rise to an unstable Riemann solution[4]. The choice between the other branches is dictated by an external criterion.

We believe that the present Riemann solver is a step in the construction of an analogous algorithm for general gas dynamics.

Acknowledgments—We wish to thank Guillermo Marshall for the suggestion to consider Riemann problems with discontinuous coefficients. Our thanks are also due to J. Blake Temple for pointing out the convenience of trying this idea for the case of isothermal gas dynamics. This work was greatly facilitated by the usage of the DOE Computer Center at the Courant Institute of Mathematical Sciences.

D.M. was partially supported by FINEP grant 4/3/82/0179/00, NSF grant MCS-8207965, DOE grant 0276ER03077 and CNPq fellowship 300204/83.

P.J.P-L. was partially supported by FINEP grant 4/3/82/0179/00 and CNPq fellowship 302482-MA.

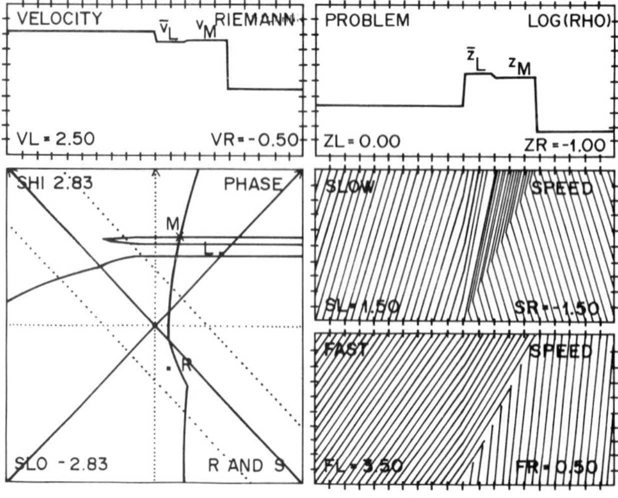

Fig. 29.

REFERENCES

1. C. M. Dafermos, Generalized characteristics and the structure of solutions of hyperbolic conservation laws. *Ind. Univ. Math. J.* **26**(6), 1097–1119 (1977).
2. Y. Lung-An and W. Ching-Hua, Global solutions of the Cauchy problem for a nonhomogeneous quasilinear hyperbolic system. *Commun. Pure Appl. Math.* **33**, 579–597 (1980).
3. C.-h. Wang, An Inhomogeneous Quasilinear Hyperbolic System. MRC Technical Summary Report No. 2137, Univ. of Wisconsin (November, 1980).
4. T. P. Liu, Transonic gas flow in a duct of varying area. *Arch. Rat. Mech. Anal.* **23**(VIII), 1–18 (1982).
5. L. Huang and T. P. Liu, A Conservative, Piecewise Steady Difference Scheme for Transonic Nozzle Flow. Univ. of Maryland preprint MD82-43-LH/TPL, TR82-36 (1982).
6. W. K. Lyons, Conservation laws with sharp inhomogeneities. *Quart. Appl. Math.* **40**(1), 385–393 (1983).
7. C. Z. Li and T. P. Liu, Asymptotic states for hyperbolic conservation laws with a moving source. *Adv. Appl. Math.* **4**, 353–379 (1983).
8. D. Marchesin and P. J. Paes–Leme, Problemas de Riemann para Equações Hiperbólicas não-homogêneas para o fluxo de fluidos, in ATAS of the 14th Colóquio Brasileiro de Matemática, Pocos de Caldas, 273–299 (July 1983).
9. P. Embid, J. Goodman and A. Majda, Multiple steady states for 1-D transonic flow. *SIAM J. Sci. Stat. Comput.* **5**(1), 21–41 (1984).
10. H. M. Glaz and T. P. Liu, The Asymptotic Analysis of Wave Interactions and Numerical Calculations of Transonic Nozzle Flow. Univ. of Maryland preprint MD82-42-HG/TPL (1982).
11. T. P. Liu, Nonlinear stability and instability of transonic flows through a nozzle. *Commun. Math. Phys.* **83**, 243–260 (1982).
12. T. P. Liu, Resonance for quasilinear hyperbolic equation. *Bull. Am. Math. Soc.* **6**(3), 463–465 (1982).
13. J. Steinhoff and A. Jameson, Multiple solutions of the transonic potential flow equation. *AIAA J.* **20**(11), 1521–1525 (1982).
14. A. Chorin, Random Choice solutions of hyperbolic systems. *J. Comput. Phys.* **22**, 517–533 (1976).
15. J. Glimm, Solutions in the large for nonlinear hyperbolic systems of equations. *Commun. Pure Appl. Math.* **18**, 695–715 (1965).
16. S. K. Godunov, Finite difference methods for numerical computation of discontinuous solutions of equations of fluid dynamics. *Mat. Sb.* **47**, 271–295 (1959).
17. T. P. Liu, Quasilinear hyperbolic systems. *Commun. Math. Phys.* **68**, 141–172 (1979).
18. J. Glimm, G. Marshall and B. Plohr, A generalized Riemann problem for Quasi-One-Dimensional gas flows. *Adv. Appl. Math.* (to appear).
19. J. B. Temple and M. Luskin, The existence of a global weak solution to the nonlinear waterhammer problem. *Commun. Pure Appl. Math.* **35**, 697–735 (1982).
20. T. Nishida, Global solution for an initial boundary value problem of a quasilinear hyperbolic system. *Proc. Japan Acad.* **44**, 642–646 (1968).
21. T. Nishida and J. Smoller, Solutions in the large for some nonlinear hyperbolic conservation laws. *Commun. Pure Appl. Math.* **26**, 183–200 (1973).
22. F. John, Partial Differential Equations. Springer-Verlag, New York (1982).

Comp. & Maths. with Appls. Vol. 12A, Nos. 4/5, pp. 457–475, 1986
Printed in Great Britain.

0886–9553/86 $3.00 + .00
© 1986 Pergamon Press Ltd.

SCATTERING PROPERTIES OF WAVE EQUATIONS WITH TIME-DEPENDENT POTENTIALS†

G. Perla Menzala

National Laboratory for Scientific Computation, LNCC/CNPq,
P.O. Box 56018, Rio de Janeiro, R. J. Brazil

Abstract—We consider finite energy solutions of the perturbed wave equation with "impurities" which depend upon space and time: $\Box u + q(x, t)u = 0$ with $x \in \mathbb{R}^n$, $t \in \mathbb{R}$ in the case $n = \text{odd} \geq 3$. We prove that under suitable assumptions on the potential $q(x, t)$, the scattering operator exist and we analyse some of its properties. Finally, we present a uniqueness result on an inverse scattering problem associated to the above equation in the case $n = 3$.

1. INTRODUCTION

In this work we shall consider finite-energy solutions of perturbed wave equations by certain "impurities" which depend upon both position and time. Our model equation will be

$$\Box u + q(x, t)u = 0 \quad \text{in } \mathbb{R}^n \times \mathbb{R} \tag{1.1}$$

Here $\Box$ denotes the d'Alembertian operator, i.e. $\Box = \partial^2/\partial t^2 - \triangle$, where $\triangle$ is the Laplacian operator. We shall consider only the case when the space dimension n is greater than or equal to 3.

We prove that under suitable assumptions on $q(x, t)$ then the scattering operator S associated with Eq. (1.1) exist. This means, roughly, that given a finite-energy solution u of (1.1) there exists two solutions u_+ and u_- of the free-wave equation (i.e. $\Box u_\pm = 0$) such that $u - u_\pm \to 0$ in energy norm as $t \to \pm\infty$. We also discuss some properties of the scattering operator. Finally, in Sec. 4 we present as an application of the previous section, a uniqueness result for the inverse scattering problem in the important physical case $n = 3$.

Although the results presented here are not surprising, the techniques involved are quite delicate. For example, we needed to find precise conditions on q such that the local energy associated with Eq. (1.1) is integrable as a function of time. We suspect that the same result should be true under less restrictive assumptions on q but have not succeeded in proving it.

This paper owes much of its existence to the important work of J. Cooper and W. A. Strauss, who in the late 1970s developed a time-dependent approach, which extends the Lax–Phillips' theory[1] for the case of wave equations outside a moving body (see [2–4]). The techniques we use are quite elementary. In order to avoid ambiguities in the notation we opted for pairs of functions (see Lemmas 3.2–3.4 and Theorem 3.5) only for convenience. Our assumption $n = \text{odd} \geq 3$ is used heavily because the Huygens' principle holds in this case for the free-wave equation. We shall use the standard notation: denote by grad u the gradient of u with respect to space variables. Let $F : \mathbb{R}^n \to \mathbb{R}^n$ be a smooth field. We shall denote by div F the divergence of F with respect to the space variables. By $C_0^\infty(\mathbb{R}^n)$ we denote the space of all C^∞ functions defined on $\mathbb{R}^n$, with compact support. We shall always assume that q is smooth enough in order to have smooth solutions of Eq. (1.1). Since (1.1) is reversible in time, we shall state and perform the estimates only for $t \geq 0$ and the same will be true for $t \leq 0$. Most of the lemmas in the subsequent sections are proved only for the case in which the initial data at $t = 0$ for (1.1) is in $C_0^\infty(\mathbb{R}^n)$. By a standard approximation procedure, the same conclusion

†This research was done while the author was visiting Brown University during 1983–1984. Work supported by CEPG-UFRJ, CNPQ (Brazil) and partially by the Lefschetz Center for Dynamical Systems at Brown University.

458   G. P. MENZALA

will be true for finite-energy solutions. All functions considered in this work will be real valued. Various positive constants will be denoted by C and they may vary from line to line.

2. PRELIMINARY LEMMAS

Let us denote by H_0 the completion of the set of pairs $f = (f_1, f_2)$ where f_1 and f_2 belong to $C_0^\infty(\mathbf{R}^n)$, with respect to the norm

$$\|f\|_{H_0}^2 = \frac{1}{2} \int_{\mathbf{R}^n} [|\text{grad } f_1|^2 + |f_2|^2] \, dx.$$

Let $v = v(x, t)$ be the solution of the free-wave equation $\Box v = 0$ in $\mathbf{R}^n \times \mathbf{R}^n$ with initial data $f = (f_1, f_2) \in C_0^\infty(\mathbf{R}^n) \times C_0^\infty(\mathbf{R}^n)$. The mapping $(f_1, f_2) \to (v, v_t)$ at time t is denoted by $U_0(t)$, which is the solution operator on H_0. Let $q = q(x, t)$ be a real-valued smooth function which is non-negative. We will be interested mainly in the case in which q has compact support in x for each t. Most of the lemmas will be proven under more general assumptions on q, but for the main result, namely the existence of the scattering operator, we shall require this assumption. For each fixed t let $H_1 = H_1(t)$ be the completion of the set of pairs $g = (g_1, g_2)$, where $g_1, g_2 \in C_0^\infty(\mathbf{R}^n) \times C_0^\infty(\mathbf{R}^n)$ with respect to the norm:

$$\|g\|_{H_1}^2 = \frac{1}{2} \int_{\mathbf{R}^n} [|\text{grad } g_1|^2 + |g_2|^2 + q(x, t)|g_1|^2] \, dx.$$

Throughout the entire paper we shall always assume that the space dimension n is greater than or equal to 3.

LEMMA 2.1

Let $q : \mathbf{R}^n \times \mathbf{R} \to \mathbf{R}, q \geq 0$. Suppose that $\|q\|_\infty = \sup_{\substack{x \in \mathbf{R}^n \\ t \in \mathbf{R}}} q(x, t) \leq C < +\infty$ and there exists $N, M > 0$ such that $q(x, t) \leq M|x|^{-2}$ for all $|x| \geq N$; then the norms $\| \; \|_{H_0}$ and $\| \; \|_{H_1}$ are equivalent.

Proof. Since $q \geq 0$ it is enough to prove that there exists a positive constant $\tilde{C}$ such that

$$\int_{\mathbf{R}^n} q(x, t)\psi^2(x) \, dx \leq \tilde{C} \int_{\mathbf{R}^n} |\text{grad } \psi|^2 \, dx \tag{2.1}$$

for all $\psi \in C_0^\infty(\mathbf{R}^n)$. The integral on the left-hand side of (2.1) can be broken in two pieces. The first one can be estimated by using Hölder's inequality and then Sobolev's inequality:

$$\int_{|x| \leq N} q\psi^2 \, dx \leq \|q\|_\infty \int_{|x| \leq N} \psi^2 \, dx \leq C\|q\|_\infty \|\psi\|_{L^{2n/n-2}(|x| \leq N)}^2$$

$$\leq C\|q\|_\infty \|\text{grad } \psi\|_{L^2(\mathbf{R}^n)}^2. \tag{2.2}$$

The second integral can be estimated by using Hardy's inequality, which says that

$$\int_{\mathbf{R}^n} \frac{\psi^2(x)}{|x|^2} \, dx \leq \left(\frac{2}{n - 2} \right)^2 \int_{\mathbf{R}^n} |\text{grad } \psi|^2 \, dx$$

for all $\psi \in C_0^\infty(\mathbf{R}^n)$ (see, for example, R. Courant and D. Hilbert[5], p. 446). Thus

$$\int_{|x| \geq N} q\psi^2 \, dx \leq M \int_{|x| \geq N} \frac{\psi^2(x)}{|x|^2} \leq M \int_{\mathbf{R}^n} \frac{\psi^2(x)}{|x|^2} \, dx \leq M \left(\frac{2}{n - 2} \right)^2 \int_{\mathbf{R}^n} |\text{grad } \psi|^2. \tag{2.3}$$

Using (2.2) and (2.3), we conclude (2.1) with

$$\tilde{C} = C\|q\|_\infty + M\left(\frac{2}{n-2}\right)^2.$$

In the following lemma a central role is played by the multiplier $M[u]$ used by J. Cooper and W. Strauss for a related problem[2].

LEMMA 2.2

Let $g \in C^2(\mathbb{R}^n)$ be a radial function. Let $u = u(x, t)$ and $q(x, t)$ be smooth functions. Define

$$M[u] = u_t + gu_r + \frac{(n-1)}{2r} gu;$$

then $M[u][\Box u + qu]$ can be written as

$$M[u][\Box u + qu] = \frac{\partial A}{\partial t} + \operatorname{div} B + D, \tag{2.4}$$

where

$$A = e(u) + gu_t\left(u_r + \frac{n-1}{2r} u\right),$$

$$B = -M[u] \operatorname{grad} u + \frac{xg}{2r} [|\operatorname{grad} u|^2 - u_r^2] + \frac{(n-1)}{4r^2}\left(g' - \frac{g}{r}\right) xu^2 + \frac{xg}{2r} qu^2,$$

$$D = g'e(u) + \left(\frac{g}{r} - g'\right)[|\operatorname{grad} u|^2 - u_r^2] - u^2\left[\left(g' - \frac{g}{r}\right)\frac{(n-1)(n-3)}{4r^2}\right.$$
$$\left. + \frac{(n-1)}{4r} g'' + \frac{1}{2} q_t + g'q + \frac{1}{2} gq_r\right].$$

Here, we denote

$$e(u) = \frac{1}{2}[u_t^2 + |\operatorname{grad} u|^2 + qu^2], \quad r = |x| \quad \text{and} \quad u_r = \frac{x}{r} \cdot \operatorname{grad} u.$$

Proof. A direct calculation shows that

$$u_t[\Box u + qu] = \frac{d}{dt} e(u) - \frac{1}{2} q_t u^2 - \operatorname{div}(u_t \operatorname{grad} u) \tag{2.5}$$

and

$$g\left[u_r + \frac{(n-1)}{2r} u\right][\Box u + qu] = \frac{\partial}{\partial t}\left(gu_t\left[u_r + \frac{(n-1)}{2r} u\right]\right) - gu_t\left[u_{rt}\right.$$
$$\left. + \frac{(n-1)}{2r} u_t\right] - g\Delta u\left[u_r + \frac{(n-1)}{2r} u\right] + g\left[u_r + \frac{(n-1)}{2r} u\right] qu \tag{2.6}$$

Now, we use the elementary identities

$$\operatorname{div}\left(\frac{xg}{2r}\,u_t^2\right) = gu_t u_{rt} + \frac{(n-1)}{2r}\,gu_t^2 + \frac{1}{2}\,u_t^2 g', \tag{2.7}$$

$$\operatorname{div}\left(\frac{x}{2r}\,|\operatorname{grad}\,u|^2\right) = \sum_{j=1}^{n}\left(\frac{\partial u}{\partial x_j}\right)_r + \frac{(n-1)}{2r}\,|\operatorname{grad}\,u|^2, \tag{2.8}$$

$$\operatorname{div}\left(\operatorname{grad}\,u\left[u_r + \frac{(n-1)}{2r}\,u\right]g\right) = g\left(u_r + \frac{(n-1)}{2r}\,u\right)\Delta u + \left[u_r + \frac{(n-1)}{2r}\,u\right]g'u_r$$
$$+ \frac{(n-1)}{2r}\,g\,|\operatorname{grad}\,u|^2 + g\sum_{j=1}^{n}\frac{\partial u}{\partial x_j}\left(\frac{\partial u}{\partial x_j}\right)_r - \frac{g}{r}\,u_r^2 - \frac{(n-1)}{2r^2}\,guu_r, \tag{2.9}$$

$$\operatorname{div}\left(\frac{gx}{2r}\,|\operatorname{grad}\,u|^2\right) = g\,\operatorname{div}\left(\frac{x}{2r}\,|\operatorname{grad}\,u|^2\right) + \frac{g'}{2}\,|\operatorname{grad}\,u|^2, \tag{2.10}$$

$$\operatorname{div}\left(\frac{xg}{2r}\,u_t^2\right) = \frac{(n-1)}{2r}\,gu_t^2 + gu_t u_{rt} + \frac{g'}{2}\,u_t^2, \tag{2.11}$$

$$\operatorname{div}\left(\frac{xgu^2q}{2r}\right) = qguu_r + \frac{(n-1)}{2r}\,gqu^2 + \frac{g}{2}\,u^2 q_r + \frac{g'}{2}\,qu^2, \tag{2.12}$$

$$\operatorname{div}\left(\frac{(n-1)}{4r^2}\left(g'-\frac{g}{r}\right)xu^2\right) = \frac{(n-1)(n-3)}{4r^2}\,g'u^2 - \frac{(n-1)(n-3)}{4r^3}\,gu^2$$
$$+ \frac{(n-1)}{2r}\,uu_r\left(g'-\frac{g}{r}\right) + \frac{(n-1)}{4r}\,g''u^2, \tag{2.13}$$

which hold for all $r > 0$ and $t \in \mathbb{R}$. Adding (2.5) and (2.6) followed by substitution of identities (2.7)–(2.13), we conclude the proof of the lemma.

Now we shall choose some convenient functions g to conclude some useful facts about the solutions of (1.1). Let $g = g(|x|)$ be a smooth radial function satisfying the following conditions: (1) $g > 0$, (2) $g'' \leq 0$, and (3) there exist $0 < \alpha < 1$ such that $0 \leq rg' \leq g \leq \alpha < 1$ for all $r \geq 0$. For example, one such choice could be $g(r) = (1 + r)/(3 + 2r)$ with $\frac{1}{2} < \alpha < 1$.

Lemma 2.3

Let us choose g as above and $q : \mathbb{R}^n \times \mathbb{R} \to \mathbb{R}$ be a smooth function such that $q \geq 0$ and

$$q_t + 2g'q + gq_r \leq \left(\frac{g}{r} - g'\right)\frac{(n-1)(n-3)}{2r^2} - \frac{(n-1)}{2r}\,g''$$

for all $|x| = r > 0$ and $t \in \mathbb{R}$. Let $u = u(x, t)$ be a solution of (1.1) with initial data in $C_0^\infty(\mathbb{R}^n)$ at time $t = 0$. Suppose that there exists a constant $R > 0$ such that $g'(R) \neq 0$; then the local energy $E_R[t]$ is integrable, i.e.

$$\int_0^\infty E_R[t]\,dt < +\infty,$$

where

$$E_R[t] = \int_{|x|\leq R} e(u)\,dx \quad \text{and} \quad e(u) = \frac{1}{2}\,[u_t^2 + |\operatorname{grad}\,u|^2 + q(x,t)u^2].$$

Proof. We can rewrite A given in Lemma 2.2 as

$$A = (1 - g)\,e(u) + g\left[e(u) + u_t\left(u_r + \frac{(n-1)}{2r}\,u\right) + \frac{(n-1)}{2r}\,uu_r + \frac{(n-1)^2}{8r^2}\,u^2\right]$$
$$+ \operatorname{div}\left(-x\,\frac{(n-1)}{4r^2}\,gu^2\right) + \frac{(n-1)}{4r}\,g'u^2 + \frac{(n-1)(n-3)}{8r^2}\,gu^2, \tag{2.14}$$

because

$$\operatorname{div}\left(\frac{x(n-1)}{4r^2}\,gu^2\right) = \frac{(n-1)^2}{8r^2}\,gu^2 + \frac{(n-1)(n-3)}{8r^2}\,gu^2$$
$$+ \frac{(n-1)}{4r}\,g'u^2 + \frac{(n-1)}{2r}\,guu_r.$$

Next, we observe that

$$\frac{1}{2}\left|u_r + \frac{(n-1)}{2r}\,u\right|^2 = \frac{1}{2}\left|\frac{x}{r}\cdot\operatorname{grad} u\right|^2 + \frac{(n-1)}{2r}\,uu_r + \frac{(n-1)^2}{8r^2}\,u^2.$$

Thus

$$\pm\left(u_r + \frac{(n-1)}{2r}\,u\right)u_t \le \frac{1}{2}\left|u_r + \frac{(n-1)}{2r}\,u\right|^2 + \frac{1}{2}\,u_t^2 \le \frac{1}{2}\,|\operatorname{grad} u|^2 + \frac{1}{2}\,u_t^2$$
$$+ \frac{(n-1)}{2r}\,uu_r + \frac{(n-1)^2}{8r^2}\,u^2 \le e(u) + \frac{(n-1)}{2r}\,uu_r + \frac{(n-1)^2}{8r^2}\,u^2, \quad (2.15)$$

because $q \ge 0$. In order to simplify the notation, let us call

$$F_\pm = e(u) + \frac{(n-1)}{2r}\,uu_r + \frac{(n-1)^2}{8r^2} \pm \left(u_r + \frac{(n-1)}{2r}\,u\right)u_t;$$

consequently by (2.15) we have that $F_\pm(t) \ge 0$. Let $T > 0$ and consider $A(T)$ written as in (2.14). Integration in space gives us

$$\int_{R^n} A(T)\,dx = \int_{R^n} (1-g)e(u)\Big|_{t=T}\,dx + \int_{R^n} gF_+(T)\,dx$$
$$+ \frac{(n-1)}{4}\int_{R^n} \frac{g'u^2}{r}\Big|_{t=T}\,dx + \frac{(n-1)(n-3)}{8}\int_{R^n} \frac{gu^2}{r^2}\Big|_{t=T}\,dx \quad (2.16)$$

Since $F_+(T)$ is non-negative, then it follows that

$$\int_{R^n} A(T)\,dx \ge 0. \tag{2.17}$$

Similarly, the expression A given in Lemma 2.2 can be written as

$$A = (1+g)e(u) - gF_- + \operatorname{div}\left(\frac{x(n-1)}{4r^2}\,gu^2\right)$$
$$- \frac{gu^2(n-1)(n-3)}{8r^2} - \frac{g'u^2(n-1)}{4r}. \quad (2.18)$$

Consider $A(0)$ and integrate (2.18) in space to obtain

$$\int_{R^n} A(0)\,dx = \int_{R^n} (1+g)e(u)\Big|_{t=0}\,dx - \int_{R^n} gF_-(0)\,dx$$
$$- \frac{(n-1)(n-3)}{8}\int_{R^n} \frac{gu^2}{r^2}\Big|_{t=0}\,dx - \frac{(n-1)}{4}\int_{R^n} \frac{g'u^2}{r}\Big|_{t=0}\,dx. \quad (2.19)$$

Since $F_-(0)$ is non-negative, $q \geq 0$ and $n \geq 3$ then (2.19) implies that

$$\int_{R^n} A(0) \, dx \leq \int_{R^n} (1 + g)e(u) \Big|_{t=0} dx = C, \qquad (2.20)$$

where C is a positive constant which depends only on the initial data at time $t = 0$ and $q(x, 0)$.

Now we use Lemma 2.2. Integration in the whole space of identity (2.4) gives us

$$\frac{d}{dt} \int_{R^n} A(t) \, dx + \int_{R^n} D(t) \, dx = 0. \qquad (2.21)$$

In case $n = 3$ we have in (2.21) an additional term which is $2\pi g(0)u^2(0, t)$. Next, we integrate (2.21) from $t = 0$ to $t = T$ to obtain

$$\int_{R^n} A(T) \, dx + \int_0^T dt \int_{R^n} D(t) \, dx = \int_{R^n} A(0) \, dx. \qquad (2.22)$$

By using (2.17) and (2.20) we obtain from (2.22) that

$$\int_0^T dt \int_{R^n} D(t) \, dx \leq \text{const} = C. \qquad (2.23)$$

Since each term in $D(t)$ is non-negative because of our assumptions on q we get from (2.23) that

$$\int_0^T dt \int_{|x| \leq R} g'e(u) \, dx \leq \int_0^T dt \int_{R^n} D(t) \, dx \leq C.$$

Therefore, since g' is decreasing and $g'(R) \neq 0$ we obtain

$$\int_0^T dt \int_{|x| \leq R} e(u) \, dx = \int_0^T E_R[u] \, dt \leq C/g'(R). \qquad (2.24)$$

Letting $T \to +\infty$ we obtain from (2.24) the desired result.

Lemma 2.4.

Let $q = q(x, t) \geq 0$ be a smooth function such that

$$q_t + \alpha q_r \leq \alpha \frac{(n - 1)(n - 3)}{2r^3}$$

for some $0 < \alpha < 1$, all $|x| = r > 0$ and $t \in R$. Let q be chosen such that if u is a solution of (1.1) with initial data in $C_0^\infty(R^n)$ at time $t = 0$ then u has finite energy, and the total energy

$$E_\infty(t) = \frac{1}{2} \int_{R^n} [u_t^2 + |\text{grad } u|^2 + qu^2] \, dx$$

is bounded for all $t \geq 0$. In fact, $E_\infty(t) \leq C_\alpha E_\infty(0)$ for all $t \geq 0$ where C_α is a non-negative constant depending only on α.

Proof. Let us choose $g(r) \equiv \alpha$. By proceeding as in the proof of Lemma 2.3 then (2.20) reads

$$\int_{R^n} A(0) \, dx \leq (1 + \alpha) \int_{R^n} e(u) \Big|_{t=0} dx = (1 + \alpha)E_\infty(0),$$

and (2.16) reads

$$\int_{R^n} A(T) \, dx \geq (1 - \alpha) \int_{R^n} e(u) \Big|_{t=T} dx = (1 - \alpha)E_\infty(t).$$

Since $D(t) \geq 0$ then (2.22) and the above inequalities imply that

$$(1 - \alpha)E_\infty(t) \leq \int_{R^n} A(T) \, dx \leq \int_{R^n} A(0) \, dx \leq (1 + \alpha)E_\infty(0).$$

Consequently

$$E_\infty(T) \leq \left(\frac{1 + \alpha}{1 - \alpha}\right) E_\infty(0)$$

for all $T > 0$, which proves the lemma.

Remarks. (1) In the particular case in which $q \in C^2$ has compact support in x for each t contained in a fixed ball $\{|x| \leq R\}$ and it is bounded in t, then Lemma 2.3 implies Lemma 2.4. In fact, since u satisfies (1.1), let us multiply (1.1) by u_t and integrate in space to obtain

$$2 \frac{d}{dt} E_\infty(t) = \int_{|x| \leq R} \left[\frac{\partial}{\partial t} (qu^2) - 2quu_t \right] dx. \tag{2.25}$$

Integration of (2.25) from $t = 0$ to $t = T$ gives us

$$2E_\infty(T) = 2E_\infty(0) + \int_{|x| \leq R} [q(x, T)u^2(x, T) - q(x, 0)u^2(x, 0)] \, dx$$
$$- 2 \int_0^T dt \int_{|x| \leq R} quu_t \, dx. \tag{2.26}$$

Let us denote by $\|q\|_\infty = \sup_{\substack{|x| \leq R \\ t \in R}} q(x, t)$. By using Schwartz's inequality we obtain that

$$\int_{|x| \leq R} quu_t \, dx \leq 2\|q\|_\infty^{1/2} E_R(t);$$

thus from (2.26) we deduce that

$$2E_\infty(T) \leq 2E_\infty(0) + 2E_R(T) + \int_{|x| \leq R} q(x, 0)u^2(x, 0) \, dx + 4\|q\|_\infty^{1/2} \int_0^T E_R(t) \, dt. \tag{2.27}$$

Since u is smooth then $E_R(t) \leq C$ for all $t \geq 0$. Using Lemma 2.3 we conclude from (2.27) that $E_\infty(T)$ is bounded for all $T > 0$.

Furthermore, by observing that the constant C which appears on the right-hand side of (2.20) is less than or equal to $2E_\infty(0)$ and $E_R(T) \leq C_R e E_\infty(0)$ (see also [6]); then it follows from (2.27) that $E_\infty(T) \leq CE_\infty(0)$ where C depends only on R (and q).

(2) We give a couple of simple examples (in case $n = 3$) of functions q satisfying the assumptions of Lemma 2.4. (a) Let $\psi = \psi(r) \in C^\infty$ such that $\alpha\psi' \leq \psi$ for some $0 < \alpha < 1$; then $q(r, t) = \psi(r)e^{-t}$ satisfies the conditions of Lemma 2.4. (b) Let $\psi : R^l \to R^l$ be a smooth function decreasing and such that $\psi(0) = 1$. Suppose that $\psi(s) = 0$ for all $s \geq m > 0$. Let $0 < \alpha < 1$ and consider $q(r, t) = \psi(r - \alpha/2t)$; then q satisfies the assumptions of Lemma 2.4.

3. THE EXISTENCE OF THE SCATTERING OPERATOR

Our goal in this section is to prove the existence of the scattering operator S associated with Eq. (1.1) under suitable assumptions on $q(x, t)$. Basically, these assumptions will be (1) $q : R^n \times R \to R$, $q \geq 0$, and smooth. (2) q has compact support in x contained in a fixed ball $\{|x| \leq R\}$ for each t. (3) The local energy is integrable on $[0, \infty)$. Furthermore, we shall prove some properties of the scattering operator.

As before, let $U_0(t)$ be the group of unitary operators on H_0 associated with the free-wave equation. For each $s \geq 0$ let us consider the pair

$$w(t) = \begin{pmatrix} w_1(t) \\ w_2(t) \end{pmatrix}$$

defined as

$$w(t) = \begin{cases} U_0(t - s) \begin{pmatrix} u(s) \\ u_t(s) \end{pmatrix} & \text{if } t \geq s, \\ \begin{pmatrix} u(t) \\ u_t(t) \end{pmatrix} & \text{if } 0 \leq t \leq s, \end{cases} \tag{3.1}$$

where u denotes the solution of (1.1) with $C_0^\infty(R^n)$ initial data at $t = 0$. We shall always assume that the space dimension n is odd ≥ 3.

Lemma 3.1

Let $q : R^n \times R \to R$ be a smooth function, $q \geq 0$. Assume that q has compact support in x, contained in a fixed ball $\{|x| \leq R\}$ for each t. Let u be a solution of the wave equation (1.1) with initial data $C_0^\infty(R^n)$. Let $s \geq 0$ and w be defined as in (3.1); then the first component of w vanishes in the set

$$\mathcal{C} = \{(x, t), \quad |x| \leq t - 2s - R, \quad t \geq 2s + R\}.$$

Proof. Here, we assume (to simplify the notation) that the initial data of u at time $t = 0$ has support contained in the ball $\{|x| \leq R\}$. Consider Fig. 1. First, we claim that the first component of w is a solution of the free-wave equation in $R^n \times (0, \infty)$ except in the set $\mathcal{C}(s) = \{(x, t), |x| \leq R, 0 \leq t \leq s\}$. This is clear because the support of q is contained in the ball $\{|x| \leq R\}$. Hence, the first component of w satisfies the free-wave equation in the set $A_1 \cup B_1$. If $t > s$ by definition, the first component of w satisfies the free-wave equation. Now by the domain of influence argument of Huygens' principle, it follows that the first component of w vanishes in the set $\mathcal{C}$.

Lemma 3.2

Let $q : R^n \times R \to R$ with the assumptions of Lemma 3.1 and let u be a solution of (1.1) with $C_0^\infty(R^n)$ initial data at $t = 0$. For each $s \geq 0$ let us define the pair

$$v_s(t) = \begin{pmatrix} u(t) \\ u_t(t) \end{pmatrix} - U_0(t - s) \begin{pmatrix} u(s) \\ u_t(s) \end{pmatrix}$$

for $t \geq s \geq 0$. Then we have the following.

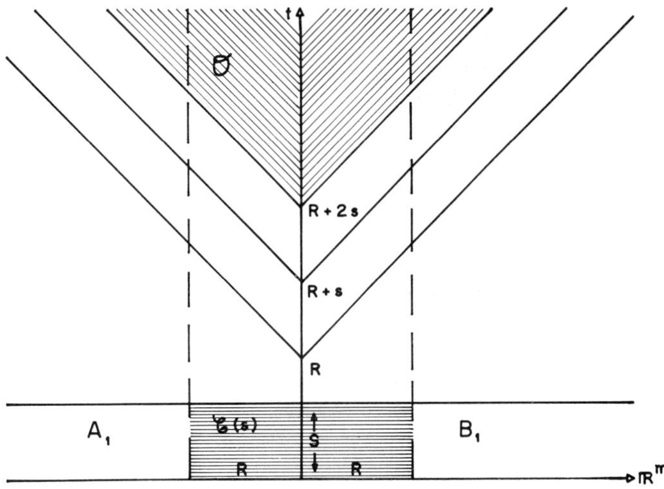

Fig. 1.

(a) Let $0 \leq s \leq \bar{s}$, then for each t such that $s \leq \bar{s} \leq t$ we have

$$v_{\bar{s}}(t) = v_s(t) + U_0(t - \bar{s})v_s(\bar{s}).$$

(b) The first component of $v_s(t)$ has support in the set

$$\{(x, t), \ |x| \leq t - s + R, t \geq s\}.$$

(c) The first component of $v_s(t)$ is a solution of (1.1) in the set

$$\{(x, t), \ |x| \leq t - 2s - R, \ t \geq 2s + R\}.$$

Proof. (a) By using the definition of $v_s(t)$ and the properties of the group $U_0(t)$ we obtain

$$
\begin{aligned}
v_{\bar{s}}(t) &= \begin{pmatrix} u(t) \\ u_t(t) \end{pmatrix} - U_0(t - \bar{s}) \begin{pmatrix} u(\bar{s}) \\ u_t(\bar{s}) \end{pmatrix} \\
&= \begin{pmatrix} u(t) \\ u_t(t) \end{pmatrix} - U_0(t - \bar{s}) \begin{pmatrix} u(\bar{s}) \\ u_t(\bar{s}) \end{pmatrix} + U_0(t - \bar{s})U_0(\bar{s} - s) \begin{pmatrix} u(s) \\ u_t(s) \end{pmatrix} - U_0(t - s) \begin{pmatrix} u(s) \\ u_t(s) \end{pmatrix} \\
&= \begin{pmatrix} u(t) \\ u_t(t) \end{pmatrix} - U_0(t - s) \begin{pmatrix} u(s) \\ u_t(s) \end{pmatrix} - U_0(t - \bar{s}) \left[\begin{pmatrix} u(\bar{s}) \\ u_t(\bar{s}) \end{pmatrix} - U_0(\bar{s} - s) \begin{pmatrix} u(s) \\ u_t(s) \end{pmatrix} \right] \\
&= v_s(t) - U_0(t - \bar{s})v_s(\bar{s}).
\end{aligned}
$$

(b) $v_s(t)$ has zero initial data at time $t = s$ and it is a solution of $\Box v_s(t) = 0$ outside the support of q; thus it follows that the support of $v_s(t)$ is contained in the set $\{(x, t), \ |x| \leq t - s + R, t \geq s\}$. (c) If $t \geq 2s + R$ we know by Lemma 3.1 that

$$U_0(t - s) \begin{pmatrix} u(s) \\ u_t(s) \end{pmatrix}$$

vanishes for $|x| \leq t - 2s - R$. Thus

$$v_s(t) = \begin{pmatrix} u(t) \\ u_t(t) \end{pmatrix}$$

in that set.

Lemma 3.3

Let $q = q(x, t) : \mathbb{R}^n \times \mathbb{R} \to \mathbb{R}$ with all of the assumptions of Lemmas 2.3 and 3.1. Let u be a solution of (1.1) with $C_0^\infty(\mathbb{R}^n)$ initial data at time $t = 0$; then there exists $f_+ \in H_0$ such that

$$\left\| \begin{pmatrix} u(t) \\ u_t(t) \end{pmatrix} - U_0(t)f_+ \right\|_{H_0} \longrightarrow 0 \quad \text{as } t \longrightarrow +\infty.$$

Proof. Let $T \geq R$ and let us define $T^* = T + 4R$. By part (b) of Lemma 3.2 we know that $v_T(T^*)$ has support in the ball $\{|x| \leq 5R\}$; thus

$$E_\infty(v_T(T^*)) = \|v_T(T^*)\|_{H_0}$$

$$= E_{5R}(v_T(T^*)) \leq E_{5R}\left(\begin{pmatrix} u(T^*) \\ u_t(T^*) \end{pmatrix} \right) + E_{5R}\left(U_0(4R) \begin{pmatrix} u(T) \\ u_t(T) \end{pmatrix} \right), \quad (3.2)$$

where we denote by

$$E_\alpha^2\left(\begin{pmatrix} u(t) \\ u_t(t) \end{pmatrix} \right) = E_\alpha^2(t) = \frac{1}{2} \int_{|x| \leq \alpha} [u_t^2 + |\text{grad } u|^2 + qu^2]\, dx.$$

By the standard energy inequality for free solutions of the wave equation we know that

$$E_{5R}\left(U_0(4R) \begin{pmatrix} u(T) \\ u_t(T) \end{pmatrix} \right) \leq E_{9R}\left(\begin{pmatrix} u(T) \\ u_t(T) \end{pmatrix} \right). \quad (3.3)$$

By Lemma 2.3 we also have that

$$\lim_{T \to \infty} E_{9R}\left(\begin{pmatrix} u(T) \\ u_t(T) \end{pmatrix} \right) = 0.$$

Consequently, there exists a sequence $\{T_j\}_{j=1}^\infty$ such that $T_1 < T_2 < \cdots$ and

$$\lim_{j \to \infty} E_{9R}\left(\begin{pmatrix} u(T_j) \\ u_t(T_j) \end{pmatrix} \right) = 0.$$

We could assume without loss of generality that $T_{j+1} \geq T_j + 4R$.

Thus, from (3.2), (3.3) and the above remarks we conclude that

$$\|v_{T_j}(T_j + 4R)\|_{H_0} \leq E_{5R}\left(\begin{pmatrix} u(T_j + 4R) \\ u_t(T_j + 4R) \end{pmatrix} \right) + E_{9R}\left(\begin{pmatrix} u(T_j) \\ u_t(T_j) \end{pmatrix} \right) \longrightarrow 0 \quad (3.4)$$

as $j \to +\infty$.

By the remark after Lemma 2.4 we know that there exists a positive constant C such that

$$\|v_{T_j}(t)\|_{H_0} = E_\infty(v_{T_j}(t)) \leq CE_\infty(v_{T_j}(T_j + 4R)) \quad (3.5)$$

for all $t \geq T_j + 4R$. Thus, from (3.4) and (3.5) we obtain that

$$\lim_{j \to \infty} \sup_{t \geq T_j + 4R} \|v_{T_j}(t)\|_{H_0} = 0. \quad (3.6)$$

We claim that the sequence $\{w_j(t)\}_{j=1}^\infty$ given by

$$w_j(t) = U_0(t - T_j)\begin{pmatrix} u(T_j) \\ u_t(T_j) \end{pmatrix}$$

is a Cauchy sequence in H_0. In fact, we have for $j > k$,

$$\|w_j(t) - w_k(t)\|_{H_0} = \left\|U_0(t - T_j)\left[\begin{pmatrix} u(T_j) \\ u_t(T_j) \end{pmatrix} - U_0(T_j - T_k)\begin{pmatrix} u(T_k) \\ u_t(T_k) \end{pmatrix}\right]\right\|_{H_0}$$

$$= \left\|\begin{pmatrix} u(T_j) \\ u_t(T_j) \end{pmatrix} - U_0(T_j - T_k)\begin{pmatrix} u(T_k) \\ u_t(T_k) \end{pmatrix}\right\|_{H_0}$$

$$= \|v_{T_k}(T_j)\|_{H_0} \longrightarrow 0 \quad \text{as } k, j \longrightarrow +\infty$$

because of (3.6). In particular, the sequence $\{w_j(0)\}_{j=1}^{\infty}$ converges in H_0.

Let us denote by $\lim_{j\to\infty} w_j(0) = f_+$. To conclude the proof of the lemma let $t \geq T_j + 4R$; then observe that

$$\left\|\begin{pmatrix} u(t) \\ u_t(t) \end{pmatrix} - U_0(t)f_+\right\|_{H_0} \leq \left\|\begin{pmatrix} u(t) \\ u_t(t) \end{pmatrix} - U_0(t - T_j)\begin{pmatrix} u(T_j) \\ u_t(T_j) \end{pmatrix}\right\|_{H_0} + \left\|U_0(t - T_j)\begin{pmatrix} u(T_j) \\ u_t(T_j) \end{pmatrix}\right.$$

$$\left. - U_0(t)f_+\right\|_{H_0} = \|v_{T_j}(t)\|_{H_0} + \|w_j(0) - f_+\|_{H_0} \longrightarrow 0 \quad \text{as } j \longrightarrow \infty,$$

which proves the lemma.

LEMMA 3.4

Under the same assumptions as in Lemma 3.3, let $f_-^1, f_-^2 \in C_0^\infty(\mathbb{R}^n)$; then there exists a solution u of (1.1) such that

$$\left\|\begin{pmatrix} u(t) \\ u_t(t) \end{pmatrix} - U_0(t)f_-\right\|_{H_0} \longrightarrow 0 \quad \text{as } t \longrightarrow -\infty,$$

where

$$f_- = \begin{pmatrix} f_-^1 \\ f_-^2 \end{pmatrix}.$$

Proof. Let $R_1 > 0$ such that supp $f_-^1 \cup$ supp f_-^2 is contained in the ball $\{|x| \leq R_1\}$. We could assume without loss of generality that $R_1 \geq R$. Let us consider the solution of the free-wave equation $U_0(t)f_-$. The support of $U_0(t)f_-$ intersected with the set $\{(x, t), |x| \leq |t| - R_1\}$ is empty because of the Huygens' principle. Consequently, $U_0(t)f_-$ vanishes in the set $\{(x, t), t \leq -2R_1, |x| \leq R_1\}$, which contains the support of q. It follows that $U_0(t)f_-$ is a solution of (1.1) for $-\infty < t \leq 2R_1$. Since the Cauchy problem for Eq. (1.1) is well posed for all time, then $U_0(t)f_-$ can be extended to a solution u of (1.1) for all time. In particular,

$$U_0(t)f_- = \begin{pmatrix} u(t) \\ u_t(t) \end{pmatrix}$$

for all $t \leq -2R_1$, so the conclusion of the lemma holds

THEOREM 3.5

Let $q : \mathbb{R}^n \times \mathbb{R} \to \mathbb{R}$ ($n = $ odd ≥ 3) satisfying the following assumptions. (1) q is smooth, $q \geq 0$. (2) For each t, $q(x, t)$ has compact support in x, contained in a fixed ball $\{|x| \leq R\}$. (3) Let us fix a radial smooth function $g = g(|x|)$ such that $g > 0$, $g'' \leq 0$ and $0 \leq rg' \leq g < \alpha < 1$ for some $0 < \alpha < 1$ and all $r = |x| \neq 0$. Choose g such that $g'(R) \neq 0$.

We require that q satisfies

$$q_t + 2g'q + gq_r \leq \left(\frac{g}{r} - g'\right)\frac{(n-1)(n-3)}{2r^2} - \frac{(n-1)}{2r}g''$$

for all $r = |x| > 0$ and $t \in \mathbb{R}$. Then, for all $f_- \in H_0$, there exists a unique solution u of (1.1) and a unique element $f_+ \in H_0$ such that

$$\|U_0(t)f_\pm - u(t)\|_{H_0} \longrightarrow 0 \quad \text{as } t \longrightarrow \pm\infty.$$

Proof. Let $f_- \in H_0$; then there exists a sequence $\{f_-^m\}_{m=1}^\infty$, $f_-^m \in C_0^\infty(\mathbb{R}^n)$ such that $f_-^m \to f_-$ as $m \to +\infty$. By Lemma 3.4, for each m there exists u^m, a solution of (1.1), such that

$$\left\|\begin{pmatrix} u^m(s) \\ u_t^m(s) \end{pmatrix} - U_0(s)f_-^m\right\|_{H_0} \longrightarrow 0 \quad \text{as } s \longrightarrow -\infty.$$

We claim that

$$\left\{\begin{pmatrix} u^m(s) \\ u_t^m(s) \end{pmatrix}\right\}_{m=1}^\infty$$

is a Cauchy sequence; in fact,

$$\left\|\begin{pmatrix} u^m(s) \\ u_t^m(s) \end{pmatrix} - \begin{pmatrix} u^k(s) \\ u_t^k(s) \end{pmatrix}\right\|_{H_0} \leq \left\|\begin{pmatrix} u^m(s) \\ u_t^m(s) \end{pmatrix} - U_0(s)f_-^m\right\|_{H_0} + \left\|U_0(s)f_-^m - U_0(s)f_-^k\right\|_{H_0}$$

$$+ \left\|\begin{pmatrix} u^k(s) \\ u_t^k(s) \end{pmatrix} - U_0(s)f_-^k\right\|_{H_0} = \left\|\begin{pmatrix} u^m(s) \\ u_t^m(s) \end{pmatrix} - U_0(s)f_-^m\right\|_{H_0}$$

$$+ \|f_-^m - f_-^k\|_{H_0} + \left\|\begin{pmatrix} u^k(s) \\ u_t^k(s) \end{pmatrix} - U_0(s)f_-^k\right\|_{H_0}. \tag{3.7}$$

For fixed m, k let $s \to -\infty$ in (3.7); then

$$\overline{\lim_{s \to -\infty}}\left\|\begin{pmatrix} u^m(s) \\ u_t^m(s) \end{pmatrix} - \begin{pmatrix} u^k(s) \\ u_t^k(s) \end{pmatrix}\right\|_{H_0} \leq \|f_-^m - f_-^k\|_{H_0}.$$

Therefore,

$$\sup_t\left\|\begin{pmatrix} u^m(t) \\ u_t^m(t) \end{pmatrix} - \begin{pmatrix} u^k(t) \\ u_t^k(t) \end{pmatrix}\right\|_{H_0} \leq C\,\overline{\lim_{s \to -\infty}}\left\|\begin{pmatrix} u^m(s) \\ u_t^m(s) \end{pmatrix} - \begin{pmatrix} u^k(s) \\ u_t^k(s) \end{pmatrix}\right\|_{H_0} \leq C\|f_-^m - f_-^k\|_{H_0}.$$

Letting $m, k \to \infty$, the claim is proved. Thus

$$\begin{pmatrix} u^m \\ u_t^m \end{pmatrix} \longrightarrow \begin{pmatrix} u \\ u_t \end{pmatrix}$$

in energy norm. Thus u is also a solution of (1.1) and

$$\sup_t\left\|\begin{pmatrix} u^m \\ u_t^m \end{pmatrix} - \begin{pmatrix} u \\ u_t \end{pmatrix}\right\|_{H_0} \leq C\|f_-^{m_l} - f_-\|_{H_0}. \tag{3.8}$$

Now we can prove the first part of the lemma; in fact,

$$\left\|\begin{pmatrix} u \\ u_t \end{pmatrix} - U_0(s)f_-\right\|_{H_0} \le \left\|\begin{pmatrix} u(s) \\ u_t(s) \end{pmatrix} - \begin{pmatrix} u^m(s) \\ u_t^m(s) \end{pmatrix}\right\|_{H_0} + \left\|\begin{pmatrix} u^m(s) \\ u_t^m(s) \end{pmatrix} - U_0(s)f_-^m\right\|_{H_0}$$

$$+ \|f_-^{m_l} - f_-\|_{H_0} \le C\|f_-^m - f_-^k\|_{H_0}$$

$$+ \left\|\begin{pmatrix} u^m(s) \\ u_t^m(s) \end{pmatrix} - U_0(s)f_-^m\right\|_{H_0} + \|f_-^{m_l} - f_-\|_{H_0} \longrightarrow 0$$

as $m \to \infty$ and $s \to -\infty$ because of (3.8). This proves the first part of the lemma. By Lemma 3.3, there exists $f_+^m \in H_0$ such that

$$\left\|\begin{pmatrix} u^m(t) \\ u_t^m(t) \end{pmatrix} - U_0(t)f_+^m\right\|_{H_0} \longrightarrow 0 \quad \text{as } t \longrightarrow +\infty.$$

We claim that $\{f_+^m\}_{m=1}^\infty$ is a Cauchy sequence. In fact,

$$\|f_+^m - f_+^k\|_{H_0} = \|U_0(t)f_+^m - U_0(t)f_+^k\|_{H_0} \le \left\|U_0(t)f_+^m - \begin{pmatrix} u^m(t) \\ u_t^m(t) \end{pmatrix}\right\|_{H_0}$$

$$+ \left\|U_0(t)f_+^k - \begin{pmatrix} u^k(t) \\ u_t^k(t) \end{pmatrix}\right\|_{H_0} + \left\|\begin{pmatrix} u^m(t) \\ u_t^m(t) \end{pmatrix} - \begin{pmatrix} u^k(t) \\ u_t^k(t) \end{pmatrix}\right\|_{H_0}. \tag{3.9}$$

Letting $t \to \infty$ and then $m, k \to \infty$, we prove the claim. Thus $\{f_+^m\}_{m=1}^\infty$ converges. Let us denote by $f_+ = \lim_{l \to \infty} f_+^m$. Let us observe that

$$\left\|\begin{pmatrix} u(t) \\ u_t(t) \end{pmatrix} - U_0(t)f_+\right\|_{H_0} \le \left\|\begin{pmatrix} u(t) \\ u_t(t) \end{pmatrix} - \begin{pmatrix} u^m(t) \\ u_t^m(t) \end{pmatrix}\right\|_{H_0}$$

$$+ \left\|\begin{pmatrix} u^m(t) \\ u_t^m(t) \end{pmatrix} - U_0(t)f_+^m\right\|_{H_0} + \|f_+^m - f_+\|_{H_0}. \tag{3.10}$$

Let $m \to +\infty$ in (3.10) and then $t \to +\infty$, we have

$$\left\|\begin{pmatrix} u(t) \\ u_t(t) \end{pmatrix} - U_0(t)f_+\right\|_{H_0} \longrightarrow 0 \quad \text{as } t \longrightarrow +\infty.$$

It remains to prove the uniqueness of u and f_+. Let us suppose that there exists another pair v and g_+ coming from the same f_-; then

$$\left\|\begin{pmatrix} u(s) \\ u_t(s) \end{pmatrix} - \begin{pmatrix} v(s) \\ v_t(s) \end{pmatrix}\right\|_{H_0} \le \left\|\begin{pmatrix} u(s) \\ u_t(s) \end{pmatrix} - U_0(s)f_-\right\|_{H_0}$$

$$+ \left\|\begin{pmatrix} v(s) \\ v_t(s) \end{pmatrix} - U_0(s)f_-\right\|_{H_0} \longrightarrow 0 \quad \text{as } s \longrightarrow -\infty. \tag{3.11}$$

By the energy boundedness (see Remark (1) after Lemma 2.4) we have

$$\left\|\begin{pmatrix} u(t) \\ u_t(t) \end{pmatrix} - \begin{pmatrix} v(t) \\ v_t(t) \end{pmatrix}\right\|_{H_0} \le C\left\|\begin{pmatrix} u(s) \\ u_t(s) \end{pmatrix} - \begin{pmatrix} v(s) \\ v_t(s) \end{pmatrix}\right\|_{H_0}$$

for all $t \geq s$, which together with (3.11) implies that $u \equiv v$. Similarly

$$\|f_+ - g_+\|_{H_0} = \|U_0(t)f_+ - U_0(t)g_+\|_{H_0} \leq \left\|\begin{pmatrix} u(t) \\ u_t(t) \end{pmatrix} - U_0(t)g_+\right\|_{H_0}$$

$$+ \left\|\begin{pmatrix} u(t) \\ u_t(t) \end{pmatrix} - U_0(t)f_+\right\|_{H_0} \longrightarrow 0 \quad \text{as } t \longrightarrow +\infty.$$

Thus $f_+ = g_+$, which proves the theorem.

Now, given q as in Theorem 3.5 and $f_- \in H_0$, let us define the scattering operator associated with Eq. (1.1) as the map $S : H_0 \rightarrow H_0$ given by

$$S(f_-) = f_+.$$

We have the following diagram:

$$\begin{array}{ccc} & S & \\ f_- & \longrightarrow & f_+ \\ & & \\ \Omega_- \searrow & & \nearrow \Omega_+^{-1}, \\ & \begin{pmatrix} u \\ u_t \end{pmatrix} & \end{array}$$

where the operators Ω_- and Ω_+^{-1} have been implicitly constructed in the proof of Theorem 3.5.

THEOREM 3.6

The scattering operator S we defined above is linear, bounded and one-to-one.

Proof. It is clear that S is linear because the whole process of solving the Cauchy problem at $-\infty$ (or $+\infty$) is linear.

Let us show that Ω_+^{-1} and Ω_- are bounded operators. In fact, for any $s \in \mathbb{R}$ and $t \geq s$,

$$\left\|\Omega_+^{-1}\begin{pmatrix} u(s) \\ u_t(s) \end{pmatrix}\right\|_{H_0} = \|f_+\|_{H_0} = \|U_0(t)f_+\|_{H_0} \leq \left\|\begin{pmatrix} u(t) \\ u_t(t) \end{pmatrix} - U_0(t)f_+\right\|_{H_0} + \left\|\begin{pmatrix} u(t) \\ u_t(t) \end{pmatrix}\right\|_{H_0}$$

$$\leq \left\|\begin{pmatrix} u(t) \\ u_t(t) \end{pmatrix} - U_0(t)f_+\right\|_{H_0} + C\left\|\begin{pmatrix} u(s) \\ u_t(s) \end{pmatrix}\right\|_{H_0}. \tag{3.12}$$

Letting $t \rightarrow +\infty$ in (3.11) we conclude that Ω_+^{-1} is a bounded operator. Similarly

$$\|\Omega_-(f_-)\|_{H_0} = \left\|\begin{pmatrix} u(s) \\ u_t(s) \end{pmatrix}\right\|_{H_0} \leq \left\|\begin{pmatrix} u(s) \\ u_t(s) \end{pmatrix} - U_0(s)f_-\right\|_{H_0} + \|f_-\|_{H_0}.$$

Letting $s \rightarrow -\infty$, we conclude that Ω_- is a bounded operator. Thus S is bounded. It remains to prove that S is one-to-one. It is enough to prove that Ω_- and Ω_+^{-1} are one-to-one. In fact, suppose that $\Omega_-(f_-) = \Omega_-(g_-)$, f_-, $g_- \in H_0$. Then

$$\|f_- - g_-\|_{H_0} = \|U_0(s)f_- - U_0(s)g_-\| \leq \|\Omega_-(f_-) - U_0(s)f_-\|_{H_0}$$

$$+ \|\Omega_-(g_-) - U_0(s)g_-\|_{H_0} \longrightarrow 0 \quad \text{as } s \longrightarrow -\infty.$$

Thus $f_- = g_-$. Similarly Ω_+^{-1} is one-to-one. This concludes the proof of the theorem.

Remark. A simple variation in the proof of Lemma 3.3 and Lemma 3.4 would prove the following result: Under the assumptions of Lemma 3.3, let f_+^1, $f_+^2 \in C_0^\infty(\mathbb{R}^n)$; then there exists a solution u of (1.1) such that

$$\left\| \begin{pmatrix} u(t) \\ u_t(t) \end{pmatrix} - U_0(t)f_+ \right\|_{H_0} \longrightarrow 0 \quad \text{as } t \longrightarrow +\infty,$$

where

$$f_+ = \begin{pmatrix} f_+^1 \\ f_+^2 \end{pmatrix}.$$

For such u there exists $f_- \in H_0$ such that

$$\left\| \begin{pmatrix} u(t) \\ u_t(t) \end{pmatrix} - U_0(t)f_- \right\|_{H_0} \longrightarrow 0 \quad \text{as } t \longrightarrow -\infty.$$

The above results combined with the techniques given in the proof of Theorem 3.5 show that the scattering operator is surjective.

4. APPLICATION

In this section we will discuss an inverse problem associated with Eq. (1.1) in the important physical case $n = 3$. Let us assume that $q = q(x, t)$ satisfies all of the assumptions in the last section, so that the scattering operator $S = S(q)$ exists. This means that we can solve "the Cauchy problem at $t = -\infty$," i.e.

$$u = u_- - \int_{-\infty}^t R(t - s)*qu \, ds. \tag{4.1}$$

Also, we can solve "the Cauchy problem at $t = +\infty$," i.e.

$$u = u_+ - \int_t^\infty R(t - s)*qu \, ds, \tag{4.2}$$

where $u_\mp = [U_0(t)f_\mp]_1$, $R(x, t)$ denotes the Riemann function associated with the free-wave equation, and $*$ denotes spatial convolution.

In terms of the scattering operator $S(q)$ we can write (4.2)–(4.2) as

$$[(S(q) - I)U_0(t)f_-]_1 = \int_{-\infty}^\infty R(t - s)*qu \, ds, \tag{4.3}$$

where $[\quad]_1$ denotes the first component of the pair $(S - I)U_0(t)f_-$.

LEMMA 4.1.

Let $q : \mathbb{R}^3 \times \mathbb{R} \to \mathbb{R}$ such that $q \geq 0$, $q_t \leq 0$ and there exists $0 < \alpha < 1$ such that $q_t + \alpha q_r \leq 0$. Let u be a smooth solution of Eq. (1.1). Then, for any $x \in \mathbb{R}^3, t \in \mathbb{R}$ we have

$$\int_{\mathbb{R}^3} q(x + y, t + |y|)u^2(x + y, t + |y|) \, dy \leq CE_\infty(0) = \text{const.}$$

Proof. Let us multiply (1.1) by u_t to obtain the identity

$$\frac{\partial}{\partial t} e(u) - \text{div}(u_t \, \text{grad } u) = q_t u^2 \leq 0, \tag{4.4}$$

where $e(u) = \frac{1}{2}[u_t^2 + |\text{grad } u|^2 + qu^2]$. Let us fix $x \in \mathbb{R}^3$ and $t \in \mathbb{R}$. Suppose for simplicity that $t > 0$. Let T such that $T \geq t$. We consider the truncated cone with vertex at (x, t):

$$K_T = \{(y, s), |y - x| \leq s - t, t \leq s \leq T\}.$$

Integration of (4.4) on k_T and use of the divergence theorem gives us

$$\int_{Ch_T} [e(u) - u_r u_t] \, dS \leq \sqrt{2} \int_{B_T} e(u) \, dx, \tag{4.5}$$

where

$$Ch_T = \{(y, s), |y - x| = s - t, t \leq s < T\},$$
$$B_T = \{(y, T), |y - x| \leq T - t\}.$$

Since $q \geq 0$ it then follows from (4.5) and Lemma 2.4 that

$$\int_{Ch_T} qu^2 \, dS \leq \sqrt{2} \int_{B_T} e(u) \, dx \leq \sqrt{2} \int_{\mathbb{R}^3} e(u) \, dx \leq C. \tag{4.6}$$

Letting $T \to +\infty$ in (4.6) we deduce the conclusion of the lemma.

Let $F : \mathbb{R}^3 \times \mathbb{R} \to \mathbb{R}$ be a continuous function such that

$$\int_{\mathbb{R}^3} |y|^{-1} F(x + y, t + |y|) \, dy < +\infty. \tag{4.7}$$

We define the operator L by

$$LF(x, t) = \int_{\mathbb{R}^3} |y|^{-1} F(x + y), t + |y|) \, dy \tag{4.8}$$

for any $x \in \mathbb{R}^3$, $t \in \mathbb{R}^+$. Let us define the auxiliary norms

$$A(q) = \int_{\mathbb{R}^3} \|q(y, \cdot)\|_{L^x(\mathbb{R})} \, dy,$$

$$B(q) = \sup_x \int_{|y| \leq 1} |y|^{-2} \|q(x + y, \cdot)\|_{L^x(\mathbb{R})} \, dy,$$

$$D(q) = \sup_x \left(\int_{|y| \leq 1} \|q^2(x + y, \cdot)\|_{L^x(\mathbb{R})} \, dy \right)^{1/2}.$$

Lemma 4.2.

Let $q : \mathbb{R}^3 \times \mathbb{R} \to \mathbb{R}$ satisfying all of the assumptions of Lemma 4.1 and u be a smooth solution of (1.1). If $A(q) < +\infty$ and $B(q) < +\infty$, then

$$\sup_{\substack{x \in \mathbb{R}^3 \\ t \in \mathbb{R}}} |L(qu)(x, t)| \leq C[A(q) + B(q)]^{1/2}.$$

Proof. Let $F = qu$; then using Holder's inequality and Lemma 4.1 we obtain for any

$x \in \mathbb{R}^3$, $t \geq 0$ that

$$|L(qu)(x, t)| \leq \left(\int_{\mathbb{R}^3} |y|^{-2} q(x + y, t + |y|) \, dt \right)^{1/2} \left(\int_{\mathbb{R}^3} (qu)^2(x + y), t + |y|) \, dy \right)^{1/2}$$

$$\leq C \left(\int_{\mathbb{R}^3} |y|^{-2} q(x + y, t + |y|) \, dy \right)^{1/2} = C \left(\int_{|y| \geq 1} + \int_{|y| \geq 1} \right)^{1/2} \leq C(A(q) + B(q))^{1/2}. \quad (4.9)$$

Similar estimates hold for the case $t \leq 0$. This proves the lemma.

LEMMA 4.3

Let q and u as in Lemma 4.2. Additionally, assume that $D(q) < +\infty$; then

$$\sup_{\substack{x \in \mathbb{R}^3 \\ t \in \mathbb{R}}} |L(qL(qu))(x, t)| \leq C[A(q) + B(q)]^{1/2}[A(q) + D(q)].$$

Proof. Clearly, by Lemma 4.2 we have

$$|L(qL(qu))(x, t)| \leq C[A(q) + B(q)]^{1/2} \int_{\mathbb{R}^3} |z|^{-1} q(x + z, t + |z|) \, dz. \quad (4.10)$$

Since

$$\int |z|^{-1} q(x + z, t + |z|) \, dz = \int_{|z| \leq 1} + \int_{|z| \leq 1} \leq A(q) + \int_{|z| \leq 1} |z|^{-1} q(x + z, t + |z|) \, dz$$

$$\leq A(q) + \left(\int_{|z| \leq 1} |z|^{-2} \, dz \right)^{1/2} \left(\int_{|z| \leq 1} q^2(x + z, t + |z|) \, dz \right)^{1/2}$$

$$\leq A(q) + CD(q),$$

which together with (4.10) proves the lemma.

LEMMA 4.4

Let $F = F(x, t)$ be a continuous function such that

$$\int_{-\infty}^{\infty} R(t - s)*(Fw) \, ds = 0 \quad (4.11)$$

for all $x \in \mathbb{R}^3$, $t \in \mathbb{R}$ and for all solutions w of the free-wave equation. Then $F \equiv 0$.

Proof. Let $(y_0, t_0) \in \mathbb{R}^m \times \mathbb{R}$. For each $\epsilon > 0$ let us consider $w^\epsilon = w^\epsilon(y, s)$ as the solution of the problem

$$\Box w^\epsilon = 0 \quad \text{in } \mathbb{R}^3 \times \mathbb{R},$$

$$w^\epsilon(y, t_0) = 0,$$

$$w_t^\epsilon(y, t_0) = \psi_\epsilon(y - y_0),$$

where $\{\psi_\epsilon\}$ is the standard delta-Dirac function, i.e. $\psi_\epsilon(y - y_0) \to \delta(y_0)$, $\psi_\epsilon \in C_0^\infty$, $\psi_\epsilon \geq 0$, $\sup \psi_\epsilon \leq \{|y - y_0| \leq \epsilon\}$ and $\int_{\mathbb{R}^3} \psi_\epsilon \, dy = 1$. Let us choose $x = y_0$ and $t = t_0 + 2\epsilon$; thus (4.11) reads

$$\int_{t_0 + \epsilon/2}^{t_0 3\epsilon/2} ds \int_{|y - y_0| = 2\epsilon - s + t_0} \frac{F(y, s)}{|y - y_0|} w^\epsilon(y, s) \, dS_y = 0 \quad (4.12)$$

because of Huygens' principle. From (4.12) we deduce that

$$F(y_0, t_0) \int_{t_0+\epsilon/2}^{t_0+3\epsilon/2} ds \int_{|y-y_0|=2\epsilon-s+t_0} \frac{w^\epsilon(y, s)}{|y - y_0|} dS_y$$

$$= -\int_{t_0+\epsilon/2}^{t_0+3\epsilon/2} ds \int_{|y-y_0|=2\epsilon-s+t_0} \frac{F(y, s) - F(y_0, t_0)}{|y - y_0|} w^\epsilon(y, s) \, dS_y \quad (4.13)$$

Next we use the continuity of F: given $\epsilon_1 > 0$ there exists $\delta > 0$ such that $|F(y, s) - F(y_0, t_0)| < \epsilon_1$ if $|(y, s) - (y_0, t_0)| < \delta$. Let us choose $\epsilon > 0$ such that $2\epsilon < \delta$. Therefore it follows from (4.13) that

$$|F(y_0, t_0)| \int_{\epsilon/2+t_0}^{3\epsilon/2+t_0} ds \int_{|y-y_0|=2\epsilon-s+t_0} \frac{w^\epsilon(y, s)}{|y - y_0|} dS_y$$

$$\leq \epsilon_1 \int_{\epsilon/2+t_0}^{3\epsilon/2+t_0} ds \int_{|y-y_0|=2\epsilon-s+t_0} \frac{w^\epsilon(y, s)}{|y - y_0|} dS_y. \quad (4.14)$$

Because of our choice of w^ϵ the integral $\int_{\epsilon/2+t_0}^{3\epsilon/2+t_0} ds \int_{|y-y_0|=2\epsilon-s+t_0} w^\epsilon(y, s)/|y - y_0| \, dS_y$ is strictly positive; therefore it follows from (4.14) that $|F(y_0, t_0)| < \epsilon_1$. Since $\epsilon_1 > 0$ was arbitrary, we conclude the proof of the lemma.

Theorem 4.5

Let $q_1, q_2 : \mathbb{R}^3 \times \mathbb{R} \to \mathbb{R}$ satisfy all the assumptions of Theorem 3.5. Assume that $\partial q_1/\partial t \leq 0$ and $\partial q_2/\partial t \leq 0$. Let $S(q_1)$ and $S(q_2)$ be the scattering operators associated with equation $\Box u + \lambda q_1 u = 0$ and $\Box v + \lambda q_2 v = 0$ respectively. Suppose that $S(q_1) = S(q_2)$; then if $0 < \lambda$ is sufficiently small we have $q_1 \equiv q_2$.

Proof. We know [by (4.3)] that $S(q_1)$ and $S(q_2)$ satisfy

$$[(S(q_1) - I)U_0(t)f_-]_1 = \lambda \int_{-\infty}^{\infty} R(t - s)^* q_1 u \, ds, \quad (4.15)$$

$$[(S(q_2)(-I)U_0(t)f_-]_1 = \lambda \int_{-\infty}^{\infty} R(t - s)^* q_2 v \, ds. \quad (4.16)$$

By subtracting (4.15) from (4.16) we obtain the identity

$$\int_{-\infty}^{\infty} R(t - s)^* [q_1 u - q_2 v] \, ds = 0. \quad (4.17)$$

Using (4.2) and (4.17) we obtain

$$\int_{-\infty}^{\infty} R(t - s)^* (q_1 - q_2) u_+ \, ds = \lambda \int_{-\infty}^{\infty} R(t - s)^* [q_1 L(q_1 u) - q_2 L(q_2 v)] \, ds$$

$$= \lambda [L(q_1 L(q_1 u)) - L(q_2 L(q_2 v))].$$

Now we use Lemma 4.3 to conclude that

$$\left| \int_{-\infty}^{\infty} R(t - s)^* (q_1 - q_2) u_+ \, ds \right| \leq \lambda [|L(q_1 L(q_1 u))| + |L(q_2 L(q_2 v))|]$$

$$\leq \lambda C[A(q_1) + B(q_1)]^{1/2}[A(q_1) + D(q_1)]$$

$$+ \lambda C[A(q_2) + B(q_2)]^{1/2}[A(q_2) + D(q_2)]. \quad (4.18)$$

Letting $\lambda \to 0$ in (4.18) we then deduce that $\int_{-\infty}^{\infty} R(t - s)^*(q_1 - q_2)u_+ \, ds = 0$. Finally, using Lemma 4.4 with $F = (q_1 - q_2)$, $w = u_+$, we have proved that $q_1 \equiv q_2$, which proves the theorem.

Acknowledgment—It is a pleasure to acknowledge the warm hospitality of both the Department of Mathematics and the Division of Applied Mathematics during my visit from 1982 to 1984. My thanks to Prof. W. A. Strauss for his continuous encouragement during my stay in Providence.

REFERENCES

1. P. Lax and R. Phillips, *Scattering Theory*. Academic Press, New York (1967).
2. J. Cooper and W. A. Strauss, Energy boundedness and decay of waves reflecting off a moving obstacle. *Indiana Univ. Math. J.* **25**, (1976), 671–696.
3. J. Cooper and W. A. Strauss, Scattering of waves by periodically moving bodies. *J. Funct. Anal.* **47**(2), 180–229 (1982).
4. W. A. Strauss, The existence of the scattering operator for moving obstacles. *J. Func. Anal.* **31**, 225–262, (1979).
5. R. Courant and D. Hilbert, *Methods of Mathematical Physics*, Vol. I. Interscience, New York (1962).
6. H. Tamura, On the decay of local energy for wave equations with time-dependent potential. *J. Math. Soc. Jpn.* **33**(4), 605–618 (1981).

Comp. & Maths. with Appls. Vol. 12A, Nos. 4/5, pp. 477–489, 1986
Printed in Great Britain.

0886–9553/86 $3.00 + .00
© 1986 Pergamon Press Ltd.

THE INFLUENCE OF NONLINEAR CONDUCTION ON SINGULARITY FORMATION IN THE INTENSE PLANE-WAVE, NONLINEAR DIELECTRIC INTERACTION PROBLEM†

Frederick Bloom

Department of Mathematical Sciences, Northern Illinois University, DeKalb, Illinois 60115, U.S.A.

Abstract—The propagation of an intense plane-wave pulse into a nonlinear dielectric half-space is considered; it is assumed that the dielectric is not a perfect nonconductor and that the conduction vector **J** is given by a simple nonlinear Ohm's law. Under these conditions we prove that singularities form in the propagating electromagnetic wave provided the initial gradients of the electromagnetic fields in the wave are positive, and pointwise sufficiently large. Our results point to the fact that nonlinear conduction currents may provide a natural dissipative mechanism in the plane-wave–nonlinear dielectric interaction problem.

1. INTRODUCTION

Let $\Omega \subseteq R^3$ be any bounded, or unbounded, open domain occupied by a nonlinear, isotropic dielectric substance which conforms to the constitutive relations

$$\mathbf{D}(\mathbf{x}, t) = \epsilon(\mathbf{E}(\mathbf{x}, t))\mathbf{E}(\mathbf{x}, t), \quad \mathbf{B}(\mathbf{x}, t) = \mu(\mathbf{H}(\mathbf{x}, t))\mathbf{H}(\mathbf{x}, t) \tag{1.1}$$

for $\mathbf{x} \in \Omega$, where $\mathbf{B}$ is the magnetic field, $\mathbf{H}$ the magnetic intensity, $\mathbf{E}$ the electric field, and $\mathbf{D}$ the electric induction field defined by $\mathbf{D} = \epsilon_0\mathbf{E} + \mathbf{P}(\mathbf{E})$; in this latter relation, ϵ_0 is the permittivity of free space and $\mathbf{P}$ is the polarization vector. Throughout Ω, for $t > 0$, it is assumed that Maxwell's equations apply, in the standard form

$$\frac{\partial \mathbf{B}}{\partial t} = -\operatorname{curl} \mathbf{E}, \quad \operatorname{div} \mathbf{B} = 0,$$

$$\mathbf{J} + \frac{\partial \mathbf{D}}{\partial t} = \operatorname{curl} \mathbf{H}, \quad \operatorname{div} \mathbf{D} = \rho, \tag{1.2}$$

where $\rho = \rho(\mathbf{x}, t)$ is the free-charge density and $\mathbf{J}$ is the conduction vector. We are going to assume, in this paper, that $\mu(\mathbf{H}) = \mu_0 = \text{const}$, where μ_0 is the permeability of free space. Although we do not assume a specific form for the scalar-valued vector constitutive function ϵ in this paper, a typical assumption, which has appeared in some of the nonlinear optics literature[1], is that

$$\epsilon(\mathbf{E}) = \epsilon_0 + \epsilon_2\|\mathbf{E}\|^2, \quad \epsilon_0 > 0, \quad \epsilon_2 > 0. \tag{1.3}$$

Our interest in this paper is in the case where Ω is the half-space $\{\mathbf{x} = (x, y, z)|x > 0\}$; we assume that a linearly polarized plane wave is propagating into Ω so that $\mathbf{E} = (0, E(x, t), 0)$, $\mathbf{D} = (0, D(x, t), 0)$, $\mathbf{B} = (0, 0, B(x, t))$ and $\mathbf{H} = (0, 0, H(x, t))$. With this latter assumption the constitutive relations (1.1) reduce to the scalar equations

$$D(x, t) = \bar{\epsilon}(E(x, t))E(x, t) \equiv \mathscr{D}(E(x, t)); \quad B(x, t) = \mu_0 H(x, t). \tag{1.4}$$

If we set $\bar{\epsilon}(\zeta) = \epsilon((0, \zeta, 0))$, $\forall \zeta \in R^1$ and assume that $(\zeta\epsilon(\zeta))' > 0$, $\forall \zeta \in R^1$, then the first

†Research supported, in part, by NSF Grant MCS 8401308.

relation in (1.4) can be written in the form

$$D(x, t) = \int_0^{E(x,t)} a(\zeta)\, d\zeta, \quad a(\zeta) \equiv (\zeta\epsilon(\zeta))' > 0. \tag{1.5}$$

Because this last integral is monotone in E, $\exists\ \mathscr{E}$ such that $E(x, t) = \mathscr{E}(D(x, t))$ and $\mathscr{E}'(D) = 1/a(E) > 0$.

In all previous work[2,3–8] on the problem of plane-wave–dielectric half-space interaction, it has been assumed *a priori* that the conduction vector $\mathbf{J} = \mathbf{0}$. In such a situation, it has been shown[2–8] that singularities form in the solutions of initial-value problems associated with the homogeneous quasilinear hyperbolic equations that result, when (1.2) is combined with our constitutive hypothesis and the assumption of plane-wave propagation, even for initial values of the electromagnetic field that are arbitrarily small, smooth, and compactly supported; examples of results of this kind, which have been obtained recently by this author in [2], will be cited below. However, no nonlinear dielectric is a perfect nonconductor and thus it may be expected that $\mathbf{J} \neq \mathbf{0}$ and that, in fact, $\mathbf{J} = \sigma(\mathbf{E})\mathbf{E}$ where σ is the conductivity; this last relationship is an example of a simple nonlinear Ohm's law. We could just as easily assume that our medium is nonhomogeneous so that $\sigma = \sigma(\mathbf{x}, \mathbf{E})$ but this leads to additional complications that will be dealt with in a future work[9]; for the sake of consistency with our assumption of homogeneity we will also assume that ρ is negligible.

Our basic interest in this paper is to begin to explore the effects, on singularity formation in the wave, which result from dropping the *a priori* assumption that $\mathbf{J} \equiv \mathbf{0}$. In referring to this problem of plane-wave–dielectric half-space interaction, Broer[3] expressed the opinion that ''in reality the solution of the physical problem may be univalent throughout. This means that there must be an effect not described by (7) and (8) [our system (2.2) below with $\tilde{\sigma} = 0$] which becomes operative whenever waves steepen. This effect must then put a bound on the steepness of the wave fronts. An analogous situation exists in the theory of non-linear acoustics . . . (where) the neglected effects are viscosity and heat conduction. In the optical case the wave propagation will certainly be rendered univalent by some dissipative process not included in the field equations, which admit an energy equation without dissipation. There is, at present, no theory of this effect.'' In this paper we will effect a partial answer to the hypothesis of Broer by showing that the inclusion of nonlinear conduction in the model leads to the conclusion that singularities form provided that initial gradients of the electromagnetic fields in the wave are (pointwise) sufficiently large; no such restriction on these initial gradients is needed for singularities to form when $\mathbf{J} \equiv \mathbf{0}$. In a forthcoming work[10] it will be shown that with $\mathbf{J} \neq \mathbf{0}$, the assumption of initial data which is both smooth and sufficiently small, in an appropriate sense, the possesses gradients which are, pointwise sufficiently small, assures the existence of a globally smooth electromagnetic field in the wave and thus precludes the formation of singularities.

2. THE SYSTEM OF EVOLUTION EQUATIONS

In all that follows we will assume that the conduction vector $\mathbf{J}$ is given by the nonlinear Ohm's law $\mathbf{J} = \sigma(\mathbf{E})\mathbf{E}$, with $\sigma(\cdot)$ of class C^1, and that ρ, the free-charge density, is negligible. If we set $\tilde{\sigma}(\zeta) = \sigma((0, \zeta, 0))$, $\zeta \in R^1$, then $\mathbf{J} = (0, \tilde{\sigma}(E(x, t)E(x, t), 0)$, and the equation expressing conservation of charge, namely $\partial\rho/\partial t + \nabla \cdot \mathbf{J} = 0$, is obviously satisfied. Combining our assumption of plane-wave propagation with our constitutive relations we find that Maxwell's equations (1.2) reduce to the scalar quasilinear system

$$\tilde{\sigma}(E)E + \frac{\partial \mathscr{D}(E)}{\partial t} = -\frac{\partial H}{\partial x}; \quad \frac{\partial B}{\partial t} = -\frac{\partial E}{\partial x}. \tag{2.1}$$

However, $\partial \mathscr{D}/\partial E = a(E)$ and $B = \mu_0 H$, so we may rewrite (2.1) in the form

$$\begin{pmatrix} E \\ H \end{pmatrix}_{,t} + \begin{pmatrix} 0 & 1/a(E) \\ 1/\mu_0 & 0 \end{pmatrix} \cdot \begin{pmatrix} E \\ H \end{pmatrix}_{,x} = \begin{pmatrix} -\tilde{\sigma}(E)E/a(E) \\ 0 \end{pmatrix}, \tag{2.2}$$

which is a strictly hyperbolic inhomogeneous quasilinear system whose characteristics are defined by the nonlinear ordinary differential equations $dx/dt = \pm 1/\sqrt{\mu_0 a(E(x, t))}$. Although (2.2) is not in conservation form, we may use the fact that $E(x, t) = \mathscr{E}(D(x, t))$ to rewrite this system as

$$\frac{\partial D}{\partial t} + \frac{1}{\mu_0} \frac{\partial B}{\partial x} = -\Sigma(D), \qquad \frac{\partial B}{\partial t} + \mathscr{E}'(D) \frac{\partial D}{\partial x} = 0, \tag{2.3}$$

where $\Sigma(D) = \bar{\sigma}(\mathscr{E}(D))\mathscr{E}(D)$; this latter system is an inhomogeneous hyperbolic conservation law of the form

$$\frac{\partial u_i}{\partial t} + \frac{\partial f_i}{\partial x}(\mathbf{u}) = \mathbf{g}(\mathbf{u})$$

with

$$\mathbf{u} = \begin{pmatrix} D \\ B \end{pmatrix}, \quad \mathbf{f}(\mathbf{u}) = \begin{pmatrix} B/\mu_0 \\ \mathscr{E}(D) \end{pmatrix}, \quad \mathbf{g}(\mathbf{u}) = \begin{pmatrix} -\Sigma(D) \\ 0 \end{pmatrix},$$

and associated characteristics defined by the equations

$$\frac{dx}{dt} = \pm \sqrt{\frac{\mathscr{E}'(D(x, t))}{\mu_0}}.$$

For $\sigma \equiv 0$, $\bar{\sigma} = 0$ and (2.3) reduces to a standard hyperbolic conservation law; it is well known that singularities will develop in solutions of initial-value problems for such systems even if the initial data is small, smooth, and compactly supported (e.g. [11] and [12]). In fact, it was shown by this author in [2] that if $D_0(x) \equiv D(x, 0)$ is periodic on R^1, $B_0(x) \equiv 0$, and $\mathscr{E}''(0) \neq 0$ (so that the problem exhibits genuine nonlinearity) then $\nabla_{(x,t)}D \equiv (D_{,t}, D_{,x})$ must blow up in finite time

$$t_{\max} \simeq \frac{\mu_0}{\max|D_0'(x)|} \times \frac{\sqrt{\mathscr{E}'(0)}}{|\mathscr{E}''(0)|},$$

while if $D_0(x)$, $B_0(x)$ both have compact support in R^1, and are of class C^1, then any C^1 solution of the associated initial-value problem must develop singularities in finite time in the first derivatives if $\mathscr{E}'(\zeta)$ is not constant on any open interval.

For $\sigma \not\equiv 0$ we easily find that (2.3) leads to the nonlinear wave equation

$$\frac{\partial^2 D}{\partial t^2} + \Sigma'(D) \frac{\partial D}{\partial t} = \frac{\partial^2 \mathscr{E}(D)}{\partial x^2}. \tag{2.4}$$

Recently, both Nishida[13] and Slemrod[14] have examined the damped quasilinear system

$$\frac{\partial w}{\partial t} - \frac{\partial v}{\partial x} = 0, \qquad \frac{\partial v}{\partial t} - \Gamma'(w) \frac{\partial w}{\partial x} = -\alpha v \quad (\alpha > 0), \tag{2.5}$$

which yields the damped nonlinear wave equation

$$\frac{\partial^2 w}{\partial t^2} + \alpha \frac{\partial w}{\partial t} = \frac{\partial^2 \Gamma(w)}{\partial x^2}. \tag{2.6}$$

Clearly the evolution equation (2.4) coincides, in form, with (2.6) iff Σ is a linear function of its argument. Nishida[13] considered the system (2.5), with associated periodic initial data $w(x, 0) = w_0(x)$, $v(x, 0) = v_0(x)$, assumed local hyperbolicity, i.e. $\Gamma'(0) > 0$, and proved an *a priori* estimate which shows that, for as long as a C^1 solution (w, v) in (x, t) exists, $\sup_x|w|$ may be held small (by choosing the L^∞ norms of the initial data $w_0(x)$, $v_0(x)$ sufficiently small) so that $\Gamma' > 0$ for as long as the smooth solution exists. By working with the Riemann invariants

$$r = v + \int_0^w \sqrt{\Gamma'(\zeta)}d\zeta, \quad s = v - \int_0^w \sqrt{\Gamma'(\zeta)}d\zeta \tag{2.7}$$

that are naturally associated with (2.5), and studying the behavior of $\partial r/\partial x$, $\partial s/\partial x$ along the characteristic curves defined by $(dx/dt = \pm\sqrt{\Gamma'(w(x, t))})$, Nishida was able to derive *a priori* estimates for

$$\left|\frac{\partial r}{\partial x}(x, t)\right| \quad \text{and} \quad \left|\frac{\partial s(x, t)}{\partial x}\right|$$

along the characteristics; these estimates were then used, in conjunction with a continuation theorem and a standard local existence theorem for initial-value problems associated with systems of the form (2.5) (e.g. [15]) to show that if $\Gamma''(0) > 0$ and the C^1 norms of $w_0'(x)$, $v_0'(x)$ are sufficiently small, a C^1 [in (x, t)] solution (w, v) of the initial-value problem for (1.1) exists for all $t > 0$. In this paper, however, our interest will not be in establishing the counterpart of Nishida's existence result for (2.5) for the system (2.3), but in establishing a counterpart of the nonexistence result proven by Slemrod for the system (2.5), namely that for r, s as defined by (2.7), $\exists t_\infty < \infty$ such that $|\partial r/\partial x(x, t)| \to \infty$ as $t \to t_\infty$ provided that $\Gamma''(0) > 0$, with $r_0'(x)$ sufficiently large and positive at some x.

In terms of the Riemann invariants r, s introduced in (2.7), the system (2.5) may be written as

$$r' = -\frac{\alpha}{2}(r + s), \quad s^\cdot = -\frac{\alpha}{2}(r + s), \tag{2.8}$$

where

$$' = \partial/\partial t - \sqrt{\hat\Gamma'(r - s)}\partial/\partial t, \quad {}^\cdot = \partial/\partial t + \sqrt{\hat\Gamma'(r - s)}\partial/\partial x,$$

and $\hat\Gamma(r - s) \equiv \Gamma(w(r - s))$. For an appropriately defined pair of Riemann invariants (which we also denote by) r, s we will be able to rewrite our system (2.3) in the form

$$r' = \Phi(r - s), \quad s^\cdot = -\Phi(r - s), \tag{2.9}$$

with r', $s^\cdot$ suitably defined derivatives of r, s along their respective characteristic curves in the (x, t) plane. Now, (2.9) is very nearly a special case of the following system in Riemann invariant form

$$r' = -\frac{\alpha}{2}(r + s) + \bar\Phi(x, t), \quad s^\cdot = -\frac{\alpha}{2}(r + s) + \bar\Phi(x, t), \tag{2.10}$$

for which Hattori[16] has established finite-time breakdown of C^1 solutions by using an argument due to Rozhdestvenskii[17], which involves showing that characteristics of the same family must cross in finite time. However, central to Hattori's analysis is the derivation and use of the *a priori* estimate

$$|r(x, t)| + |s(s, t)| \le \sup_x |r_0(x)| + \sup_x |s_0(x)| + \left(\frac{4}{\alpha}\right)\sup_x |\bar\Phi(x, t)|, \tag{2.11}$$

where $r_0(x) = r(x, 0)$, $s_0(x) = s(x, 0)$; clearly, (2.11) has no relevance *vis á vis* the system (2.9) for which $\alpha = 0$, and thus Hattori's nonexistence results cannot be carried over to establish singularity development in smooth solutions of the inhomogeneous quasilinear hyperbolic system (2.3) which is of interest in this paper.

In order to establish that (2.3) may be rewritten in the form (2.9), for appropriately defined r, s, and Φ we introduce the functions

$$r(x, t) = -\frac{1}{\mu_0} B(x, t) + \int_0^{D(x,t)} \sqrt{\frac{\mathscr{E}'(\zeta)}{\mu_0}} \, d\zeta,$$

$$s(x, t) = -\frac{1}{\mu_0} B(x, t) - \int_0^{D(x,t)} \sqrt{\frac{\mathscr{E}'(\zeta)}{\mu_0}} \, d\zeta,$$

$$(2.12)$$

and set

$$\lambda = -\sqrt{\mathscr{E}'(D(x, t))/\mu_0}, \quad v = \sqrt{\mathscr{E}'(D(x, t))/\mu_0}; \tag{2.13a}$$

$$' = \frac{\partial}{\partial t} + \lambda(x, t) \frac{\partial}{\partial x}, \qquad ` = \frac{\partial}{\partial t} + v(x, t) \frac{\partial}{\partial x}. \tag{2.13b}$$

By (2.12) and (2.13a),

$$\frac{\partial r}{\partial t}(x, t) = -\frac{1}{\mu_0} \frac{\partial B}{\partial t}(x, t) - \lambda(x, t) \frac{\partial D}{\partial t}(x, t), \tag{2.14a}$$

$$\frac{\partial r}{\partial x}(x, t) = -\frac{1}{\mu_0} \frac{\partial B}{\partial x}(x, t) - \lambda(x, t) \frac{\partial D}{\partial x}(x, t), \tag{2.14b}$$

with similar results for $(\partial s/\partial t)(x, t)$ and $(\partial s/\partial x)(x, t)$.

Combining the equations in (2.3) and using (2.14a,b) we find that

$$\frac{\partial r}{\partial t}(x, t) + \lambda(x, t) \frac{\partial r}{\partial x}(x, t) = -\sqrt{\frac{\mathscr{E}'(D(x, t))}{\mu_0}} \, \Sigma(D(x, t)) \tag{2.15}$$

or

$$r'(x, t) = -\sqrt{\frac{\mathscr{E}'(D(x, t))}{\mu_0}} \, \Sigma(D(x, t)). \tag{2.16}$$

In a similar manner we obtain from (2.3) and the expressions for $\partial s/\partial t$, $\partial s/\partial x$, which are analogous to (2.14a,b),

$$s`(x, t) = +\sqrt{\frac{\mathscr{E}'(D(x, t))}{\mu_0}} \, \Sigma(D(x, t)). \tag{2.17}$$

Now, by using (2.12),

$$r(x, t) - s(x, t) = 2 \int_0^{D(x,t)} \sqrt{\frac{\mathscr{E}'(\zeta)}{\mu_0}} \, d\zeta \equiv \eta(x, t).$$

As $\eta = \hat{\eta}(D(x, t))$ satisfies

$$\frac{d\hat{\eta}}{dD} = 2\sqrt{\frac{\mathscr{E}'(D)}{\mu_0}} > 0,$$

$\hat{\eta}$ is monotonic in D and we may define an inverse $\hat{\eta}^{-1}$ such that $D(x, t) = \hat{\eta}^{-1}(r(x, t) - s(x, t))$; thus

$$\lambda(x, t) = -\sqrt{\mathscr{E}'(\hat{\eta}^{-1}(r(x, t) - s(x, t)))/\mu_0} \equiv \hat{\lambda}(r(x, t) - s(x, t)). \tag{2.18}$$

As $v(x, t) = -\lambda(x, t)$, $\hat{v}(= -\hat{\lambda})$ such that $v(x, t) = \hat{v}(r(x, t) - s(x, t))$. Finally, if we define

$$\Phi(\kappa) = -\sqrt{\frac{\mathcal{E}'(\hat{\eta}^{-1}(\kappa))}{\mu_0}} \, \Sigma(\hat{\eta}^{-1}(\kappa)), \quad \kappa \in R^1, \tag{2.19}$$

then Eqs. (2.16), (2.17) assume the form

$$\begin{aligned} r'(x, t) &= \Phi(r(x, t) - s(x, t)), \\ s'(x, t) &= -\Phi(r(x, t) - s(x, t)). \end{aligned} \tag{2.20}$$

The functions $r(x, t)$, $s(x, t)$ are, of course, the Riemann invariants naturally associated with our system (2.5) and along the characteristic curves $x_1(t; \beta_1)$, $x_2(t; \beta_2)$, defined by the solutions of the initial-value problems

$$\begin{aligned} \frac{dx_1}{dt} &= \hat{\lambda}(r(x_1, t) - s(x_1, t)); \; x_1(0, \beta_1) = \beta_1, \\ \frac{dx_2}{dt} &= \hat{v}(r(x_2, t) - s(x_2, t)); \; x_2(0, \beta_2) = \beta_2, \end{aligned} \tag{2.21}$$

the system of partial differential equations (2.20) reduces to the pair of ordinary differential equations

$$\begin{aligned} \frac{d}{dt} r(x_1(t, \beta_1), t) &= \Phi(r(x_1(t, \beta_1), t) - s(x_1(t, \beta_1), t)), \\ \frac{d}{dt} s(x_2(t, \beta_2), t) &= -\Phi(r(x_2(t, \beta_2), t) - s(x_2(t, \beta_2), t)). \end{aligned} \tag{2.22}$$

Associated with the system (2.20) we have initial conditions of the form

$$r(x, 0) = r_0(x), \quad s(x, 0) = s_0(x), \tag{2.23}$$

where $r_0(\cdot)$, $s_0(\cdot)$ are assumed to be periodic in x and of class C^1 for $x > 0$. In the standard fashion we now extend $r(\cdot, t)$, $s(\cdot, t)$, $r_0(\cdot)$, and $s_0(\cdot)$ to all of R^1, in such a manner that the extended initial data is periodic and of class C^1 for $x \in R^1$, and we consider the initial-value problem (2.20), (2.23) for $-\infty < x < \infty$, $t > 0$.

3. A PRIORI BOUNDS ON C^1 SOLUTIONS

Up till now we have required only that our constitutive theory conform to the hypothesis that $\forall \zeta \in R^1$, $(\zeta \epsilon(\zeta))' > 0$ which, in turn, implies that $\mathcal{E}'(\zeta) > 0$, $\forall \zeta \in R^1$. We now make the further assumption that $\mathcal{E}''(0) > 0$ and that $\mathcal{E}(\cdot)$ and $\tilde{\sigma}(\cdot)$ jointly satisfy the condition

$$\sup_{\eta} \left| \frac{d}{d\eta} (\sqrt{\mathcal{E}'(\hat{\eta}^{-1}(\eta))} \, \Sigma(\hat{\eta}^{-1}(\eta)) \right| = \mathcal{J} < \infty. \tag{3.1}$$

Now, as $\hat{\eta}(0) = 0$, we clearly have

$$\Phi(0) = -\sqrt{\frac{\mathcal{E}'(0)}{\mu_0}} \, \Sigma(0) \equiv -\sqrt{\frac{\mathcal{E}'(0)}{\mu_0}} \, \tilde{\sigma}(\mathcal{E}(0)) \mathcal{E}(0) = 0, \tag{3.2}$$

so that $\Phi(r - s) = \int_0^{r-s} d/d\eta\,\Phi(\eta)\,d\eta$. Therefore, in view of (3.1),

$$|\Phi(r - s)| \le \sup_{\eta} \left|\frac{d\Phi(\eta)}{d\eta}\right| (|r| + |s|) \le \frac{\mathcal{J}}{\sqrt{\mu_0}} (|r| + |s|).$$

For future reference we define $M = 2\mathcal{J}/\sqrt{\mu_0}$ so that the above estimate assumes the form

$$|\Phi(r - s)| \le \frac{M}{2} (|r| + |s|). \tag{3.3}$$

The last estimate is the basis for the following.

LEMMA

Let (r, s) be a C^1 solution of the initial-value problem (2.20), (2.23) for $0 \le t \le T$, $T < \infty$. Then for all $t \le T$, and $x \in (-\infty, \infty)$,

$$|r(x, t)| + |s(x, t)| \le (|r_0| + |s_0|)e^{MT}, \tag{3.4}$$

where $|r_0| = \sup_x|r_0(x)|$, $|s_0| = \sup_x|s_0(x)|$.

Proof. We consider the characteristic curves which are defined by the solutions of the initial-value problems (2.21), denoting the solutions of these problems, respectively, by $x_1(t; \beta_1)$ and $x_2(t; \beta_2)$; along these curves the system (2.20) reduces to the set of nonlinear ordinary differential equations (2.22). By integrating Eqs. (2.22) along their respective characteristics we then obtain

$$r(x_1, t) = r_0(\beta_1) + \int_0^t \Phi(r(x_1, \tau) - s(x_1, \tau))\,d\tau, \tag{3.5a}$$

$$s(x_2, t) = s_0(\beta_2) - \int_0^t \Phi(r(x_2, \tau) - s(x_2, \tau))\,d\tau. \tag{3.5b}$$

Thus, for $t \le T$,

$$|r(x_1, t)| \le |r_0(\beta_1)| + \int_0^t |\Phi(r(x_1, \tau) - s(x_1, \tau))|\,d\tau, \tag{3.6a}$$

$$|s(x_2, t)| \le |s_0(\beta_1)| + \int_0^t |\Phi(r(x_2, \tau) - s(x_2, \tau))|\,d\tau. \tag{3.6b}$$

But, in view of (3.3),

$$|\Phi(r(x_1, \tau) - s(x_1, \tau))| \le \frac{M}{2} (|r(x_1, \tau)| + |s(x_1, \tau)|),$$

$$|\Phi(r(x_2, \tau) - s(x_2, \tau))| \le \frac{M}{2} (|r(x_2, \tau)| + |s(x_2, \tau)|);$$

and therefore, (3.6a,b) imply that for $t \le T$,

$$|r(x_1, t)| \le |r_0(\beta_1)| + \frac{M}{2} \int_0^t |r(x_1, \tau)|\,d\tau + \frac{M}{2} \int_0^t |s(x_1, \tau)|\,d\tau, \tag{3.7a}$$

$$|s(x_2, t)| \le |s_0(\beta_2)| + \frac{M}{2} \int_0^t |s(x_1, \tau)|\,d\tau + \frac{M}{2} \int_0^t |r(x_2, \tau)|\,d\tau. \tag{3.7b}$$

If we set $R(t) = \sup_x |r(x, t)|$, $S(t) = \sup_x |s(x, t)|$, then

$$|r(x_1, t)| \leq |r_0| + \frac{M}{2} \int_0^t (R(\tau) + S(\tau))\, d\tau, \tag{3.8a}$$

$$|s(x_2, t)| \leq |s_0| + \frac{M}{2} \int_0^t (R(\tau) + S(\tau))\, d\tau. \tag{3.8b}$$

Because of the assumed periodicity of the initial data, $r(x, t)$ and $s(x, t)$ are periodic in x for each $t \leq T$; thus for $t = \hat{t}$, $\hat{x}_1$, $\hat{x}_2$ such that $R(\hat{t}) = |r(\hat{x}_1, \hat{t})|$, $S(\hat{t}) = |s(x_2, \hat{t})|$. We now choose $\omega_1 = \omega_1(t)$, $\omega_2 = \omega_2(t)$ such that $\hat{x}_1 = x_1(\omega_1(\hat{t}), \hat{t})$, $\hat{x}_2 = x_2(\omega_2(\hat{t}), \hat{t})$. Then for each $t = \hat{t}$, $R(\hat{t}) = |r(x_1(\omega_1(\hat{t}), \hat{t}), \hat{t})|$ and $S(\hat{t}) = |s(x_2(\omega_2(\hat{t}), \hat{t}), \hat{t})|$. Thus, by choosing $x_1 = x_2(\omega_1(t), t)$, $x_2 = x_2(\omega_2(t), t))$ for each $t \leq T$, on the left-hand sides of (3.7a,b), respectively, we obtain from these estimates the bounds

$$R(t) \leq |r_0| + \frac{M}{2} \int_0^t (R(\tau) + S(\tau))\, d\tau, \tag{3.9a}$$

$$S(t) \leq |s_0| + \frac{M}{2} \int_0^t (R(\tau) + S(\tau))\, d\tau. \tag{3.9b}$$

By defining $W(t) = R(t) + S(t)$ and adding (3.9a,b) we then obtain the estimate

$$W(t) \leq |r_0| + |s_0| + M \int_0^t W(\tau)\, d\tau. \tag{3.10}$$

This last result implies (by Gronwall's inequality) that

$$W(t) \leq (|r_0| + |s_0|)e^{Mt}, \quad t \leq T \tag{3.11}$$

But, for each $t \leq T$, $x \in (-\infty, \infty)$, $W(t) \geq (|r(x, t)| + |s(x, t)|)$ by virtue of the definitions of $R(\cdot)$, $S(\cdot)$, and $W(\cdot)$, and (3.4) follows.   ∎

4. SINGULARITY FORMATION

In this section we begin by assuming that there exists a C^1 solution $(r(x, t), s(x, t))$ of the initial-value problem (2.20), (2.23) for $0 \leq t \leq T$, $T > 0$ arbitrary; we want to derive a contradiction to this assumption by proving that for $|r_0|$, $|s_0|$ sufficiently small, and $(\partial r/\partial x)(x, 0)$ sufficiently large, for some x, $(\partial r/\partial x)(x, t) \to \infty$ as $t \to t_\infty < \infty$. To this end we first rewrite the first equation in the set (2.20) in the form

$$\frac{\partial r}{\partial t} - \hat{v}(r - s)\frac{\partial r}{\partial x} = \Phi(r - s), \tag{4.1}$$

where

$$\hat{v}(r - s) \equiv \sqrt{\frac{\mathscr{E}'(\hat{\eta}^{-1}(r - s))}{\mu_0}},$$

and $\Phi(r - s) = -v(r - s)\,\Sigma(\hat{\eta}^{-1}(r - s)) \equiv \lambda(r - s)\,\Sigma(\hat{\eta}^{-1}(r - s))$. Differentiating (4.1) through with respect to x, we obtain

$$\frac{\partial^2 r}{\partial t \partial x} - \hat{v}\frac{\partial^2 r}{\partial x^2} - \frac{\partial \hat{v}}{\partial x} = \Phi'(r - s)\left(\frac{\partial r}{\partial x} - \frac{\partial s}{\partial x}\right), \tag{4.2}$$

where

$$\Phi'(r - s) = \left.\frac{d\Phi(\eta)}{d\eta}\right|_{\eta = r - s}.$$

As $' = \partial/\partial t - \hat{v}\partial/\partial x$ we may put (4.2) in the form

$$\left(\frac{\partial r}{\partial x}\right)' + \frac{\partial \hat{v}}{\partial x}\frac{\partial r}{\partial x} = \Phi'(r - s)\left(\frac{\partial r}{\partial x} - \frac{\partial s}{\partial x}\right). \tag{4.3}$$

But

$$\frac{\partial \hat{v}}{\partial x} = \frac{\partial \hat{v}}{\partial r}\frac{\partial r}{\partial x} + \frac{\partial \hat{v}}{\partial s}\frac{\partial s}{\partial x} \equiv \frac{\partial \hat{v}}{\partial r}\left(\frac{\partial r}{\partial x} - \frac{\partial s}{\partial x}\right)$$

and thus if we set $\Psi(r - s) = \Phi'(r - s)$,

$$\left(\frac{\partial r}{\partial x}\right)' - \frac{\partial \hat{v}}{\partial r}\left(\frac{\partial r}{\partial x} - \frac{\partial s}{\partial x}\right)\frac{\partial r}{\partial x} = \Psi(r - s)\left(\frac{\partial r}{\partial x} - \frac{\partial s}{\partial x}\right) \tag{4.4}$$

or

$$\left(\frac{\partial r}{\partial x}\right)' + \frac{\partial \hat{v}}{\partial r}\frac{\partial s}{\partial x}\frac{\partial r}{\partial x} = \frac{\partial \hat{v}}{\partial r}\left(\frac{\partial r}{\partial x}\right)^2 + \Psi(r - s)\left(\frac{\partial r}{\partial x} - \frac{\partial s}{\partial x}\right). \tag{4.5}$$

By combining Eqs. (2.20) we have $r' - s` = 2\Phi$. However, $s' = \partial s/\partial t - \hat{v}\partial s/\partial x$, and, thus, as $s` = \partial s/\partial t + \hat{v}\partial s/\partial x$ it follows that

$$\frac{\partial s}{\partial x} = -\frac{\Phi}{\hat{v}} + \frac{(r' - s')}{2\hat{v}}.$$

Substituting this last result into (4.5) and simplifying we obtain

$$\left(\frac{\partial r}{\partial x}\right)' + \frac{\partial \hat{v}}{\partial r}\cdot\frac{(r' - s')}{2\hat{v}}\frac{\partial r}{\partial x} - \frac{\Phi}{\hat{v}}\frac{\partial \hat{v}}{\partial r}\frac{\partial r}{\partial x} = \frac{\partial \hat{v}}{\partial r}\left(\frac{\partial r}{\partial x}\right)^2$$
$$+ \Psi\frac{\partial r}{\partial x} - \Psi\left(\frac{r' - s'}{2\hat{v}}\right) + \frac{\Phi\Psi}{\hat{v}}. \tag{4.6}$$

We now note that

$$(\log \hat{v})' = \frac{\hat{v}'}{\hat{v}} = \frac{(\partial \hat{v}/\partial r)r' + (\partial \hat{v}/\partial s)s'}{\hat{v}},$$

or

$$(\log \hat{v})' = \frac{\partial \hat{v}/\partial r}{\hat{v}}(r' - s');$$

using this fact we may rewrite (4.6) in the form

$$\left(\frac{\partial r}{\partial x}\right)' + \frac{1}{2}(\log \hat{v})'\frac{\partial r}{\partial x} = \frac{\partial \hat{v}}{\partial r}\left(\frac{\partial r}{\partial x}\right)^2 + \left(\Psi + \frac{\Phi}{\hat{v}}\right)\frac{\partial r}{\partial x} - \Psi\left(\frac{r' - s'}{2\hat{v}}\right) + \frac{\Phi}{\hat{v}}\Psi. \tag{4.7}$$

Multiplying the last equation through by $\hat{v}^{1/2}$ and using the fact that $\hat{v}^{1/2}(\log \hat{v})' = \hat{v}^{-1/2}\hat{v}'$ we obtain from (4.7)

$$\left(v^{1/2}\frac{\partial r}{\partial x}\right)' = \hat{v}^{1/2}\frac{\partial \hat{v}}{\partial r}\left(\frac{\partial r}{\partial x}\right)^2 + \left(\Psi + \frac{\Phi}{\hat{v}}\frac{\partial \hat{v}}{\partial r}\right)\hat{v}^{1/2}\frac{\partial r}{\partial x}$$
$$- \frac{\Psi}{2}\hat{v}^{-1/2}(r' - s') + \hat{v}^{-1/2}\Phi\Psi. \quad (4.8)$$

We now set $\chi = \hat{v}^{1/2}\partial r/\partial x$, thus reducing (4.8) to

$$\chi' = \hat{v}^{-1/2}\frac{\partial \hat{v}}{\partial r}\chi^2 + \left(\Psi + \frac{\Phi}{\hat{v}}\frac{\partial v}{\partial r}\right)\chi - \frac{\Psi}{2}\hat{v}^{-1/2}(r' - s') + \hat{v}^{-1/2}\Phi\Psi. \quad (4.9)$$

At this stage of the analysis it is natural to introduce the definition

$$\mathcal{F}(r - s) = -\frac{1}{2}\int_0^{r-s}\Psi(\zeta)\hat{v}^{-1/2}(\zeta)\,d\zeta, \quad (4.10)$$

so that (4.9) becomes

$$\chi' = \hat{v}^{-1/2}\frac{\partial v}{\partial r}\chi^2 + \left(\Psi + \frac{\Phi}{\hat{v}}\frac{\partial \hat{v}}{\partial r}\right)\chi + \hat{v}^{-1/2}\Phi\Psi + \mathcal{F}'. \quad (4.11)$$

In order to simplify the subsequent analysis we now introduce the notation

$$\phi = \Psi + \frac{\Phi}{\hat{v}}\frac{\partial \hat{v}}{\partial r}; \quad \psi = \hat{v}^{-1/2}\Phi\Psi. \quad (4.12)$$

We note that both ϕ, ψ are functions of $r - s$. In view of the definitions (4.12), (4.11) may be rewritten as

$$\chi' = \hat{v}^{-1/2}\frac{\partial \hat{v}}{\partial r}\chi^2 + \phi\chi + \psi + \mathcal{F}'. \quad (4.13)$$

We make the natural definition $\theta \equiv \chi - \mathcal{F}$ and (4.13) becomes

$$\theta' = \hat{v}^{-1/2}\frac{\partial \hat{v}}{\partial r}\theta^2 + \mathcal{H}\theta + \mathcal{G}, \quad (4.14)$$

where

$$\mathcal{H} \equiv 2v^{-1/2}\frac{\partial \hat{v}}{\partial r}\mathcal{F} + \phi \quad \text{and} \quad \mathcal{G} \equiv \hat{v}^{-1/2}\frac{\partial \hat{v}}{\partial r}\mathcal{F}^2 + \phi\mathcal{F} + \psi$$

are both functions of $r - s$. We now pause to examine the coefficient $\hat{v}^{-1/2}\partial\hat{v}/\partial r$ of the quadratic term in (4.14); we compute directly from the relations

$$r - s \equiv \eta = \hat{\eta}(D) \equiv 2\int_0^D \sqrt{\frac{\mathscr{E}'(\zeta)}{\mu_0}}\,d\zeta$$

and

$$\hat{v}(r - s) = \sqrt{\frac{\mathscr{E}'(\hat{\eta}^{-1}(r - s))}{\mu_0}} \equiv \sqrt{\frac{\mathscr{E}'(D)}{\mu_0}}.$$

that

$$\frac{\partial D}{\partial r} = \frac{\partial D}{\partial \eta} = \frac{1}{\dfrac{\partial \eta}{\partial D}} = \frac{\sqrt{\mu_0}}{2} \cdot \frac{1}{\sqrt{\mathscr{E}'(D)}} \quad \text{or} \quad \frac{\partial D}{\partial r} = \frac{\sqrt{\mu_0}}{2} \frac{1}{\sqrt{\mathscr{E}'(\hat{\eta}^{-1}(r-s))}}.$$

Also,

$$\frac{\partial \hat{v}}{\partial r} = \frac{\partial \hat{v}}{\partial D} \cdot \frac{\partial D}{\partial r}$$

so $\partial \hat{v}/\partial r \equiv \frac{1}{4}(\mathscr{E}'(D))^{-1}\mathscr{E}''(D)$, where $D = \hat{\eta}^{-1}(r-s)$. Therefore

$$\hat{v}^{-1/2} \frac{\partial \hat{v}}{\partial r} = \frac{1}{4}\mu_0^{1/4} \frac{\mathscr{E}''(D)}{(\mathscr{E}'(D))^{5/4}}; \quad D = \hat{\eta}^{-1}(r-s). \tag{4.15}$$

In view of our assumption that $\mathscr{E}''(0) > 0$ it follows that $\exists \Lambda > 0$ such that $\mathscr{E}''(\zeta) > 0$ for $|\zeta| \leq \Lambda$. But for $t \leq T$, $|r - s| \leq (|r_0| + |s_0|)e^{MT} \leq \Lambda$ provided we choose $r_0(\cdot)$, $s_0(\cdot)$ such that $|r_0| + |s_0| \leq \Lambda e^{-MT}$; with such a choice of the initial data, it follows that for $t \leq T$, $-\infty < x < \infty$ $\mathscr{E}''(\eta^{-1}(r(x, t) - s(x, t))) > 0$ and thus by (4.15) that $\hat{v}^{-1/2}\partial \hat{v}/\partial r > 0$, $0 \leq t \leq T$. If we define

$$\Gamma = \inf \hat{v}^{-1/2} \frac{\partial \hat{v}}{\partial r}; \quad h = \sup|\mathscr{H}|, \quad g = \inf \mathscr{G}, \tag{4.16}$$

where the respective inf and sup are, in each case, taken over the bounded set of arguments $\{\eta \equiv r - s | \eta| \leq \Lambda, 0 \leq t \leq T\}$, then Eq. (4.14) implies the ordinary differential inequality

$$\theta' \geq \Gamma\theta^2 + \mathscr{H}\theta + \mathscr{G}. \tag{4.17}$$

However,

$$\mathscr{H}\theta \geq -\frac{\Gamma}{2}\theta^2 - \frac{1}{2\Gamma}\mathscr{H}^2 \geq -\frac{\Gamma}{2}\theta^2 - \frac{1}{2\Gamma}h^2,$$

so for $0 \leq t \leq T$,

$$\theta' \geq \frac{\Gamma}{2}\theta^2 - \frac{1}{2\Gamma}h^2 + g. \tag{4.18}$$

We may, without loss of generality, assume that $g \leq 0$ [if $g > 0$, (4.18) is then strengthened by dropping g]. Thus

$$\theta' \geq \frac{\Gamma}{2}(\theta^2 - \mathscr{J}^2), \quad 0 \leq t \leq T; \quad \mathscr{J} \equiv \left[\left(\frac{h}{\Gamma}\right)^2 - \frac{2}{\Gamma}g\right]^{1/2}. \tag{4.19}$$

With (4.19) we have associated initial data

$$\theta(x, 0) \equiv \chi(x, 0) - \mathscr{H}(x, 0)$$

$$= \hat{v}^{1/2}(r_0(x) - s_0(x)) \frac{\partial r}{\partial x}(x, 0) + \frac{1}{2}\int_0^{r_0(x) - s_0(x)} \Psi(\zeta)\hat{v}^{-1/2}(\zeta) \, d\zeta. \tag{4.20}$$

The solution of (4.19), (4.20) is now compared with the solution of the initial-value problem

$$\hat{\theta}' = \frac{\Gamma}{2}(\hat{\theta}^2 - \mathscr{J}^2), \quad 0 \leq t \leq T; \quad \hat{\theta}(x, 0) = \theta(x, 0). \tag{4.21}$$

By a standard comparison theorem[18], $\theta(x, t) \geq \hat{\theta}(x, t)$ for $0 \leq t \leq T$. However, the solution of (4.21) is easily seen to be given by

$$\frac{1}{\hat{\theta}(x, t) + \mathcal{J}} = \frac{e^{\mathcal{J}\Gamma t}}{\theta(x, 0) + \mathcal{J}} + \frac{1}{2\mathcal{J}}(1 - e^{\mathcal{J}\Gamma t}). \tag{4.22}$$

By (4.2) it follows that $\exists t_\infty < \infty$ such that $\theta(x, t) \to \infty$ as $t \to t_\infty$ iff for such t_∞

$$\lim_{t \to t_\infty} \left[\frac{e^{\mathcal{J}\Gamma t}}{\theta(x, 0) + \mathcal{J}} + \frac{1}{2\mathcal{J}}(1 - e^{\mathcal{J}\Gamma t}) \right] = 0. \tag{4.23}$$

But, (4.23) is satisfied iff

$$\frac{2\mathcal{J}}{\theta(x, 0) + \mathcal{J}} = 1 - e^{-\mathcal{J}\Gamma t_\infty} \tag{4.24}$$

for some $t_\infty < \infty$. As $0 < 1 - e^{-\mathcal{J}\Gamma t} < 1$ for all $t > 0$, it follows that (4.24) is satisfied iff $\theta(x, 0)$ may be chosen so as to satisfy

$$0 < \frac{2\mathcal{J}}{\theta(x, 0) + \mathcal{J}} < 1, \tag{4.25}$$

in which case

$$t_\infty = - \frac{1}{\mathcal{J}\Gamma} \ln \left[1 - \frac{2\mathcal{J}}{\theta(x, 0) + \mathcal{J}} \right] > 0. \tag{4.26}$$

However, in view of (4.20), $\theta(x, 0)$ will certainly satisfy (4.25) if at some x, $\partial r/\partial x(x, 0)$ is positive and sufficiently large. Thus, for $\partial r/\partial x(x, 0)$ positive and sufficiently large $\hat{\theta}(x, t)$ (and, then, $\theta(x, t)$ as well) approaches ∞ as $t \to t_\infty$, where $t_\infty > 0$ is given by (4.26). Moreover, t_∞, as given by (4.26), will satisfy $t_\infty < T$ provided $\partial r/\partial x(x, 0) > 0$ is chosen so large that

$$\theta(x, 0) > \frac{2\mathcal{J}}{1 - e^{-\mathcal{J}\Gamma T}} - \mathcal{J} \tag{4.27}$$

But

$$\theta(x, t) = \nu^{1/2}(r(x, t) - s(s, t)) \frac{\partial r}{\partial x}(x, t) + \frac{1}{2} \int_0^{r(x,t) - s(x,t)} \Psi(\zeta)\hat{\nu}^{1/2}(\zeta)\, d\zeta, \tag{4.28}$$

and thus $\partial r/\partial x(x, t) \to \infty$ as $t \to t_\infty$, for $\partial r/\partial x(x, 0) > 0$ and sufficiently large at x, where $t_\infty < T$ is given by (4.26); this contradicts our assumption that $r(x, t) \in C^1$ (in (x, t)) for all t, $0 \leq t < \infty$. As

$$\frac{\partial r}{\partial x}(x, t) = - \frac{1}{\mu_0} \frac{\partial B(x, t)}{\partial x} + \sqrt{\frac{\mathcal{E}'(D(x, t))}{\mu_0}} \frac{\partial D}{\partial x}(x, t) \tag{4.29}$$

we have established the following result on singularity formation in the plane-wave–nonlinear dielectric interaction problem.

THEOREM

   Consider the inhomogeneous quasilinear system (2.3) with periodic initial data $B(x, 0) = B_0(x)$, $D(x, 0) = D_0(x)$. Suppose that $\mathcal{E}'(\zeta) > 0$, $\forall \zeta \in R^1$, $\mathcal{E}''(0) > 0$, and that $\epsilon(\cdot)$, $\tilde{\sigma}(\cdot)$ jointly satisfy (3.1). Then a C^1 solution $(B(x, t), D(x, t))$ cannot exist for all $t > 0$ if $\sup_x |B_0(x)|$,

$\sup_x |D_0(x)|$ are chosen sufficiently small while

$$\sqrt{\frac{\mathscr{E}'(D_0(x))}{\mu_0}} \, D_0'(x) \; - \; \frac{1}{\mu_0} \, B_0'(x)$$

is chosen so as to be positive and sufficiently large at some x. In fact, given $T > 0$, $\exists \rho_1(T) > 0$, $\rho_2(T) > 0$ such that for $\max(\sup_x |B_0(x)|, \sup|D_0(x)|) < \rho_2(T)$ and

$$\sqrt{\frac{\mathscr{E}'(D_0(x))}{\mu_0}} \, D_0'(x) \; - \; \frac{1}{\mu_0} \, B_0'(x) > \rho_2(T), \quad \exists t_\infty < T$$

such that

$$\lim_{t \to t_\infty} \left(\left[\frac{\partial B(x, t)}{\partial x} \right]^2 + \left[\frac{\partial D(x, t)}{\partial x} \right]^2 \right)^{1/2} = \infty.$$

Remarks

It is clear from the conditions cited in the theorem that singularities also form, for $\sup_x |B_0(x)|$ and $\sup_x |D_0(x)|$ chosen sufficiently small, if $D_0'(x) > 0$, $-\infty < x < \infty$, while $B_0'(x) < 0$, $-\infty < x < \infty$, with $|B_0'(x)|$ sufficiently large at some x.

REFERENCES

1. N. Bloembergen, *Nonlinear Optics*. W. A. Benjamin, New York (1965).
2. F. Bloom, Nonexistence of smooth electromagnetic fields in nonlinear dielectrics II: Shock development in a half-space. *J. Math. Anal. Applic.*, (1986) in press.
3. F. DeMartini, C. H. Townes, T. K. Gustafson and P. L. Kelley, Self-steepening of light pulses. *Phys. Rev.* **164**, 312–323 (1967).
4. L. J. F. Broer, Wave propagation in non-linear media. *ZAMP* **16**, 11–26 (1965).
5. A. Jeffrey, Non-dispersive wave propagation in nonlinear dielectrics. *ZAMP* **16**, 741–745 (1968).
6. A. Jeffrey and V. P. Korobeinikov, Formation and decay of electromagnetic shock waves. *ZAMP* **20**, 440–447 (1969).
7. I. G. Kataev, *Electromagnetic Shock Waves*. Iliffe Pub., London (1966).
8. A. Donato and D. Fusco, Some applications of the Riemann method to electromagnetic wave propagation in nonlinear media. *ZAMP* **60**, 537–613 (1980).
9. F. Bloom, Shock formation in an inhomogeneous nonlinear dielectric in the presence of a nonlinear conduction current (in preparation).
10. F. Bloom, Smoothing of plane electromagnetic waves in nonlinear dielectrics (in preparation).
11. P. D. Lax, Development of singularities of solutions in nonlinear hyperbolic partial differential equations. *J. Math. Phys.* **S**, 611–613 (1964).
12. S. Klainerman and A. Majda, Formation of singularities for wave equations including the nonlinear vibrating string. *CPAM* **XXXIII** (1980).
13. T. Nishida, *Nonlinear Hyperbolic Equations and Related Topics in Fluid Dynamics*. Publications Mathematiques D'Orsay 78.02, Universite de Paris-Sud, Department de Mathematique (1978).
14. M. Slemrod, Instability of steady shearing flows in a non-linear viscoelastic fluid. *Arch. Rat. Mech. Anal.* **68**, 211–255 (1978).
15. A. Douglis, Some existence theorems for hyperbolic systems of partial differential equations in two independent variables. *Comm. Pure Appl. Math.* **5**, 119–154 (1952).
16. H. Hattori, Breakdown of smooth solutions in dissipative nonlinear hyperbolic equations. Ph.D. thesis, R.P.I. (1981).
17. Rozhdestviskii, *Translations of the A.M.S.* **101** (1973).
18. V. Lakshmikantham and S. Leela, *Differential and Integral Inequalities*, Vol. 1. Academic Press, New York (1969).

Comp. & Maths. with Appls. Vol. 12A, Nos. 4/5, pp. 491–512, 1986
Printed in Great Britain.

0886–9553/86 $3.00 + .00
© 1986 Pergamon Press Ltd.

ON THE STABILITY OF THE CELL-SIZE DISTRIBUTION
II: TIME-PERIODIC DEVELOPMENTAL RATES

O. DIEKMANN
Centrum voor Wiskunde en Informatica, Kruislaan 413, 1098 SJ Amsterdam, The Netherlands

H. J. A. M. HEIJMANS
Centrum voor Wiskunde en Informatica, Kruislaan 413, 1098 SJ Amsterdam, The Netherlands

and

H. R. THIEME
Sonderforschungsbereich 123, Universität Heidelberg, Im Neuenheimer Feld 293, D-6900 Heidelberg,
F.R.G.

Abstract—A deterministic model for the growth of a size-structured proliferating cell population is analyzed. The developmental rates are allowed to vary with time. For periodically varying rates stability of the cell-size distribution is shown under similar conditions for the growth rate of individual cells as found before in the time-homogeneous case. Strongly positive quasicompact linear operators on Banach lattices serve as powerful abstract tools. Finally, the autonomous case is revisited and the conditions for stability found in [1] are relaxed.

INTRODUCTION

The growth of a size-structured population of cells which reproduce by fission into two equal parts can be described by the partial differential equation

$$\partial_t n(t, x) + \partial_x(g \cdot n)(t, x) + (\mu + b) \cdot n(t, x) = 4b \cdot n(t, 2x). \tag{1a}$$

Here $\partial_x(g \cdot n)(t, x)$ denotes $(\partial/\partial x)(g(t, x)n(t, x))$, $(\mu + b) \cdot n(t, x)$ denotes $(\mu(t, x) + b(t, x))n(t, x)$. The independent variables t and x denote, respectively, time and cell size (e.g. the length, volume, weight, protein content, etc. of a cell). At fixed time t, $n(t, x)$ describes the size density of the population. The development of an individual cell is governed by three processes: growth (i.e. increase in size), death (or dilution), and reproduction by splitting into two parts of equal size. $g(t, x)$, $\mu(t, x)$ and $b(t, x)$ indicate the respective rates in dependence on time t and cell size x. Equation (1a) relates these individual changes to the change of the population density. For a further explanation and a derivation of (1a) see [1, Sec. 2 and Appendix].

We assume that cells can only divide between a minimum size $a > 0$ and a maximum size which has been normalized to be 1, i.e.

$$b \cdot n(t, x) = 0 \quad \text{if } x \notin [a, 1), \tag{1b}$$

with $a < 1$. Consequently there are no cells with size less than $a/2$, so we require

$$n(t, a/2) = 0. \tag{1c}$$

Usually we consider (1) as an initial-value problem, i.e. (1a–c) are assumed to hold for $t > t_0$ with $t_0 \in \mathbb{R}$, and

$$n(t_0, x) = n_0(x) \tag{2}$$

with a given initial cell density n_0.

In [1] we impose conditions on the splitting rate b which guarantee that all cells die or divide before reaching the maximum size 1. But this is an unnecessary restriction. We confine the size interval to $[a/2, 1]$ in (1a), however, and so assume that cells with a size exceeding the maximum splitting size 1 do not affect the further development of the population. This assumption, as well as the linear character of Eq. (1), is justified if the environment of the population is unlimited (this can be artificially achieved in a laboratory) such that density-dependent feedback mechanisms can be ignored.

In [1] we considered (1) for time-independent rates g, μ, b and found conditions under which a cell population (with an arbitrary cell-size distribution at the beginning) asymptotically ($t \to \infty$) exhibits exponential growth with a stationary cell-size distribution, i.e.

$$n(t, x) \sim Ce^{\sigma t}\bar{n}(x), \quad t \to \infty. \tag{3}$$

Here σ and $\bar{n}$ do not depend on the initial state of the population, whereas the positive scalar C depends on the initial function in a linear and strictly positive way. The conditions mainly concerned the individual growth rate g: (3) holds, for example, if $g(2x) < 2g(x)$ for $x \in [a/2, \frac{1}{2}]$, whereas (3) does not hold if $g(2x) = 2g(x)$ for $x \in [a/2, \frac{1}{2}]$.

We generalize the results of [1] to the case of time-periodic rates g, μ, b (with the same period for b, g, μ) in presenting similar conditions under which

$$n(t, x) \sim Ce^{\sigma t}\bar{n}(t, x), \quad t \to \infty, \tag{4}$$

with C, σ, $\bar{n}$ having the same characteristics as in the time-homogeneous case and $\bar{n}(t, x)$ being periodic in t (having the same period as g, b, μ) (see Sec. 6). In Sec. 7 we revisit the time-homogeneous case and prove the conjecture at the end of Sec. 8 in [1], namely that $g(2x) \neq 2g(x)$ on an open subinterval of $[a/2, \frac{1}{2}]$ is sufficient for (3) to hold. In [1] we already showed that (3) does not hold for arbitrary initial values if $g(2x) = 2g(x)$ on $[a/2, \frac{1}{2}]$, e.g. in the case of exponential individual cell growth. So a complete characterization of those growth rates g has been achieved which cause convergence to steady-state exponential growth from arbitrary initial states. This improvement of our former result is of particular importance because the stronger assumptions in [1, Sec. 8] are not satisfied by data found for g by Anderson *et al.* (see [2, Fig. 4.B]) whose work[2–5] (besides the work of Sinko and Streifer[6,7], see also [8]) has been the main motivation for our study. See [1] for some more references. We mention that splitting into two unequal parts of fixed ratio (as it is considered in Sinko and Streifer's work[7] on planarian worms) can be dealt with in essentially the same way. Heijmans[9] deals with a model in which the ratio of mother size and daughter size is described by a probability distribution.

Recently, related models of proliferating cell populations have been studied by Lasota and Mackey[10] and by Tyson and coworkers[11–16]. Hannsgen, Tyson and Watson[16] examine the stationary size distribution for populations growing under steady-state conditions and they find, among other things, that such a distribution does not exist if growth is proportional to size and division is governed by the (purely age-dependent) transition probability model. In [13] Tyson and Hannsgen analyse the "Tandem Model," which is obtained from the transition probability model by adding a critical size requirement. Lasota and Mackey[10] consider the size distribution at birth in successive generations (so they are not concerned with the evolution in time of the size distribution of extant cells). Their assumptions about the dynamics of individual cells resemble ours and they prove that the birth-size distribution converges to a unique globally stable distribution when the generation number tends to infinity. In [15] Tyson and Hannsgen derive a similar result for the case that the probability of division is governed by age (and not size) and individual cell growth is linear. Moreover, they show that such a result does not hold if one assumes that individual cell growth is exponential instead of linear. In [11,12] finally, Tyson makes a comparison of the generation time distribution and the division-size distribution predicted by various models and observed for a population of fission yeast cells.

The organisation of our paper is as follows: In Sec. 1 we study the characteristic curves associated with the first-order partial differential equation (1a). These are important tools to transform (1), (2) into an integral equation, the solutions of which can be considered weak solutions of (1), (2) (see Sec. 2). In Sec. 3 we prove uniqueness and existence of solutions to the integral equation and derive some properties. In Sec. 4 we study the solution operators corresponding to the integral equations, in particular their positivity and compactness properties. In Sec. 5 we investigate their spectral properties, if the rates g, μ, b are time-periodic. Here we make substantial use of the theory of strongly positive linear operators on Banach lattices (see, e.g. [17,18]). In Secs. 6 and 7 we formulate and prove our results, first for the periodic and, under less restrictive assumptions, for the time-homogeneous case. In the appendix we present some material from the theory of Banach lattices.

Finally, we mention that the extensions and improvements of the results in [1], which we achieved in this paper, lead to corresponding extensions and improvements of the results in [19] for a rather special nonlinear variant of the model.

1. THE CHARACTERISTIC CURVES

The growth of an individual cell

In dealing with PDEs of first order, integration along characteristic curves plays an important role (see, e.q., [20,21]). In our case these characteristic curves describe the growth of an individual cell.

Throughout this paper we impose the following conditions on the growth rate g.

ASSUMPTION 1.1

g is a continuous nonnegative function on $\mathbb{R} \times [a/2, 1]$ with the following properties:

(a) g is bounded and bounded away from zero, i.e.

$$0 < g_{\min} \leq g(t, x) \leq g_{\max} < \infty \quad \text{for } t \in \mathbb{R}, x \in [a/2, 1].$$

(b) The partial derivatives $\partial_t g(t, x)$, $\partial_t^2 g(t, x)$, $\partial_x g(t, x)$ exist and are continuous and bounded on $\mathbb{R} \times [a/2, 1]$.

We recall that even if cells should grow beyond the maximum splitting size 1, only cells of size $x \in [a/2, 1)$ affect the further development of the population [see (1b, c)]. So it is sufficient to know g on $\mathbb{R} \times [a/2, 1]$. For convenience we extend g to $\mathbb{R}^2$ by

$$\begin{aligned}
g(t, x) &= g(t, 1) &&\text{for } x \geq 1, \\
g(t, x) &= g(t, a/2) &&\text{for } x \leq a/2.
\end{aligned} \tag{5}$$

The derivatives $\partial_t g$ and $\partial_t^2 g$ now exist and are continuous on $\mathbb{R}^2$, $\partial_x g$ still exists in a generalized sense and is bounded.

So g is smooth enough such that the following ODE initial-value problems can be solved:

$$\begin{aligned}
\partial_y T(t, x, y) &= 1/g(T(t, x, y), y), \\
T(t, x, x) &= t,
\end{aligned} \tag{6}$$

$$\begin{aligned}
\partial_t X(t, s, x) &= g(t, X(t, s, x)), \\
X(s, s, x) &= x,
\end{aligned} \tag{7}$$

for $t, s, x, y \in \mathbb{R}$.

T and X can be interpreted biologically: A cell with size x at time s has size $X(t, s, x)$ at time t; a cell with size x at time t has size y at time $T(t, x, y)$. The following lemma lists a number of properties of the unique solutions T and X to (6) and (7). These will serve as paramount tools in our analysis of problem (1). A proof can be established by standard methods or can be found in textbooks dealing with differential equations (see, e.g. [22, Chap. VI]).

LEMMA 1.2

There exist unique continuously differentiable solutions T of (6) and X of (7) on $\mathbb{R}^3$. They have the following properties:

(a) $T(t, x, y)$ and $X(t, s, y)$ strictly increase as t and y increase and x and s decrease.

(b) $T(T(t, x, y), y, z) = T(t, x, z)$, $X(t, r, X(r, s, x)) = X(t, s, x)$.

(c) $T(s, x, X(t, s\ x)) = t$, $T(t, X(t, s, x), x) = s$, $T(t, X(t, s, x), y) = T(s, x, y)$.

(d) $X(s, T(s, x, y), y) = x$, $X(T(s, y, x), s, y) = x$, $X(t, T(s, x, y), y) = X(t, s, x)$.

(e) $\partial_t T(t, x, y) = \exp\left(-\int_x^y \dfrac{\partial_1 g(T(t, x, z), z)}{(g(T(t, x, z), z))^2}\, dz\right).$

(f) $\partial_x X(t, s, x) = \exp\left(\int_s^t \partial_2 g(r, X(r, s, x))\, dr\right).$

(g) $\partial_t T(t, x, y) + g(t, x)\partial_x T(t, x, y) = 0.$

(h) $\partial_s X(t, s, x) + g(s, x)\partial_x X(t, s, x) = 0.$

As an exercise the reader might verify (a)–(d) from the biological interpretation of T and X. Because of their permanent use these properties will not explicitly be quoted in the sequel.

Since we extended g from $\mathbb{R}^2 \times [a/2, 1]$ to $\mathbb{R}^2$ we should be aware of the domains in which T and X only depend on the values of g on $\mathbb{R} \times [a/2, 1]$: Eq. (6) tells us that this is the case for $T(t, x, y)$ iff $x, y \in [a/2, 1]$. It follows from (7) that, if $x \in [a/2, 1]$, $X(t, s, x)$ depends on the values of g on $\mathbb{R} \times [a/2, 1]$ iff $X(t, s, x) \in [a/2, 1]$, i.e. iff $t \in [T(s, x, a/2), T(s, x, 1)]$ or, equivalently, iff $s \in [T(t, a/2, x), T(t, 1, x)]$.

2. TRANSFORMATION OF THE PROBLEM

Weak solutions

As we shall see in Sec. 3 one cannot find classical solutions of (1) for $t > t_0$ if the initial values $n(t_0, \cdot)$ and the rates μ and b are not differentiable. Since we want to include initial values and rates which are continuous only, we look for a reformulation of (1) which is equivalent to the original one for differentiable data and solutions, but makes sense for continuous non-differentiable data and solutions as well.

Throughout this paper we make the following assumptions on the splitting and mortality rates b and μ.

ASSUMPTION 2.1

(a) μ is a continuous nonnegative function on $\mathbb{R} \times [a/2, 1]$.

(b) b is a nonnegative function on $\mathbb{R}^2$ with the following properties:

    (i) b is continuous on $\mathbb{R} \times [a, 1)$.

    (ii) $b(t, x) > 0$ if $x \in (a, 1)$, $b(t, x) = 0$ if $x \notin [a, 1)$.

    (iii) There exists a continuous function b_0 on $(a, 1)$ and some $c > 0$ such that

$$b_0(x) \leq b(t, x) \leq cb_0(x).$$

By assumption (b) cells can divide at any size in $(a, 1)$, but at no size outside $[a, 1)$. If $\int_a^1 b_0(x)\, dx = \infty$, then every cell dies or divides before reaching the maximum size 1.

In a first step we transform (1) to a simpler equation. We define

$$m = g \cdot n/E \tag{8}$$

with

$$E(t, x) = \exp\left(\int_{a/2}^x \left[\frac{\partial_1 g}{g^2} - \frac{\mu + b}{g}\right](T(t, x, z), z)\, dz\right). \tag{9}$$

If $(\mu + b)(s, z)$ is differentiable in s, by Lemma 1.2(g),

$$(\partial_t + g(t, x)\partial_x)E = \left[\frac{\partial_1 g}{g} - (\mu + b)\right] E;$$

hence Eq. (1) is equivalent to the following equations for m:

$$(\partial_t m/g)(t, x) + \partial_x m(t, x) = D(t, x)m(t, 2x), \quad x \in (a/2, 1); \tag{10a}$$

$$D(t, x)m(t, 2x) = 0, \quad x \notin [a/2, \tfrac{1}{2}); \tag{10b}$$

$$m(t, a/2) = 0; \tag{10c}$$

with

$$D(t, x) = 4\,\frac{b(t, 2x)E(t, 2x)}{g(t, 2x)E(t, x)}, \quad a/2 \leqslant x \leqslant \tfrac{1}{2}. \tag{11}$$

The initial function takes the form

$$m(t_0, x) = (g \cdot n/E)(t_0, x) = :\Phi(x). \tag{12}$$

The following properties of D follow from Assumptions 1.1 and 2.1.

LEMMA 2.2
   (a) D is continuous on $\mathbb{R} \times [a/2, \tfrac{1}{2})$.
   (b) $D(t, x) > 0$ if $x \in (a/2, \tfrac{1}{2})$, $D(t, x) = 0$ if $x \notin [a/2, \tfrac{1}{2})$.
   (c) There exists a continuous function D_0 on $[a/2, \tfrac{1}{2}]$ such that $D(t, x) \leqslant D_0(x)$, $\int_{a/2}^{1/2} D_0(x)\,dx < \infty$.

Note that Lemma 2.2 even holds if $\int_a^1 b_0(x)\,dx = \infty$. This "reduction in the singularity" will be very useful in the next section and is an extra motivation for the transformation (8).

The transformation from (1) to (10) does not yet settle the problems we mentioned at the beginning of this section. So, in a next step, we integrate (10a) along the characteristic curves T. To this end we define

$$u(s, z, x) = m(T(s, z, x), x). \tag{13}$$

By (10a) and (6),

$$\partial_x u(s, z, x) = D(T(s, z, x), x).m(T(s, z, x), 2x).$$

Hence, since $T(s, z, z) = s$,

$$u(s, z, x) = \int_z^x D(T(s, z, y), y).m(T(s, z, y), 2y)\,dy + m(s, z). \tag{14}$$

In order to return to an equation for m, we use that $m(t, x) = u(s, z, x)$ with $s = T(t, x, z)$ by (13). But, dealing with an initial-value problem, we have to avoid that $s < t_0$. So, if $x < X(t, t_0, a/2)$ we choose $z = a/2$ and $s = T(t, x, a/2) > t_0$.

If $x \geqslant X(t, t_0, a/2)$, we choose $z = X(t_0, t, x) \geqslant a/2$ and $s = T(t, x, z) = t_0$. In this way we arrive at the following integral equation for m on which we will focus in the sequel:

$$m(t, x) = \int_{a/2}^{1/2} K(t, x, y)m(T(t, x, y), 2y)\,dy + m_0(t, x) \tag{15a}$$

for $t \geq t_0$, $x \in [a/2, 1)$, with

$$K(t, x, y) = D(T(t, x, y), y), \quad \text{if } X(t_0, t, x) \leq y \leq x, \text{ and}$$
$$K(t, x, y) = 0 \qquad\qquad\qquad \text{otherwise.} \qquad\qquad (15b)$$

m_0 contains the information about the initial function Φ:

$$m_0(t, x) = \Phi(X(t_0, t, x)) \quad \text{if } X(t, t_0, a/2) \leq x < 1, \quad m_0(t, x) = 0 \quad \text{otherwise,} \quad (16)$$

with Φ being the initial values of m at $t = t_0$.

Let us summarize what we have done so far. We have shown that, if $(\mu + b)(s, x)$ is differentiable in s, solutions n of (1) correspond to solutions m of (10) via the transformation (8) and conversely. Furthermore, any solution of (10) with initial function $m(t_0, \cdot) = \Phi$ at $t = t_0$ solves (15), (16). Conversely, if m is a solution of (15), (16) and $m(t, x)$, $D(t, x)$ are differentiable in t and $\Phi(x)$ is differentiable in x, then m is a solution of (10) by Lemma 1.2 (g, h). But Eq. (15) also makes sense, if m and m_0 are continuous (without the special form of m_0 in (16)) and D has the properties listed in Lemma 2.2. So (15), (16) can be considered a weak version of (10), (12) or (1), (2) respectively, and we define the following.

Definition 2.3. A continuous solution m of (15), (16) is called a "weak solution of (10) with initial function Φ at $t = t_0$". If $m(t_0, \cdot) = \Phi$ is given by (12), $n = m \cdot E/g$ is called a weak solution of (1) with initial function $n(t_0, \cdot)$ at $t = t_0$.

Remarks
   (a) Since

$$K(t, x, y) \leq D_0(y), \quad \int_{a/2}^{1/2} D_0(y)\,\mathrm{d}y < \infty \qquad (17)$$

by (15b) and Lemma 2.2(c), it will be appropriate to handle (15) with m, m_0 being continuous on $[t_0, T] \times [a/2, 1]$, $T > t_0$. This involves that, in order to obtain weak solutions n of (1) for initial functions $n(t_0, \cdot)$, we must assume $(g \cdot n/E)(t_0, \cdot)$ to be continuous on $[a/2, 1]$. As a kind of tradeoff we obtain that $g \cdot n/E$ is continuous in $[t_0, t] \times [a/2, 1]$, $t > t_0$.

   (b) There are two other ways of defining and/or handling weak solutions of (10). The first one approximates the initial function and the other data of the equation by sufficiently smooth ones, and finds "strong" solutions of the approximating equations which have a limit: the weak solution. This can be done, for example, by studying Eq. (15). The second way studies (10) as a temporally inhomogeneous evolution equation (see, e.g. [23, XIV.4] or [24,25]) of the form

$$m'(t) = A(t)m(t), \; m(t_0) = \Phi$$

and joins an "evolutionary system" $U(t, s)$ with the operators $A(t)$. Since $U(t, t_0)\Phi$ provides "strong" solutions of (10) for sufficiently smooth data and initial values Φ, $U(t, t_0)\Phi$ can be considered a weak solution of (10). Whereas the first alternative only provides an additional characterization of weak solutions, the second also suggests a different mathematical approach.

In this paper we concentrate on (15). In a second step we show that the solution operators associated with (15), (16) form an evolutionary system. In the temporally homogeneous case [1] we subsequently refer to the underlying evolution equation and consider the infinitesimal generator A and its spectrum in order to derive conclusions about the asymptotic behaviour of the semigroup. Here the corresponding approach would consist in considering the generating operators $A(t)$ and deriving information about the qualitative behaviour of $U(t, t_0)$ from spectral properties of $A(t)$. Since the theory of abstract evolutionary systems has not yet been developed so far as the theory of semigroups, we will not do so. Instead we draw the required information about the spectrum of $U(t, t_0)$ more or less directly from (15), (16) by means of positivity and compactness arguments.

3. EXISTENCE AND UNIQUENESS OF WEAK SOLUTIONS

In the preceding section we revealed a one-to-one correspondence between weak solutions of (1), (2) and solutions of (15), (16). In this section we look for solutions m of (15) in the space $Y = C([t_0, t_1] \times [a/2, 1])$ of real-valued continuous functions with $m_0 \in Y$ being given. By Lemma 2.2 and (17) the integral in the right-hand side of (15a) provides a bounded linear operator F on Y. We show that the spectral radius of F is zero by finding equivalent norms $\|\cdot\|_\lambda$, $\lambda > 0$, on Y such that, for the associated operator norms $\|F\|_\lambda \to 0$ for $\lambda \to \infty$. We take

$$\|m\|_\lambda = \sup\{e^{-\lambda t}|m(t, x)|; \ t_0 \leq t \leq t_1, \ a/2 \leq x \leq 1\}.$$

Then, by (15) and (17),

$$\left|e^{-\lambda t}(Fm)(t, x)\right| \leq \|m\|_\lambda \int_{a/2}^{x} e^{\lambda(T(t,x,y) - t)} D_0(y) \ \mathrm{d}y.$$

We have to show that the integrals $I_\lambda(t, x)$ on the right-hand side of this equation converge to zero for $\lambda \to \infty$ uniformly in t. Since $T(t, x, y) < t$ for $y < x$ and D_0 is integrable, pointwise decreasing convergence follows from Lebesgue's theorem of dominated convergence. As $I_\lambda(t, x)$ continuously depends on t and x, uniform convergence follows from Dini's lemma. Thus (15) takes the abstract form

$$m = Fm + m_0$$

with the bounded linear operator F having zero spectral radius. Hence $(I - F)^{-1}$ exists (with I being the identity operator on Y) and can be represented by the Neumann series ΣF^j with convergence in the operator topology. Note that the operator F preserves continuity and non-negativity and so does $(I - F)^{-1}$. So we obtain the following.

THEOREM 3.1

Let m_0 be a continuous function from $[t_0, \infty) \times [a/2, 1]$ to $\mathbb{R}$. Then there exists a unique continuous solution m of (15) on $[t_0, \infty) \times [a/2, 1]$. m can be represented as

$$m(t, x) = \sum_{j=0}^{\infty} m_j(t, x) \tag{18}$$

with

$$m_j(t, x) = \int_{a/2}^{1/2} K(t, x, y)m_{j-1}(T(t, x, y), 2y) \ \mathrm{d}y \tag{19}$$

for $j = 1, 2, \ldots$. The convergence of the series in (18) is uniform on $[t_0, t_1] \times [a/2, 1]$ for any $t_1 > t_0$. If m_0 is nonnegative, so is m.

We have not yet stated that m depends continuously on m_0. To make this precise let $\| \ \|_{0,t}$ be the sup-norm on $[t_0, t] \times [a/2, 1]$ for $t < \infty$.

PROPOSITION 3.2

Let m_0 and m be as in Theorem 3.1. Then

$$\|m\|_{0,t} \leq c(t)\|m_0\|_{0,t} \quad \text{for } t \geq t_0$$

with constants $c(t)$ depending continuously on t, but not depending on m_0.

This boundedness result can be sharpened. To this end we define

$$v_0(t, x) = \int_{a/2}^{x} D(T(t, x, y), y) \ \mathrm{d}y, \quad a/2 \leq x \leq 1, \tag{20}$$

with D from (11), $D(t, y) = 0$ for $y > \frac{1}{2}$. It follows from assumption 2.1(b) (iii) that

$$v_0(t, x) \le \int_{a/2}^{x} D_0(y) \, dy \tag{21}$$

for $t \in \mathbb{R}$, $x \in [a/2, 1]$, with D_0 from Lemma 2.2(c). Hence we obtain the following estimate from proposition 3.2 and (15a).

COROLLARY 3.3

Let m_0 and m be as in Theorem 3.1. Then

$$|m(t, x)| \le c(t)\|m_0\|_{0,t} \cdot v_0(t, x) + |m_0(t, x)|$$

with constant $c(t)$ depending continuously on t, but not on m_0.

In order to let the reader appreciate this estimate we recall that $m_0(t, \cdot) \equiv 0$, if m_0 is given by (16) and $t \ge T(t_0, a/2, 1)$.

Remark 3.4

(a) As in [1], Sec. 4, m_j in (18) can be considered the jth generation of cells and for any $t_1 > t_0$ it can be shown that there exists $j_0 \in \mathbb{N}$ with $m_j(t, x) = 0$ for $j \ge j_0$, $t \in [t_0, t_1]$, $x \in [a/2, 1]$. So at any time only finitely many generations are present.

(b) The last observation helps to find conditions under which a solution m of (15), (16), i.e. a weak solution of (10) with initial function Φ, actually is a strong solution of (10) and so provides a strong solution of (1). As we mentioned in Sec. 2 the crucial step consists in proving the differentiability of $m(t, x)$ in t. If m_0 is given by (16), it is differentiable if Φ is differentiable and $\Phi(a/2) = \Phi'(a/2) = 0$. The operator F defined by (19) preserves differentiability except at $x = \frac{1}{2}$, if D is differentiable in t, i.e. if b and μ are differentiable with respect to time. Thus all m_j are differentiable and so is m because the series (18) is locally finite.

4. PROPERTIES OF THE SOLUTION OPERATORS

In studying the asymptotic behaviour of weak solutions n to (1), (2), or, equivalently, of solutions m to (15), (16) in the case of time-periodic developmental rates, we want to apply the spectral theory of strongly positive, quasicompact operators on Banach lattices (see, e.g. [18]). To this end it is convenient to consider the solution operators belonging to Eqs. (15), (16).

Let Z be the Banach space of continuous real-valued functions u on $[a/2, 1]$ with $u(a/2) = 0$. The norm $\|\cdot\|$ on Z is provided by the supremum-norm. Note that the continuity of u at $x = 1$, in view of Eq. (1), involves for the initial values of n that $(g \cdot n/E)(t_0, x)$ converges for $x \uparrow 1$.

For $\Phi \in Z$ there exists a unique solution m of (15), (16) by Theorem 3.1. The solution operators $U(t, t_0)$, $t \ge t_0$ are now defined by

$$U(t, t_0)\Phi = m(t, \cdot). \tag{22}$$

It turns out that the operators $U(t, t_0)$, $t \ge t_0$, form an evolutionary system.

PROPOSITION 4.1

(a) $U(t, t_0)$ is a bounded linear operator on Z.

(b) $U(t, s)U(s, t_0) = U(t, t_0)$, $t_0 \le s \le t$, $U(t, t)u = u$, $u \in Z$.

(c) For $u \in Z$, $U(t, t_0)u$ continuously depends on t, $t \ge t_0$.

(a) and (c) follow from Theorem 3.1 immediately. (b) can also be formulated in this way: if m is a weak solution of (10) with initial values at $t = t_0$, then

$$m(t, \cdot) = U(t, s)m(s, \cdot), \ t \ge s \ge t_0. \tag{23}$$

Since solutions of (15) are unique by Theorem 3.1, (23) follows by showing that $m(t, x)$ is a weak solution of (10) for $t > s$ with initial values $m(s, \cdot)$ at $t = s$. This is easily done by using the properties of T and X in Lemma 1.2.

In order to prepare the application of the theory of strongly positive operators we recall that Z is a Banach lattice. The cone Z_+ is formed by the nonnegative functions, the ordering '$\leq$' is the point-wise ordering and the modulus (or absolute value) of $u \in Z$ is given by $|u|(x) = |u(x)|$, $x \in [a/2, 1]$ (see the Appendix and [18]).

First we note that the boundedness of $U(t, t_0)$ holds in a stricter, order-theoretic way if t is large enough.

PROPOSITION 4.2

For $t \geq T(t_0, a/2, 1)$ we have $|U(t, t_0)u| \leq c\|u\|v_0$ for $u \in Z$ with c depending on t and t_0, and v_0 given by (20).

In other words $U(t, t_0)$ continuously maps Z into the Banach space $Z_{v_0(t)}$ (see Appendix 2), if $t \geq T(t_0, a/2, 1)$. For then $m_0(t, \cdot) = 0$, if m_0 is given by (16), and so the proposition easily follows from Corollary 3.3. Finding conditions under which the operators $U(t, t_0)$ are strongly positive is much more involved.

Strong positivity of the solution operators

We look for conditions under which, after a sufficiently long time t, cells of every size in $(a/2, 1)$ are present in the population, i.e.

$$m(t, x) > 0 \quad \text{for } x \in (a/2, 1]. \tag{24}$$

We claim that the following assumption will work.

ASSUMPTION 4.3

For any $x \in [\frac{1}{2}, 1)$ there exist $\epsilon > 0$, $y, z \in [a, 1]$, $y \geq x$, $z \geq \frac{1}{2}$ such that

$$T(T(s, x, y), y/2, x) + \epsilon \leq T(T(s, \tfrac{1}{2}, z), z/2, \tfrac{1}{2})$$

for all $s \in \mathbb{R}$.

Assumption 4.3 roughly states the following: consider a cohort (group) of cells all having the same size $x \in [\frac{1}{2}, 1)$ and a cohort of cells having size $\frac{1}{2}$, at time s. Then *some* daughters of the first cohort have reached size x again before *all* daughters of the second cohort (either have divided or) have reached $\frac{1}{2}$.

In order to show that Assumption 4.3 actually implies (24) for large t, we define $\tilde{T}(s, x)$ to be the latest possible time at which a daughter of a cell having size x at time s can reach size x again, formally

$$\tilde{T}(s, x) = \sup\{T(T(s, x, y), y/2, x); x, a \leq y \leq 1, 2x\}. \tag{25}$$

Remember that the mother cell can split at any size y with x, $a < y < 1$. We also define $\tilde{X}(t, s, x)$ and $\underset{\sim}{X}(t, s, x)$ to be the maximum and the minimum size which daughters of a cell having size x at time s can reach up to time t, formally

$$\tilde{X}(t, s, x) = \sup\{X(t, T(s, x, y), y/2); x, a \leq y \leq X(t, s, x), 1\} \tag{26}$$

for $t \geq s$, $t \geq T(s, x, a)$. $\underset{\sim}{X}(t, s, x)$ is the corresponding infimum. Equation (26) only makes sense if the mother cell can reach the minimum splitting size a up to time t, i.e. for $t > T(s, x, a)$, or equivalently, $a \leq X(t, s, x)$. We mention that $\tilde{X}$, $\underset{\sim}{X}$ are continuous and that $\tilde{X}(t, s, x)$ and $\underset{\sim}{X}(t, s, x)$ strictly monotone increase if t, x increase and s decreases. More precisely, there is some $\bar{\epsilon} > 0$ such that

$$\tilde{X}(t_2, s, x) - \tilde{X}(t_1, s, x) \geq \bar{\epsilon}(t_2 - t_1) \quad \text{for all } t_2 \geq t_1 \geq s, x \in [a/2, 1] \text{ with } t_1 \geq T(s, x, a).$$

Furthermore, for any $r > 0$, $1 \geq x_2 > x_1 > a/2$ there exists $\epsilon > 0$ such that

$$\tilde{X}(t, s, x_2) - \tilde{X}(t, s, x_1) \geq \epsilon, \quad \text{if } 0 \leq t - s \leq r, t \geq T(s, x_1, a).$$

The same statements hold for $\underline{X}$. They follow from the Assumptions 1.1 and from Lemma 2.2.

Since Assumption 4.3 states that daughters of a cell having size $x \in [\frac{1}{2}, 1)$ at time s can reach size x again before time $\tilde{T}(s, \frac{1}{2})$, it follows that $\tilde{T}(s, \frac{1}{2}) \geq T(s, x, a)$ and

$$\tilde{X}(\tilde{T}(s, \tfrac{1}{2}), s, x) \geq x + \delta \quad \text{for } x \in [\hat{x}, 1), \tag{27}$$

where $a/2 < \hat{x} < \frac{1}{2}$ and $\delta > 0$ can be chosen independently of $s \in \mathbb{R}$ and of x in any compact subset of $[\hat{x}, 1)$. Equation (27) can be derived from Assumption 4.3 rigorously by exploiting Lemma 1.2. Similarly, it is intuitively clear and can rigorously be derived from Lemma 1.2 that

$$\underline{X}(\tilde{T}(s, \tfrac{1}{2}), s, \tfrac{1}{2}) = \tfrac{1}{2}. \tag{28}$$

Before we actually dive into the proof of (24) we state a couple of useful lemmas.

The following statement is intuitively evident from the interpretation of X.

Lemma 4.4

Let m be a weak solution of (10) for $t > t_0$. If $s \geq t_0$, $a/2 < x_1 \leq x_2 \leq 1$, and $m(s, y) > 0$ for $x_1 \leq y \leq x_2$, then $m(t, x) > 0$ for $t \geq s$, $X(t, s, x_1) \leq x \leq X(t, s, x_2)$, $x \leq 1$.

In order to prove Lemma 4.4 take into account that $m(t, x)$ is a weak solution of (10) with initial function $m(s, \cdot)$ at $t = s$ (see Proposition 4.1 and the subsequent remarks). Hence, by (15) and (16), $m(t, x) \geq m(s, X(s, t, x))$ if $X(t, s, a/2) \leq x \leq 1$.

The next lemma states that the presence of cells of size x, $a < x < 1$, implies the presence of cells of size $x/2$ by splitting. It is an obvious consequence of the continuity of m and of (15a).

Lemma 4.5

Let m be as in Lemma 4.4, $s > t_0$, $x \in (a, 1]$. If $m(s, x) > 0$, then $m(s, x/2) > 0$.

Combining Lemmas 4.4 and 4.5 yields the following statement, which is intuitively evident from the interpretation of $\tilde{X}$ and $\underline{X}$.

Lemma 4.6

Let m be as in Lemma 4.4. Let $t > s > t_0$, $a/2 < x_1 \leq x_2 \leq 1$, $t \geq T(s, x_1, a)$. If $m(s, x) > 0$ for $x_1 \leq x \leq x_2$, then $m(t, z) > 0$ for $\underline{X}(t, s, x_1) \leq z \leq \tilde{X}(t, s, x_1)$, $z \leq 1$.

Actually combining Lemmas 4.4 and 4.5 leads to the following conclusion: if $s > t_0$, $a/2 < x < 1$ and $m(s, x) > 0$, then $m(t, z) > 0$ for all $z = X(t, r, X(r, s, x)/2)$ with $z \leq 1$, $s \leq r \leq t$, $a \leq X(r, s, x) \leq 1$, or, equivalently for all $z = X(t, T(s, x, y), y/2)$ with $z \leq 1$, x, $a \leq y \leq 1$. Lemma 4.6 now follows from the definitions of $\tilde{X}$ and $\underline{X}$ in (26).

From (27) and (28) we can now derive Lemma 4.7.

Lemma 4.7

Let m be as in Lemma 4.4, $s > t_0$, $t = \tilde{T}(s, \frac{1}{2})$, $x \in [\frac{1}{2}, 1)$. If $m(s, y) > 0$ for $\frac{1}{2} \leq y \leq x$ then $m(t, z) > 0$ for $z \in [\frac{1}{2}, \tilde{x}] \cap [\frac{1}{2}, 1]$ with $\tilde{x} = \tilde{X}(t, s, x)$.

Remark

Note that $\tilde{x} > x + \delta$ and that $\delta > 0$ can be chosen independently of $s \in \mathbb{R}$ and of x in compact subsets of $[\frac{1}{2}, 1)$.

After these preparations we are ready for the proof of (24).

First we note from Lemmas 4.4 and 4.5 that, if the initial value $\Phi \in Z_+$ at time $t = t_0$ satisfies $\Phi \neq 0$, then

$$m(s_0, \tfrac{1}{2}) > 0$$

with some $s_0 \in (t_0, T(t_0, a/2, 1)]$. Guided by Lemma 4.5 we define

$$s_{j+1} = \tilde{T}(s_j, \tfrac{1}{2}),$$
$$x_{j+1} = \min(1, \tilde{X}(s_{j+1}, s_j, x_j)),$$
$$x_0 = \tfrac{1}{2},$$

It follows that

$$x^* = \lim_{j \to \infty} x_j = 1,$$

for $j = 0, 1, \ldots,$ and find a strictly increasing sequence $x_j \leqslant 1$ with

$$m(s_j, x) > 0 \quad \text{for } x \in [\tfrac{1}{2}, x_j].$$

for, if $x^* < 1$, by Lemma 4.7 and the subsequent remark, $x_{j+1} \geqslant x_j + \delta$ with $\delta > 0$ for all $j \in \mathbb{N}$, in contradiction to the convergence of x_j. Now, by Lemma 4.5,

$$m(s_j, x) > 0 \quad \text{for } x \in (\underset{\sim}{x}, x_j/2] \cup [\tfrac{1}{2}, x_j] \tag{29}$$

with $\underset{\sim}{x} = \max(a/2, \tfrac{1}{4})$ and $x_j \to 1$ for $j \to \infty$. By (27) there exists $\underset{\wedge}{x} \in [\hat{x}, \tfrac{1}{2})$ such that

$$\tilde{X}(\tilde{T}(s, \tfrac{1}{2}), s, \underset{\wedge}{x}) \geqslant \tfrac{1}{2}$$

for all $s \in \mathbb{R}$, in particular

$$\tilde{X}(s_{j+1}, s_j, \underset{\wedge}{x}) \geqslant \tfrac{1}{2}.$$

By the remarks following the definition of $\tilde{X}$, $\underset{\sim}{X}$ in (26) we obtain

$$\underset{\sim}{X}(s_{j+1}, s_j, \underset{\wedge}{x}) \leqslant \underset{\sim}{X}(s_{j+1}, s_j, \tfrac{1}{2}) - \delta = \tfrac{1}{2} - \delta$$

with $\delta > 0$ not depending on j. Since, for large j, $x_j/2 > \underset{\wedge}{x}$ and $x_{j+1}/2 > \tfrac{1}{2} - \delta$ we obtain from (29) and Lemma 4.6 that

$$m(s_{j+1}, x) > 0 \quad \text{for } x \in (\underset{\wedge}{x}, x_{j+1}],$$

for sufficiently large j, with $x_j \to 1$ for $j \to \infty$. If j is large enough, $T(s_j, x_j, 1) < T(s_j, \underset{\wedge}{x}, \tfrac{1}{2})$; hence

$$m(\tilde{s}, x) > 0 \quad \text{for } \tfrac{1}{2} \leqslant x \leqslant 1,$$

with some large $\tilde{s} > s_0$. By Lemma 4.5,

$$m(\tilde{s}, x) > 0 \quad \text{for } a/2 < x \leqslant 1.$$

The continuity of m and Lemma 4.5 yield

$$m(s, x) > 0 \quad \text{for } a/2 < x \leqslant 1, s \geqslant \tilde{s}.$$

We formulate this result in terms of the solution operators $U(t, t_0)$. $v \in Z_+$ is a quasi-interior point of Z_+ (see the Appendix, point 4) iff v is continuous on $[a/2, 1]$ and $v(x) > 0$ for $x \in (a/2, 1]$.

PROPOSITION 4.8

Let the assumption 4.3 be satisfied. Then, for $t_0 \in \mathbf{R}$, there exists $t_1 > t_0$ such that $U(t, t_0)\Phi$ is a quasiinterior point of Z_+ if $t \geq t_1$, $\Phi \in Z_+$, $\Phi \neq 0$.

Note that t_1 is independent of Φ.

After having established this positivity property of the operators $U(t, t_0)$ for $t - t_0$ being large we now turn to compactness.

Compactness of the solution operators

In order to show that the operators $U(t, t_0)$ are compact on Z if $t - t_0$ is large enough, we consider an arbitrary bounded subset M of Z and consider the weak solutions m of (10) with initial function $\Phi \in M$ for $t = t_0$. By the Arzela–Ascoli theorem we have to show that $m(t, \cdot)$ is bounded and continuous uniformly in $\Phi \in M$ for $t - t_0$ being large. Boundedness is obvious from Theorem 3.1 and (16). For the proof of equicontinuity we set

$$v(s, x) = m(T(s, a/2, x), x) \quad \text{for } s \geq t_0.$$

It follows from (14) and Theorem 3.1 that $v(s, x)$ is continuous in x uniformly for s ranging in a bounded interval and $\Phi \in M$. Transforming (14) into an equation for v we obtain

$$v(s, x) = \int_{a/2}^{\min(x, 1/2)} D(T(s, a/2, y), y)v(f(s, y), 2y) \, dy \tag{30}$$

for $s \geq T(t_0, a/2, 1)$, with

$$f(s, y) = T(T(s, a/2, y), 2y, a/2). \tag{31}$$

Since $m(t, x) = v(T(t, x, a/2), x)$ for $t \geq T(t_0, a/2, x)$ we are done, if v can be shown to be continuous in (s, x) uniformly in $\Phi \in M$. This follows from the uniform (with respect to s and $\Phi \in M$) continuity in x, if a change of variables $f(s, y) = r$ can be performed in the right-hand side of (30). To this end we differentiate f with respect to y and obtain, with $\bar{s} = T(s, a/2, y)$ from Lemma 1.2(g),

$$\partial_y f(s, y) = 2\partial_1 T(\bar{s}, 2y, a/2) \cdot \left(\frac{1}{2g(\bar{s}, y)} - \frac{1}{g(\bar{s}, 2y)} \right). \tag{32}$$

Guided by this formula we make the following assumption.

ASSUMPTION 4.9

There exist at most finitely many points $x_i \in [a/2, \tfrac{1}{2}]$ such that $2g(s, x_i) = g(s, 2x_i)$ for some $s \in \mathbf{R}$.

By this assumption the interval $[a/2, \tfrac{1}{2}]$ can be divided into intervals $[x_i, x_{i+1}]$ such that $2g(s, y) - g(s, 2y)$ is either strictly positive on (x_i, x_{i+1}) or strictly negative.

We want to solve the equation

$$f(s, h_i(s, r)) = r \tag{33}$$

with $h_i(s, r) \in (x_i, x_{i+1})$. To this end we differentiate (33) with respect to r and obtain the following differential equation for h_i:

$$\partial_r h_i(s, r) = \frac{1}{\partial_2 f(s, h_i(s, r))}. \tag{34}$$

Since $\partial_1^2 g$ exists and is continuous by Assumption 1.1, $\partial_2 f(s, y)$ is Lipschitz continuous in (s, y) by Lemma 1.2(e) and (32). So we find a continuous solution $h_i(s, r)$ of (34) and thus of (33) for $s \in \mathbb{R}$, r in the interval with endpoints $f(s, x_i)$, $f(s, x_{i+1})$. We now split up the integral (30) into integrals

$$\int_{\min\{x, x_i\}}^{\min\{x, x_{i+1}\}} D(T(s, a/2, y), y)v(f(s, y), 2y) \, \mathrm{d}y$$

$$= \int_{f(s, \min\{x, x_i\})}^{f(s, \min\{x, x_{i+1}\})} \frac{D(T(s, a/2, h_i(s, r)), h_i(s, r))}{\partial_2 f(s, h_i(s, r))} v(r, 2h_i(s, r)) \, \mathrm{d}r.$$

Since the first integral exists if we take absolute values of the integrand, so does the second. We conclude that the integrals are continuous in (s, x) uniformly for $\Phi \in M$.

Summarizing our preceding considerations we conclude that $v(s, x)$ is continuous in (s, x), $s \geq T(t_0, a/2, 1)$, $a/2 \leq x \leq 1$, uniformly for $\Phi \in M$. Since $m(t, x) = v(T(t, x, a/2), x)$ for $t \geq T(t_0, a/2, x)$, $m(t, x)$ is continuous in (t, x) uniformly for $\Phi \in M$, provided that $T(t, x, a/2) \geq T(t_0, a/2, 1)$, $a/2 \leq x \leq 1$, thus in particular if $t \geq \phi(\phi(t_0))$ with $\phi(s) = T(s, a/2, 1)$. By the Arzela–Ascoli theorem (see, e.g. [22, IX, Sec. 4]) we obtain the following result.

PROPOSITION 4.10

Let Assumptions 4.9 be satisfied. Then the operators $U(t, t_0)$ are compact on Z for $t \geq \phi(\phi(t_0))$.

One can presumably prove that the operators $U(t, t_0)$ are compact on Z for $t \geq \phi(t_0)$ by studying the generation expansion (18), (19) (see [1], Sec. 5). Recall that at time $\phi(t_0)$ all cells from the initial population (i.e. the cells of the zero generation) have divided or died.

Furthermore, Assumption 4.9 can be relaxed, e.g. by assuming that the set $\{(s, y); s \in \mathbb{R}, a/2 \leq y \leq \frac{1}{2}, 2g(s, y) = g(s, 2y)\}$ is contained in the union of finitely many graphs of continuous functions $x_i \colon \mathbb{R} \to [a/2, \frac{1}{2}]$.

5. THE SOLUTION OPERATORS UNDER PERIODICITY

In this section we study the operator

$$B = U(t_0 + p, t_0) \tag{35}$$

the properties of which are of crucial importance for the asymptotic behaviour of weak solutions of (10), if the developmental rates are time-periodic with period p. We will use the language of Banach lattices which is summarized in the Appendix.

From now on we make the following assumption.

ASSUMPTION 5.1

$g(t, x)$, $\mu(t, x)$, $b(t, x)$ are periodic in $t \in \mathbb{R}$ with the same period $p > 0$ for g, b, μ.

This periodicity assumption has the following consequence for the solution operators.

PROPOSITION 5.2

$U(t + p, t_0 + p) = U(t, t_0)$ for all $t, t_0 \in \mathbb{R}$, $t \geq t_0$.

Proof. Let m be the weak solution of (10) with initial function $\Phi \in Z$ at $t = t_0$. By the uniqueness of solutions it is sufficient to show that $\tilde{m}(t, \cdot) = m(t - p, \cdot)$ is a weak solution of (10) with initial values Φ at $t = t_0 + p$. This easily follows from Lemma 5.3.

LEMMA 5.3

$T(t + p, x, y) = T(t, x, y) + p$, $X(t + p, s + p, x) = X(t, s, x)$.

The lemma follows from the uniqueness of solutions to the differential equations (6), (7). Proposition 4.1 now implies the following.

PROPOSITION 5.4

    (a) B is a positive bounded linear operator on Z.

    (b) For $k \in \mathbb{N}$, $t \in [0, p]$, $U(t_0 + kp + t, t_0) = U(t_0 + t, t_0)B^k$.

Assumptions 4.3 now take the following seemingly weaker form (see Lemma 5.3).

ASSUMPTION 5.5

    For any $x \in [\frac{1}{2}, 1)$ there exist $y, z \in [a, 1]$, $y \geqslant x$, $z \geqslant \frac{1}{2}$ such that $T(T(s, x, y), y/2, x)$ $< T(T(s, \frac{1}{2}, z), z/2, \frac{1}{2})$ for all $s \in \mathbb{R}$.

Recall the interpretation we gave after Assumption 4.3. Propositions 4.2, 4.8 and 4.10 imply the following properties of B.

PROPOSITION 5.6

    (a) If Assumption 4.9 holds then B^j is a compact operator on Z for sufficiently large j.

    (b) For large enough j, B^j maps Z continuously into the Banach lattice $Z_{v_0(t_0)}$ with $v_0(t_0)$ being the quasi-interior point of Z_+ defined in (20).

    (c) If Assumption 5.5 holds then B^j maps $Z_+ \backslash \{0\}$ into the quasi-interior points of Z_+ for large enough j. In particular B is strongly positive on Z (see Appendix, points 4, 7).

The following spectral properties of B now follow easily from the theory of power compact strongly positive operators (see [18, Chap. V]).

PROPOSITION 5.7

    Let Assumptions 4.9 and 5.5 be valid. Then the following holds:

    (a) The spectral radius $r_0 = \mathrm{spr}\, B$ of B is different from zero.

    (b) r_0 is an algebraically simple eigenvalue of B and B'.

    (c) There is an eigenvector w_0 of B belonging to r_0 which is a quasi-interior point of Z_+.

    (d) There is a strictly positive eigenfunctional $w_0' \in Z'$ of B' belonging to r_0, $w_0' = 0$ on $(r_0 I - B)Z$.

    (e) All spectral values of B different from r_0 lie in a circle around $0 \in \mathbb{C}$ with radius strictly smaller than r_0.

    (f) r_0 is the unique eigenvalue of B with an eigenvector in Z_+.

See the Appendix, point 5. Since r_0 is a pole of the resolvent of B, we can split up the space Z into a direct sum

$$Z = \mathrm{span}\{w_0\} \oplus \check{Z},$$

with the eigenvector w_0 being provided by Proposition 5.7(c) and the B-invariant closed subspace $\check{Z} = (r_0 I - B)Z$ (see, e.g. [26, Chap. 1]). The spectral radius of B restricted to $\check{Z}$ is strictly smaller than r_0 by Proposition 5.7(e). In (d) it is stated that $w_0' = 0$ if restricted to $\check{Z}$. So we obtain the following result.

PROPOSITION 5.8

    Let w_0' be normalized such that $w_0' w_0 = 1$. Then there exists a bounded linear projection P on Z with the following properties:

    (a) $PB = BP$.

    (b) $Pw_0 = 0$, $w_0' P = 0$.

    (c) $r_1 := \mathrm{spr}\, BP < r_0 := \mathrm{spr}\, B$.

    (d) Any $w \in Z$ has the unique representation $w = w_0'(w)w_0 + Pw$.

We conclude this section by characterizing the nonzero eigenvalues and eigenvectors of B. Let $Bw = qw$, $q \neq 0$ and let m be the weak solution of (10) with initial value w at $t = t_0$, i.e. $m(t, \cdot) = U(t, t_0)w$. Let $\lambda \in \mathbb{C}$ be such that $e^{\lambda p} = q$ and set

$$\tilde{m}(t, \cdot) = e^{-\lambda t}m(t, \cdot).$$

Then

$$\tilde{m}(t + p, \cdot) = e^{-\lambda(t+p)}U(t + p, t_0)w = e^{-\lambda t}U(t + p, t_0 + p)w = e^{-\lambda t}U(t, t_0)w = \tilde{m}(t, \cdot)$$

by Proposition 5.4(b). We extend $\tilde{m}$ for $t < t_0$ in a p-periodic way. If $t \geq T(t_0, a/2, 1)$

$$\tilde{m}(t, x) = \int_{a/2}^{\min\{x,1/2\}} D(T(t, x, y), y)e^{\lambda(T(t,x,y)-t)}\tilde{m}(T(t, x, y), 2y)\, dy \qquad (36)$$

by (15a). Since $\tilde{m}(t, x)$ is p-periodic in t, this equality holds for all $t \in \mathbb{R}$ (see Lemma 5.3). Moreover, $\tilde{m}$ is continuous.

Conversely, if $\tilde{m}$ is a continuous p-periodic solution of (36) on $\mathbb{R} \times [a/2, 1]$, then $m(t, x) = e^{\lambda t}\tilde{m}(t, x)$ is a weak solution of (10) with initial value $w = e^{\lambda t_0}\tilde{m}(t_0, \cdot)$ at $t = t_0$ satisfying $Bw = e^{\lambda p}w$. (Use Lemma 1.2.)

Thus we have obtained the following relation between eigenvalues of B and p-periodic solutions of (36).

PROPOSITION 5.9

Let $q, \lambda \in \mathbb{C}, q = e^{\lambda p}$. Then q is an eigenvalue of B iff there exists a p-periodic continuous solution $\tilde{m}$ on $\mathbb{R} \times [a/2, 1]$ of (36). The eigenvectors w of B belonging to q are related to the periodic solutions $\tilde{m}$ of (36) by

$$U(t, t_0)w = e^{\lambda t}\tilde{m}(t, \cdot) \quad \text{for } t \geq t_0.$$

6. ASYMPTOTIC BEHAVIOUR OF SOLUTIONS

In order to prove our main result on the asymptotic behaviour of weak solutions to (10) or to (1) respectively, we continue the considerations of the last section.

Let $\Phi \in Z$ and let m be the weak solution to (10) with initial value Φ at $t = t_0$. Furthermore, let $t > t_0 + kp$, $k \in \mathbb{N}$. Then, by Propositions 5.4 and 5.8,

$$m(t, \cdot) = U(t, t_0)\Phi = U(t - kp, t_0)B^k\Phi = w_0'(\Phi)U(t, t_0)w_0 + U(t - kp, t_0)(PB^k\Phi),$$

and $\|PB^k\| \leq cr^k$ for any $r \in (r_1, r_0)$ with $c > 0$ depending on r, but not on k.

By Proposition 5.9 $\tilde{m}(t, \cdot) = e^{-\lambda t}U(t, t_0)w_0$ is a p-periodic continuous solution of (36) on $\mathbb{R} \times [a/2, 1]$. Since w_0 is strictly positive on $(a/2, 1]$ by Proposition 5.7(a), $\tilde{m}(t, x) > 0$ for $t \in \mathbb{R}, x \in (a/2, 1]$ by Proposition 4.8. Now (36), (20), and the periodicity of $\tilde{m}$ imply that

$$\epsilon v_0(t, x) \leq \tilde{m}(t, x) \leq c v_0(t, x)$$

for $t \in \mathbb{R}, x \in [a/2, 1]$ with $\epsilon, c > 0$ not depending on t and x. On the other hand, $|U(t - kp, t_0)PB^k\Phi| \leq cr^k\|\Phi\|v_0(t)$ by Proposition 4.2, if $t - kp \geq T(t_0, a/2, 1)$ with c depending on r. Combining these observations and Corollary 3.3 yields the following.

THEOREM 6.1

Let Assumptions 1.1, 2.1, 4.9, 5.1 and 5.5 be satisfied. Then the following holds:

(a) There exists a unique $\lambda \in \mathbb{R}$ such that (36) admits a continuous nonnegative time-periodic (with period p) solution $\tilde{m} \neq 0$ on $\mathbb{R} \times [a/2, 1]$. $\tilde{m}$ is uniquely determined up to a scalar factor.

(b) If m is a weak solution of (10) with continuous initial function Φ on $[a/2, 1]$ at $t = t_0$, $\Phi(a/2) = 0$, then, for $t \to \infty$,

$$m(t, x) = e^{\lambda t}\tilde{m}(t, x).(\alpha + e^{-\epsilon t}\mathcal{O}(1))$$

with some $\epsilon > 0$. The scalar α depends in a linear and strictly positive way on the initial function Φ.

More precisely, one can say that $\alpha(\Phi)$ is a strictly positive bounded linear functional on Z. The Landau symbol $\mathcal{O}(1)$ represents a bounded function $m(t, x)$. Actually $m(t, \cdot) = \tilde{U}(t, t_0)\Phi$ with bounded linear operators $\tilde{U}(t, t_0)$ on Z and $\|\tilde{U}(t, t_0)\|$ is bounded on $[t_0, \infty)$.

We now translate this result to weak solutions of (1) via (8), (9). We define

$$\bar{E}(t, x) = \exp\left(-\int_{a/2}^{x} (b/g)(T(t, x, z), z)\, \mathrm{d}z\right).$$

COROLLARY 6.2

Under the assumptions of Theorem 6.1 the following holds.

Let n be a weak solution of (1) for $t > t_0$ such that $n(t_0, x)/\bar{E}(t_0, x)$ is continuous in $x \in [a/2, 1]$ and $n(t_0, a/2) = 0$. Then, for $t \to \infty$,

$$n(t, x) = e^{\lambda t}\bar{n}(t, x)(\alpha + e^{-\epsilon t}\mathcal{O}(1))$$

with some $\epsilon > 0$. In this expression λ and the time-periodic (with period p) function $\bar{n}$ do not depend on the initial function $n(t_0, \cdot)$. The scalar α, however, depends linearly and strictly positive on the initial function.

More precisely, $\alpha(\Psi)$, $\Psi = n(t_0, \cdot)$, is a strictly positive bounded linear functional on the Banach space $Z_{\bar{E}(t_0, \cdot)}$. $\mathcal{O}(1)$ stands for a bounded function $n(t, x)$. Actually $n(t, \cdot) = \tilde{U}(t, t_0)\Psi$ with $\tilde{U}(t, t_0)$ now being bounded linear operators from $Z_{\bar{E}(t_0, \cdot)}$ to Z and $\|\tilde{U}(t, t_0)\| \leqslant$ const for all $t \geqslant t_0$.

Though Assumption 5.5 has a clear biological meaning (see the interpretation following Assumption 4.3) one would like to have an assumption in terms of the growth rate g as well. Unfortunately a condition of that kind which is not too complicated can only be given in the very special case

$$g(t, x) = \gamma(t)g(x).$$

Then, by (6),

$$\int_{t}^{T(t,x,y)} \gamma(s)\, \mathrm{d}s = \int_{x}^{y} \frac{\mathrm{d}z}{g(z)},$$

and Assumption 5.5 takes the following form.

For any $x \in [\tfrac{1}{2}, 1)$ there exist $y, z \in [a, 1]$, $y \geqslant x$, $z \geqslant \tfrac{1}{2}$ such that

$$\int_{y/2}^{y} \frac{\mathrm{d}\zeta}{g(\zeta)} < \int_{z/2}^{z} \frac{\mathrm{d}\zeta}{g(\zeta)}, \quad \text{i.e.} \quad \int_{z/2}^{y/2} \left\{\frac{1}{g(2\zeta)} - \frac{1}{2g(\zeta)}\right\} < 0.$$

But this already follows from Assumption 4.9. So we obtain the following result.

COROLLARY 6.3

Let Assumptions 1.1, 2.1 and 5.1 be satisfied, $g(t, x) = \gamma(t)g(x)$ with $g(2x) \neq 2g(x)$ for all but finitely many $x \in [a/2, \tfrac{1}{2}]$. Then the statements (a) and (b) of Theorem 6.1 and Corollary 6.2 are valid.

7. THE TIME-HOMOGENEOUS CASE REVISITED

For time-independent developmental rates one expects Theorem 6.1 and Corollary 6.2 to hold with $\bar{m}(t, x)$ and $\bar{n}(t, x)$ not depending on t. In fact, we established the theorems in this form in [1] under one of the following conditions on the growth rate g:

(i) $g(2x) < 2g(x)$, $x \in [a/2, \tfrac{1}{2}]$.

(ii) $a \geqslant \tfrac{1}{2}$, $g(x) = x$ for $x \in [a/2, \beta]$; $g(x) < x$ for $x \in (\beta, 1]$ with $\beta < 1$.

Unfortunately neither condition is satisfied by the experimental data for g of Anderson *et al.* [2]; in particular they found $a < \frac{1}{2}$. This situation is especially unsatisfactory because Anderson *et al.* gathered their data from cell populations which, after the experiment had run for some time, showed exponential growth with a stationary size distribution. So our theorem should hold for their data. In [1] we guessed that a substantial generalization of conditions (i) or (ii) would be very laborious or even impossible. Fortunately, this fear is unfounded. Let us assume that

$$g(2x) \neq 2g(x) \quad \text{for } x \in J, \tag{37}$$

J being a nonempty open subinterval of $[a/2, \frac{1}{2}] =: I$. Since g is continuous, (37) is a necessary condition for asymptotic exponential growth with stationary size distribution, as we pointed out at the beginning of [1, Sec. 8]. Replacing Assumption 4.9 by (37) we in general lose the compactness of the operators $U(t, t_0)$ even if $t - t_0$ is large. The hope, however, is that (37) is strong enough to imply that the radius of the essential spectrum of $U(t, t_0)$ is strictly smaller than the spectral radius of $U(t, t_0)$. The spectral values in the complement of the essential spectrum essentially behave like the spectral values of a compact operator. We proceed as in Sec. 8 of [1] where the reader can also find details and references concerning the essential spectrum. (The essential spectrum consists of all elements of the spectrum which are not poles of the resolvent with a residue of finite rank.)

A solution m of (15) can be split up as follows:

$$m(t, x) = \overline{m}(t, x) + \hat{m}(t, x), \tag{38}$$

with

$$\overline{m}(t, x) = \int_{I \setminus J} K(t, x, y)\overline{m}(T(t, x, y), 2y) \, dy + m_0(t, x), \tag{39}$$

$$\hat{m}(t, x) = \int_I K(t, x, y)\hat{m}(T(t, x, y), 2y) \, dy + \int_J K(t, x, y)\overline{m}(T(t, x, y), 2y) \, dy. \tag{40}$$

With m_0 being given by (16) we define solution operators $\overline{U}(t, t_0)$ and $\hat{U}(t, t_0)$ by

$$\overline{U}(t, t_0)\Phi = \overline{m}(t, \cdot),$$
$$\hat{U}(t, t_0)\Phi = \hat{m}(t, \cdot). \tag{41}$$

In particular

$$U(t, t_0) = \overline{U}(t, t_0) + \hat{U}(t, t_0). \tag{42}$$

Observe that $\overline{U}(t, t_0)$ defines an evolutionary system.

We claim that (37) implies compactness of the operators $\hat{U}(t, t_0)$ for $t - t_0$ being large enough. We proceed as in the proof of the compactness of the operators $U(t, t_0)$ for large $t - t_0$ in Sec. 4.

We find that

$$\overline{v}(s, x) = \overline{m}(T(s, a/2, x), x)$$

is continuous in x uniformly for s ranging in a bounded interval in $[t_0, \infty)$ and for initial values Φ in a bounded subset M of Z. Let $\tilde{m}(t, x)$ denote the second integral on the right-hand side of (40). Then

$$\tilde{m}(T(s, a/2, x), x) = \int_{J \cap [0, x]} D(T(s, a/2, y), y)\overline{v}(f(s, y), 2y) \, dy$$

for $s \ge T(t_0, a/2, 1)$, with f in (31). By (37) we can perform the transformation $f(s, y) = r$ and find that $\bar{m}(t, x)$ is jointly continuous in $t \ge \phi(\phi(t_0))$, $x \in [a/2, 1]$ uniformly in $\Phi \in M$. Here, as before,

$$\phi(t) = T(t, a/2, 1). \tag{43}$$

By (40) $\hat{m}(t, x)$ is jointly continuous in $t \ge \phi(\phi(\phi(t_0)))$, $x \in [a/2, 1]$, uniformly in $\Phi \in M$. Note that $\hat{m}$ can be given as a convergent series as in Theorem 3.1. By the Arzela–Ascoli theorem we obtain the following.

Lemma 7.1

The operators $\hat{U}(t, t_0)$ are compact on Z for $t \ge \phi(\phi(\phi(t_0)))$.

For Lemma 7.1 to hold we did not need that the developmental rates are time independent. This is different for the next lemma.

Lemma 7.2

Let g, μ, b be time independent. If $t \ge \phi(\phi(\phi(t_0)))$, the spectral radius of $\bar{U}(t, t_0)$ is strictly smaller than the spectral radius of $U(t, t_0)$, in particular the radius of the essential spectrum of $U(t, t_0)$ is strictly smaller than the spectral radius of $U(t, t_0)$.

In order to prove Lemma 7.2 we first note that, if the developmental rates do not depend on t,

$$D(t, y) = D(y) \tag{44}$$

is independent of t [see (11)]. Furthermore,

$$T(t, x, y) - t =: T_0(x, y) \tag{45}$$

does not depend on t. We note that (39) can be written in the form of (15) if D is replaced by D^J:

$$D^J(y) = D(y) \quad \text{for } a/2 \le y \le \tfrac{1}{2}, y \notin J; \tag{46}$$
$$D^J(y) = 0 \qquad \text{otherwise.}$$

Though the positivity and compactness results in Secs. 4, 5 may cease to be valid under this modification, Propositions 4.1, 4.2 and 5.9 equally hold for $\bar{U}$.

In Sec. 5 we found p-periodic solutions of (36) from the properties of the operators $U(t, t_0)$. We now go this way backwards and try to obtain information on the operators $U(t, t_0)$, $\bar{U}(t, t_0)$ by looking for special periodic, namely t-independent solutions of (36). This is possible because of (44) and (45). To this end we define operators V_λ^J on Z by

$$(V_\lambda^J u)(x) = \int_{a/2}^{\min(x, 1/2)} D^J(y) e^{\lambda T_0(x, y)} u(2y) \, \mathrm{d}y \tag{47}$$

for $\lambda \in \mathbb{R}$, $u \in Z$. Here we admit any open subinterval J of $[a/2, \tfrac{1}{2}]$, in particular $J = \emptyset$, i.e. $D^J = D$. Recall that $D^J(y) > 0$ if a $a/2 < y < \tfrac{1}{2}$, $y \notin J$ (see Lemma 2.2). We now study the eigenvalues and the positive eigenvectors of V_λ^J in dependence on λ. Since Proposition 5.9 also holds if D is replaced by D^J, any pair $\lambda \in \mathbb{C}$, $u \in Z$ with $u = V_\lambda^J u$ provides an eigenvalue $e^{\lambda p}$ of $U^J(t_0 + p, t_0)$ with eigenvector u. Here $U^J(t, t_0)$, $t \ge t_0$, denote the solution operators associated with Eq. (39).

Lemma 7.3

(a) Let J be an open subinterval of $(a/2 + \epsilon, \tfrac{1}{2} - \epsilon)$ for some $\epsilon > 0$. Then there exists $\lambda = \lambda_J \in \mathbb{R}$ such that the spectral radius of V_λ^J is equal to 1. Furthermore, for $\lambda = \lambda_J$, there are $v_J \in Z_+$, $v_J \ne 0$ and $v_J' \in Z_+'$, $v_J' \ne 0$ such that $V_\lambda^J v_J = v_J$ and $(V_\lambda^J)' v_J' = v_J'$.

(b) Let, in addition, the length of J be strictly smaller than

$$\frac{1}{n+1} - \frac{1}{2n}$$

for any $n \in \mathbb{N}$ with $a < 1/n$. Then v_J is comparable with v_0 in (20) and $v_J'(v) > 0$ for every $v \in Z_+$ with $V_\lambda^J(v) \neq 0$. If $J = \emptyset$, i.e. $D^J = D$, v_J' is strictly positive.

Remark

The notation λ_J, v_J, v_J' may be misleading insofar as it suggests that these entities are uniquely determined (in the case of v_J, v_J' up to normalization). In general this might not be the case. It will be the case, however, if J satisfies both the conditions of Lemma 7.3(a) and (b). Note that v_0 in (20) is now time independent. The case $J = \emptyset$ has already been dealt with by one of us in [27].

Proof of Lemma 7.3. Drop the index J for convenience. Let r_λ denote the spectral radius of V_λ. It can readily be seen that $\|V_\lambda\| \to 0$ for $\lambda \to \infty$. Recall that $T_0(x, y) < 0$ for $x > y$. Thus $r_\lambda \to 0$ for $\lambda \to \infty$. In order to see that $r_\lambda \to \infty$ for $\lambda \to -\infty$ we choose $v \in Z_+$, $v(x) = 0$ for $x \leq \frac{1}{2}$, $v(1) > 0$. Then

$$(V_\lambda v)(x) \geq \int_{a/2}^{1/2} D^J(y) e^{\lambda T_0(x,y)} v(2y)\, \mathrm{d}y > 0$$

for $\frac{1}{2} \leq x \leq 1$, in particular $V_\lambda v \geq c_\lambda v$ with $c_\lambda > 0$ for $\lambda \in \mathbb{R}$ and $c_\lambda \to \infty$ for $\lambda \to -\infty$. This implies that $r_\lambda \geq c_\lambda > 0$ for $\lambda \in \mathbb{R}$ and $c_\lambda \to \infty$ for $\lambda \to -\infty$. Note that V_λ continuously depends on λ in the uniform operator topology. The formula $r_\lambda = \inf_n \|V_\lambda^n\|^{1/n} = \lim_{n\to\infty} \|V_\lambda^n\|^{1/n}$ reveals that r_λ is continuous from above, i.e. $\varlimsup_{\bar\lambda\to\lambda} r_{\bar\lambda} \leq r_\lambda$. Since every nonzero spectral value of the compact operator V_λ is an isolated point of the spectrum, the perturbation result in Theorem 3.16 in [28, Chap. IV] implies the continuity of r_λ from below, i.e. $\varliminf_{\bar\lambda\to\lambda} r_{\bar\lambda} \geq r_\lambda$. The intermediate value theorem now implies the existence of some $\lambda \in \mathbb{R}$ with $r_\lambda = 1$. The existence of v_J and v_J' follows from the Krein–Rutman theorem (see, for example, [29]).

Part (b) follows from the fact that there is some $j \in \mathbb{N}$ such that $(V_\lambda^j v)(x) > 0$ for $a/2 < x \leq 1$, if $v \in Z_+$ and $V_\lambda v \neq 0$. In fact, if $V_\lambda v \neq 0$, $v \in Z_+$, then $(V_\lambda v)(x) > 0$ for $\frac{1}{2} \leq x \leq 1$. Let us suppose that we have already proved

$$(V_\lambda^{j-1} v)(x) > 0 \quad \text{for} \quad \frac{a}{2} < \frac{1}{j} \leq x \leq 1,\, j \geq 2.$$

Then

$$D^J(y)(V_\lambda^{j-1} v)(2y) > 0 \quad \text{for } \tfrac{1}{2}j,\, a/2 \leq y \leq \tfrac{1}{2};\, y \in J.$$

If $\tfrac{1}{2}j \leq a/2$, $D^J(y)(V_\lambda^{j-1} v)(2y) > 0$ for $a/2 \leq y \leq a/2 + \epsilon$; hence $(V_\lambda^j v)(x) > 0$ for $a/2 < x \leq \frac{1}{2}$. If $\tfrac{1}{2}j > a/2$, $D^J(y)(V_\lambda^{j-1} v)(2y)$ cannot vanish a.e. on $[\tfrac{1}{2}j,\, 1/(j+1)]$ because, by assumption, the length of J is strictly smaller than the length of $[\tfrac{1}{2}j,\, 1/(j+1)]$; hence $(V_\lambda^j v)(x) > 0$ for $1/(j+1) < x < \frac{1}{2}$. Repeating this step several times we find some j such that $(V_\lambda^j v)(x) > 0$ for $a/2 < x \leq 1$. ∎

We now prove the crucial result which will imply that the spectral radius of $\overline{U}(t, t_0)$ is strictly smaller than the spectral radius of $U(t, t_0)$, $t > t_0$.

LEMMA 7.4

Let J be an interval satisfying the assumptions in Lemma 7.3(a, b). Let λ, $\lambda_J \in \mathbb{R}$ be such that the spectral radius of V_λ^0 and $V_{\lambda_J}^J$ equal 1. Then $\lambda > \lambda_J$.

Proof. We suppose that $\lambda \leq \lambda_J$. Choose $v_J \in Z_+$, $v_J \neq 0$ such that $V_{\lambda_J}^J v_J = v_J$. Since v_J is strictly positive by Lemma 7.3(b), $V_\lambda^0 v_J > V_\lambda^J v_J \geq V_{\lambda_J}^J v_J = v_J$. Recall that $T_0(x, y) \leq 0$ for

$y \leqslant x$. Now choose a strictly positive functional $v' = V_\lambda^\emptyset v'$ according to Lemma 7.3(b). Then

$$1 = \frac{(V_\lambda^\emptyset v')(v_J)}{v'(v_J)} = \frac{v'(V_\lambda^\emptyset v_J)}{v'(v_J)} > \frac{v'(v_J)}{v'(v_J)} = 1,$$

a contradiction.   ∎

Let us now fix the nonempty open subinterval J of $I = [a/2, 1]$ such that J satisfies both the assumptions in Lemma 7.3 and in (37). Choose an arbitrary $p > 0$. Proposition 5.9 and Lemmas 7.3, 7.4 now imply that $e^{p\lambda}$ and $e^{p\lambda_J}$ are positive eigenvalues of $B = U(t_0 + p, t_0)$ and $\overline{B} = \overline{U}(t_0 + p, t_0)$ respectively with eigenvectors v, $\overline{v}$ which are comparable to v_0 in (20), and that $e^{p\lambda} > e^{p\lambda_J}$. Thus $Z_{v_0} = Z_v = Z_{\overline{v}}$ as Banach spaces. Obviously B and $\overline{B}$ map Z_v and $Z_{\overline{v}}$ continuously into themselves and have the spectral radii $e^{p\lambda}$ and $e^{p\lambda_J}$ on these spaces (see the Appendix, point 8). It follows from proposition 4.2 that B and $\overline{B}$ map Z continuously into Z_{v_0} for $p \geqslant T_0(a/2, 1)$. The next lemma will imply that B and $\overline{B}$ have the spectral radii $e^{p\lambda} > e^{p\lambda_J}$ on Z so that Lemma 7.2 follows from Lemma 7.1.

LEMMA 7.5

Let W, Z be Banach spaces, W a linear subspace of Z. Let B be a bounded linear operator on Z such that $BZ \subseteq W$ and B is also a bounded operator on W. Then, with the possible exception of $0 \in \mathbb{C}$, B has the same spectrum on Z and on W.

By the open mapping theorem the proof of the lemma reduces to showing that, for $q \neq 0$, $qI - B$ is a bijection on Z iff it is a bijection on W. But proving this is almost trivial and left to the reader.

In order to conclude the consideration of the time-homogeneous case we remind that (37) implies Assumption 5.5, as we have seen in the remarks preceding Corollary 6.3. Thus Propositions 5.6(b, c) hold. The power compactness of B stated in proposition 5.6(a) was only needed to guarantee that the spectral radius r_0 of $B = U(t_0 + p, t_0)$ is a pole of the resolvent of B. In order to realize that r_0 keeps this property under the present conditions we first note that r_0 is a spectral value of the positive operator B (see [18, V.4.1]). By Lemma 7.2 r_0 has the same properties as a nonzero spectral value of a compact operator, in particular r_0 is an eigenvalue and even a pole of the resolvent of B (see [1, Sec. 8]). Thus Propositions 5.7 and 5.8 are valid (see [18, Chap. V]) and so are Theorem 6.1 and Corollary 6.2, if we replace Assumptions 4.8, 5.1 and 5.5 by (37) and the time independence of g, b, μ. Note that $\tilde{m}$ in Theorem 6.1 is now a time-independent solution of $\tilde{m} = V_\lambda^\emptyset \tilde{m}$. Thus, by (47), $\tilde{m}$ is differentiable for $a/2 < x < 1$, $x \neq \frac{1}{2}$ and satisfies the differential equation

$$\left(\frac{\lambda}{g(x)} + \frac{\mathrm{d}}{\mathrm{d}x}\right) \tilde{m}(x) = D(x)\tilde{m}(2x) \quad \text{for } a/2 < x < 1, \, x \neq \tfrac{1}{2}, \tag{48a}$$

$$D(x)\tilde{m}(2x) = 0, \quad x \geqslant \tfrac{1}{2}, \tag{48b}$$

and the boundary condition

$$\tilde{m}(a/2) = 0. \tag{49}$$

So Theorem 6.1 takes the following form.

THEOREM 7.6

Let Assumptions 1.1, 2.1 be satisfied. Let g, μ, b be time independent and $g(2x) \neq 2g(x)$ for some $x \in (a/2, \tfrac{1}{2})$. Then the following holds.

(a) There exists a unique $\lambda \in \mathbb{R}$ with a nonnegative solution $\tilde{m} \neq 0$ of (48), (49). $\tilde{m}$ is uniquely determined up to a scalar factor.

(b) If m is a weak solution of (10) with a continuous initial function Φ on $[a/2, 1]$ at $t = t_0$, $\Phi(a/2) = 0$, then for $t \to \infty$,

$$m(t, x) = e^{\lambda t}\tilde{m}(x) \cdot (\alpha + e^{-\epsilon t} \mathcal{O}(1))$$

with $\epsilon > 0$ and α as in Theorem 6.1.

Again the dependence of α and $\mathscr{C}(1)$ on the initial values u can be described in more detail, as we did in the sequel of Theorem 6.1.

When Theorem 7.6 is translated to Eq. (1), we note that $\hat{n}(t, x)$ in Corollary 6.2 is time independent, continuous in x, differentiable in $x \neq a, \frac{1}{2}$ and a solution of the differential equation

$$\lambda \tilde{n}(x) + \frac{\mathrm{d}}{\mathrm{d}x}(g \cdot \tilde{n})(x) + (\mu + b) \cdot \tilde{n}(x) = 4b \cdot \tilde{n}(2x), \quad x \in (a/2, 1), \ x \neq a, \tfrac{1}{2}, \quad (50a)$$

$$b \cdot \tilde{n}(x) = 0 \quad \text{if } x \notin [a, 1), \tag{50b}$$

and the boundary condition

$$\tilde{n}(a/2) = 0. \tag{51}$$

Furthermore, $\tilde{n}(x)/\tilde{E}(x)$ is bounded in $x \in (a/2, 1)$ with

$$\tilde{E}(x) = \exp\left(-\int_{a/2}^{x} (b/g)(z)\,\mathrm{d}z\right). \tag{52}$$

So Corollary 6.2 takes the following form.

COROLLARY 7.7

Let the assumptions of Theorem 7.6 be satisfied. Then the following holds.

(a) There exists a unique $\lambda \in \mathbb{R}$ with a nonnegative solution $\tilde{n} \neq 0$ of (50), (51) such that $\tilde{n}(x)/\tilde{E}(x)$ is continuous on $[a/2, 1]$. $\tilde{n}$ is uniquely determined up to a scalar factor.

(b) Let n be a weak solution of (1) for $t > t_0$ such that $n(t_0, x)/\tilde{E}(x)$ is continuous in $x \in [a/2, 1]$. Then, for $t \to \infty$,

$$n(t, x) = e^{\lambda t}\tilde{n}(x) \cdot (\alpha + e^{-\epsilon t}\,\mathscr{C}(1))$$

with $\epsilon > 0$ and α as in Corollary 6.2.

We remark that these results for the time-homogeneous model can also be obtained using the semigroup theory we applied in [1]. The proof of the strong positivity of the solution operators could then be replaced by a more thorough analysis of the operators V_λ in (47), in particular by the considerations in [27, Sec. 7]. Though $g(2x) < 2g(x)$ for all $x \in [a/2, 1]$ is supposed there, the proof of Theorem 7.2 reveals that (37) is sufficient because the eigenfunction Φ_0 is strictly positive. For time-dependent developmental rates a direct proof of the strong positivity of $U(t, t_0)$ seems unavoidable.

Acknowledgment—The work of H. Thieme has been supported by the Deutsche Forschungs-gemeinschaft (DFG) and, through a visitors grant, by the Netherlands Organization for the Advancement of Pure Research (ZWO).

REFERENCES

1. O. Diekmann, H. J. A. M. Heijmans and H. R. Thieme, On the stability of the cell size distribution. *J. Math. Biol.* **19**, 227–248 (1984).
2. E. C. Anderson, G. I. Bell, D. F. Petersen and R. A. Tobey, Cell Growth and Division IV. Determination of volume growth rate and division probability. *Biophys. J.* **9**, 246–263 (1969).
3. E. C. Anderson and D. F. Petersen, Cell Growth and Division II. Experimental studies of cell volume distributions in mammalian suspension cultures. *Biophys. J.* **7**, 353–364 (1967).
4. G. I. Bell, Cell Growth and Division III. Conditions for balanced exponential growth in a mathematical model. *Biophys. J.* **8**, 431–444 (1968).
5. G. I. Bell and E. C. Anderson, Cell Growth and Division I. A mathematical model with applications to cell volume distributions in mammalian suspension cultures. *Biophys. J.* **7**, 329–351 (1967).
6. J. W. Sinko and W. Streifer, A new model for the age–size structure of a population. *Ecology* **48**, 910–918 (1967).
7. J. W. Sinko and W. Streifer, A model for populations reproducing by fission. *Ecology* **52**, 330–335 (1971).
8. W. Streifer, *Realistic Models in Population Ecology, Advances in Ecology Research*, (Edited by A. MacFadyen), Vol. 8. Academic Press (1974).

9. H. J. A. M. Heijmans, On the stable size distribution of populations reproducing by fission into two unequal parts. *Math. Biosc.* **72**, 19–50 (1984).

10. A. Lasota and M. C. Mackey, Global asymptotic properties of proliferating cell populations. *J. Math. Biol.* **19**, 43–62 (1984).

11. J. J. Tyson, The coordination of cell growth and division—intentional or incidental? *Bio Essays* **2**, 72–77 (1985).

12. J. J. Tyson, The coordination of cell growth and division: a comparison of models, in *Temporal Order* (Edited by L. Rensing and N. I. Jaeger), pp. 291–295. Springer-Verlag (1985).

13. J. J. Tyson and K. B. Hanssgen, The distributions of cell size and generation time in a model of the cell cycle incorporating size control and random transitions. *J. Theor. Biol.* **113**, 29–62 (1985).

14. J. J. Tyson and O. Diekmann, Sloppy size control of the cell division cycle, J. Theor. Biol. (in press).

15. J. J. Tyson & K. B. Hannsgen, Global asymptotic stability of the size distribution in probabilistic models of the cell cycle, *J. Math. Biol.* **22**, 61–68 (1985).

16. K. B. Hannsgen, J. J. Tyson and L. T. Watson, Steady-state size distributions in probabilistic models of the cell division cycle. *SIAM J. Appl. Math.* **45**, 523–540 (1985).

17. M. A. Krasnoselskii, *Positive Solutions of Operator Equations*. Groningen, Noordhoff (1964).

18. H. H. Schaefer, *Banach Lattices and Positive Operators*. Springer-Verlag (1974).

19. O. Diekmann, H. A. Lauwerier, T. Aldenberg and J. A. J. Metz, Growth, fission and the stable size distribution. *J. Math. Biol.* **18**, 135–148 (1983).

20. R. Courant and D. Hilbert, *Methods of Mathematical Physics,* Vol. II. Interscience Publishers, (1962).

21. R. R. Garabedian, *Partial Differential Equations*. John Wiley (1964).

22. S. Lang, *Analysis II*. Addison-Wesley (1969).

23. K. Yosida, Functional Analysis. 4th ed., Springer-Verlag (1974).

24. A. Pazy, *Semigroups of Linear Operators and Applications to Partial Differential Equations*. Springer-Verlag, Berlin (1983).

25. H. Tanabe, *Equations of Evolution*. Pitman (1979).

26. H. R. Dowson, *Spectral Theory of Linear Operators*. Academic Press (1978).

27. H. J. A. M. Heijmans, An eigenvalue problem related to cell growth. *J. Math. Anal. Appl.* **111**, 253–280 (1985).

28. T. Kato, *Perturbation Theory for Linear Operators*, 2nd Ed. Springer-Verlag, (1976).

29. H. H. Schaefer, *Topological Vector Spaces*. Macmillan, (1966).

APPENDIX: SOME VOCABULARY FROM THE THEORY OF BANACH LATTICES

1. A Banach lattice Z is a Banach space with norm $\|\cdot\|$ and a vector lattice with cone Z_+, ordering "$\leq$" and absolute value (or modulus) $|\cdot|$ with these two structures being interlinked by

$$|u| \leq |v| \quad \text{implies} \quad \|u\| \leq \|v\| \tag{A1}$$

for all $u, v \in Z$ (see [18]).

2. Let $v \in Z_+$, $v \neq 0$. $u \in Z$ is called *v-bounded* iff

$$|u| \leq cv \quad \text{for some } c > 0. \tag{A2}$$

The set Z_v of v-bounded elements is a linear subspace of Z and becomes a Banach space itself by the v-norm $\|u\|_v$, which is by definition the smallest c such that (A2) is satisfied (see [17, Secs. 1.2, 1.3]). The Banach space Z_v becomes a Banach lattice by restricting the lattice structure of Z to Z_v, in particular $Z_{v,+} = Z_v \cap Z_+$. If Z is a function space, then $\|u\|_v = \sup\{|(u/v)(x)|; v(x) \neq 0\}$. The cone $Z_{v,+}$ has interior points in the Z_v-topology.

3. Let $u, v \in Z_+$. u is called *v-positive* iff $u \geq \epsilon v$ for some $\epsilon > 0$. u and v are called *comparable* (or order-equivalent) iff u is v-bounded and v-positive. u and v are comparable iff Z_v and Z_u are equal as Banach lattice (in particular $Z_u = Z_v$ as sets and $\|\cdot\|_v$ and $\|\cdot\|_u$ are equivalent norms); furthermore, u and v are comparable iff u is an interior point in Z_v and *vice versa*.

4. $v \in Z_+$ is called a *quasi-interior point* iff Z_v is dense in Z, or equivalently, iff $v'v > 0$ for any $v' \in Z'_+$, $v' \neq 0$ (see [18, Thm. II.6.3]).

5. A functional v' on Z is called *positive* iff $v'(u) \geq 0$ for all $u \in Z_+$. v' is called *strictly positive* iff $v'(u) > 0$ for all $u \in Z_+$, $u \neq 0$.

6. A linear operator A from one Banach lattice into another is called *positive* iff it maps one cone into the other. Positive linear operators between Banach lattices are automatically bounded (see [18, Sec. II.5.3]).

7. A positive linear operator A is called *strongly positive* iff for any $u \in Z_+$, $u \neq 0$, $A^n u$ is a quasi-interior point of Z_+ for some $n \in \mathbb{N}$.

8. A positive operator A on Z maps Z_v, $v \in X_+$, into itself iff Av is v-bounded (i.e. $Av \in Z_v$). The operator norm $\|A\|_v$ of A on X_v satisfies $\|A\|_v = \|Av\|_v$. If $Av = rv$ for some $r > 0$, then $\|A\|_v = v$ and r is the spectral radius of A on Z_v.

Comp. & Maths. with Appls. Vol. 12A, Nos. 4/5, pp. 513–526, 1986
Printed in Great Britain.

0886–9553/86 $3.00 + .00
© 1986 Pergamon Press Ltd.

PERIODIC McKENDRICK EQUATIONS FOR AGE-STRUCTURED POPULATION GROWTH

J. M. Cushing
Department of Mathematics and Program on Applied Mathematics, University of Arizona,
Building #89, Tucson, Arizona 85721, U.S.A.

Abstract—With the averaged net reproductive rate used as a bifurcation parameter, the existence of a local parameterized branch of time-periodic solutions of the McKendrick equations is proved under the assumption that the death and fertility rates suffer small-amplitude time periodicities. The required linear theory is developed and the results are illustrated by means of a simple example in which fertility varies cosinusoidaliy in time.

INTRODUCTION

The equations

$$\rho_t + \rho_a + D\rho = 0, \quad t > 0, \quad 0 < a < A < +\infty, \tag{1.1}$$

$$\rho(t, 0) = \int_0^A F\rho(t, a) \, \mathrm{d}a, \quad t > 0, \tag{1.2}$$

for the age-specific density $\rho = \rho(t, a)$ of a single age-structured species at time t describe the death and birth processes respectively in terms of a per unit death rate D and fertility rate F. The real A is a maximum age for any individual in the population and it is required that $\rho(t, A) \equiv 0$ for all $t > 0$. These equations are now usually referred to as the McKendrick equations.

In using the system of Eqs. (1.1)–(1.2) as a model for population growth most studies allow the vital rates D and F to be dependent upon age a and also, in modelling density-dependent growth, upon the density ρ. The vast majority of population growth models are autonomous, i.e. they assume that these vital rates do not depend explicitly on time t. Under such an assumption the important fundamental questions concerning asymptotic states as $t \to +\infty$, around which theoretical population dynamics centers, deal with the existence and stability of equilibrium (i.e. time-independent) solutions and there is a rapidly growing literature on these topics for the general McKendrick Eqs. (1.1)–(1.2) as well as many specialized cases derived from them.

Despite the widely recognized biological fact that death and fertility rates for real populations are rarely constant in time the amount of literature dealing with model equations in which vital rates are explicitly time dependent is comparatively very small. This is true even for the simpler classical models of non-age-structured populations. In recent years considerable attention has been paid to the important and difficult question of the effects due to stochastic fluctuations of model parameters in population-growth models. While it is true that vital rates and other model parameters can be expected to suffer significant stochastic variations in time for some populations under certain conditions, it is also true that some parameters for other populations or for populations under other conditions may well exhibit regular recurring fluctuations in time (e.g., see [1]). For example, physical environmental conditions such as temperature and humidity and the availability of food, water and other resources (just to mention a few) usually vary in time, often fairly regularly, with the yearly seasons (or with daily or monthly periods or even cycles with other periods). Natural birth rates are often markedly seasonal, as are death rates. These can be due to such things as exposure to seasonal weather patterns and resource availabilities, susceptibility to diseases or exposure to predators and competitors, etc.

A natural simplifying mathematical assumption to make in considering such regular fluctuations in model parameters is that they are exactly periodic. This leads to the study of

514   J. M. CUSHING

nonautonomous equations with periodic coefficients. For example, in the general model (1.1)–(1.2) D and F might be assumed to depend explicitly on time t in a periodic manner.

In general, nonautonomous periodic differential equations do not have equilibrium solutions and the familiar techniques available for studying the existence and stability properties of equilibrium solutions are not applicable. Instead one has the more difficult challenge of dealing with the existence and the stability of periodic solutions, as well as the challenge of analysing the properties of these solutions. Such problems have been considered in recent papers for a variety of non-age-structured population-growth models[2–13].

With regard to the autonomous McKendrick equations (1.1)–(1.2) there is a growing body of literature dealing with equilibrium solutions (e.g. [15–19]), but little in the literature concerning periodic solutions of periodic equations[20]. The purpose of this paper is to consider the existence of positive periodic solutions of (1.1)–(1.2) when the vital rates D and F are explicitly periodic in time t, using a bifurcation theory approach analogous to that used in [15,16] for equilibrium solutions of the autonomous case.

The primary assumption will be that the time periodicities in D and F are of small amplitude α. This assumption permits the use of perturbation techniques and the calculation of lower-order approximations to solutions. It is, however, a restrictive assumption and precludes the consideration of large ''catastrophic'' fluctuations in such things as environmental conditions, availability of resources, population densities, etc. Nonetheless, as is often pointed out with regard to oscillatory phenomena in many kinds of models and applications, small-amplitude parameter oscillations can lead to significant effects and to ignore them in favor of averaged values can be misleading. Moreover, the effects of small-amplitude periodicities often persist in specific models for larger-amplitude periodicities. In any case, the study of small-amplitude time periodicities in the vital rates D and F in (1.1)–(1.2) certainly contributes fundamentally to the general understanding of time-periodic fluctuations in these vital rates.

In order to use a bifurcation-theory approach to Eqs. (1.1)–(1.2) it is necessary to distinguish a bifurcation parameter. In [15,16] this parameter was taken to be the biologically meaningful ''inherent net reproductive rate'' n defined by

$$n = \int_0^A F_0 \exp\left(-\int_0^a D_0 \, ds \right) da,$$

where D_0 and F_0 denote the age-specific vital rates D and F evaluated at $\rho \equiv 0$. Then the fertility rate F is written as $F = nf$, where f is appropriately normalized. When F is periodic in time we will use, more generally, a time-averaged inherent net reproductive rate for the bifurcation parameter.

Suppose $D = D(\rho)(t, a)$ and $F = F(\rho)(t, a)$ depend on the density ρ, age a and periodically on time t. Let av$[f]$ denote the time t average of f over one period and set

$$n = \text{av}\left[\int_0^A F(0)(t, a) \exp\left(-\int_0^a D(0)(t, s) \, ds \right) da \right].$$

If we write $F = nf$, where $f = f(\rho)(t, a)$ is normalized so that

$$\text{av}\left[\int_0^A f(0)(t, a) \exp\left(-\int_0^a D(0)(t, s) \, ds \right) da \right] = 1,$$

then Eqs. (1.1)–(1.2) become

$$\rho_t + \rho_a + D(\rho)(t, a)\rho = 0, \quad t > 0, \quad 0 < a < A < +\infty, \tag{1.3}$$

$$\rho(t, 0) = n \int_0^A f(\rho)(t, a)\rho(t, a) \, da, \quad t > 0, \tag{1.4}$$

for $\rho = \rho(t, a) \geq 0$. Of interest here is the existence of nontrivial time t periodic solutions $\rho \geq 0$ ($\not\equiv 0$) satisfying $\rho(t, A) \equiv 0$, or more specifically a determination of those values of n for which such solutions exist.

In [16] it was shown under only mild continuity conditions that when D and f are independent of t a global branch of nontrivial equilibrium solutions $\rho = \rho(a) \geq 0$ exists and bifurcates from (and only from) the critical point $(n, \rho) = (1, 0)$. This continuum branch of equilibria (n, ρ) connects to the boundary of the domain upon which the problem is posed in certain Banach spaces. In [15] this bifurcation phenomenon is studied in more detail locally in a neighborhood of the bifurcation point $(1, 0)$.

In Sec. 3 below a local bifurcation theorem for the existence of nontrivial periodic solutions is proved for (1.3)–(1.4) when D and f are periodic in t, under the assumption that these periodicities have "small amplitude." It establishes the local existence of a parameterized branch of nontrivial periodic solutions which bifurcates from a critical point $(n, \rho) = (n_0, 0)$, where n_0 is an eigenvalue of the linearized problem whose existence is established in Sec. 2. This result extends, indeed generalizes, the existence result of [15], to the case of time-periodic vital rates D and f. The lemmas and theorem of Secs. 2 and 3 allow for the computation of any number of lower-order terms in perturbation expansions for n_0 and the nontrivial solution (n, ρ). These lower-order terms can of course be used as approximations to these solutions in specific applications. A simple example is given in Sec. 4.

The stability of the branch solutions given in the theorem of Sec. 3 is not studied here. For the autonomous case stability was shown in [15] to depend on the "direction of bifurcation," i.e. the branch solutions ρ are stable near bifurcation only if these solutions correspond to $n > n_0 = 1$. Moreover, the trivial solution $\rho \equiv 0$ is stable for $n < 1$ and unstable for $n > 1$. A natural conjecture is that these stability properties remain in force for the periodic, small-amplitude case considered here.

2. THE LINEAR THEORY

Let R and R^+ denote the set of reals and the set of nonnegative reals respectively. Denote by Δ the set of continuous functions μ: $[0, A) \to R$ satisfying

$$\lim_{a \to A^-} M(a) = +\infty, \quad M(a) := \int_0^a \mu(s) \, ds.$$

Set

$$\rho_0(a) = \exp(-M(a)) \quad \text{for } 0 \leq a < A, \ \rho_0(A) = 0.$$

The linear space of continuous functions $h: R \times [0, A] \to R$ for which $h(t, a)/\rho_0(a)$ is continuous and for which h is periodic of period 1 in t is denoted by $P(\mu)$. This space $P(\mu)$ is a Banach space under the norm

$$\|h\|_\mu := \max_{R \times [0,A]} |h(t, a)|/\rho_0(a).$$

The subspace of $h \in P(\mu)$ which are independent of t is the space B_μ used in [15,16] to study equilibrium solutions of autonomous Eqs. (1.3)–(1.4). Note that $\rho_0 \in P(\mu)$ and that $h(t, A) \equiv 0$, $t \in R$, for any $h \in P(\mu)$.

The linear space of functions h for which in addition $h(t, a)/\rho_0(a)$ has continuous first partial derivatives is a Banach space under the norm

$$\|h\|_{\mu, 1} := \|h\|_\mu + \max_{R \times [0,A]} \left| \frac{\partial}{\partial a} (h/\rho_0) \right| + \max_{R \times [0,A]} \left| \frac{\partial}{\partial t} (h/\rho_0) \right|,$$

and is denoted by $P^1(\mu)$. The Banach space of functions h for which $h(t, a)/\rho_0(a)$ has a continuous

516   J. M. Cushing

first partial in t is denoted by $P^{1-}(\mu)$ and is given the norm

$$\|h\|_{\mu,1-} := \|h\|_\mu + \max_{R \times [0,A]} \left| \frac{\partial}{\partial t}(h/\rho_0) \right|.$$

For $h \in P(\mu)$ define the average $\mathrm{av}[h] := \int_0 h(t, a)\, dt$ and denote by $P_0(\mu)$ the subspace of functions $h \in P(\mu)$ with zero average (for all $a \in [0, A]$); similarly for $P_0^1(\mu)$ and $P_0^{1-}(\mu)$.

Finally, let P^1 be the Banach space of continuously differentiable, periodic functions h: $R \to R$ under the usual supremum norm $\|h\|_1 := \|h\|_0 + \|h'\|_0$, $\|h\|_0 := \max_{0 \le t \le 1} |h(t)|$. P_0^1 is the subspace of functions with zero average. The cross-product space $P^1 \times P^{1-}(\mu)$ is given the norm $\|\cdot\|_1 + \|\cdot\|_{\mu,1-}$. Similar norms are taken for other cross-product spaces.

Consider the nonhomogeneous linear system

$$\rho_t(t, a) + \rho_a(t, a) + \mu(a)\rho(t, a) = h_2 \in P^{1-}(\mu), \quad 0 < a < A < +\infty, \tag{2.1a}$$

$$\rho(t, 0) - \int_0^A \beta(a)\rho(t, a)\, da = h_1 \in P^1, \tag{2.1b}$$

and its associated homogeneous system

$$\begin{aligned}
\rho_t(t, a) + \rho_a(t, a) + \mu(a)\rho(t, a) &= 0, \quad 0 < a < A, \\
\rho(t, 0) - \int_0^A \beta(a)\rho(t, a)\, da &= 0.
\end{aligned} \tag{2.2}$$

We will also refer to the related systems

$$\begin{aligned}
\rho_a(a) + \mu(a)\rho(a) &= h_2 \in B_\mu, \quad 0 < a < A, \\
\rho(0) - \int_0^A \beta(a)\rho(a)\, da &= h_1 \in R,
\end{aligned} \tag{2.3}$$

and

$$\begin{aligned}
\rho_a(a) + \mu(a)\rho(a) &= 0, \quad 0 < a < A < +\infty, \\
\rho(0) - \int_0^A \beta(a)\rho(a)\, da &= 0.
\end{aligned} \tag{2.4}$$

We assume throughout that $\mu \in \Delta$, $\beta \in C^0 = C^0([0, A]; R)$. It is required in all equations that ρ vanish identically when $a = A$, a condition which is met by finding solutions in $P^1(\mu)$.

In [15] a Fredholm-type alternative was proved for (2.3)–(2.4), which formed the basis of the study of equilibrium solutions of nonlinear equations. The goal of this section is to derive a similar result from the t-periodic systems (2.1)–(2.2). First, however, a lemma concerning the integral equation

$$B(t) - \int_0^A \beta(a)\rho_0(a)B(t - a)\, da = h \in P^1, \tag{2.5}$$

and its associated homogeneous equation

$$B(t) - \int_0^A \beta(a)\rho_0(a)B(t - a)\, da = 0 \tag{2.6}$$

is needed. Let Z denote the set of all integers and N denote the subspace of all solutions $B \in P^1$ of (2.6). Also, let $N^\perp = \{h \in P^1 : \mathrm{av}[hB] = 0 \ \forall\, B \in N\}$.

LEMMA 2.1

Assume $\mu \in \Delta$ and $\beta \in C^0$.

(a) dim N is finite.
(b) Equation (2.5) has a solution $B \in P^1$ if and only if $h \in N^\perp$, in which case (2.5) has a unique solution $B \in N^\perp$ and the operator $L: N^\perp \to N^\perp$ defined by $B = Lh$ is linear and bounded.

Proof. $B \in P^1$ has a Fourier series $B = \Sigma \, c_m e^{2\pi imt}$, which when substituted into (2.5) yields the equations

$$\left(1 - \int_0^A \beta(a)\rho_0(a)e^{-2\pi ima}\,da\right)c_m = h_m \tag{2.7}$$

for the complex coefficients c_m. Here the h_m are the Fourier coefficients of $h \in P^1$.

(a) For the homogeneous Eq. (2.6) all $h_m = 0$ and a nontrivial solution exists in P^1 if and only if

$$\int_0^A \beta(a)\rho_0(a)e^{-2\pi ima}\,da = 1$$

for at least one $m \in Z$. The space N is spanned by the real and imaginary parts of $e^{2\pi imt}$, $m \in Z_0$, where

$$Z_0 = \left\{ m \in Z: \int_0^A \beta(a)\rho_0(a)e^{-2\pi ima}\,da = 1 \right\}.$$

By the Riemann–Lebesgue theorem

$$\int_0^A \beta(a)\rho_0(a)e^{-2\pi ima}\,da \to 0 \quad \text{as } |m| \to +\infty, \tag{2.8}$$

and hence Z_0 is finite.

(b) Clearly it is necessary for the solution of (2.7) that $h_m = 0$ for all $m \in Z_0$. This shows that $h \in N^\perp$ is necessary. Conversely, suppose $h \in N^\perp$. Then (2.7) can be solved to yield the solution

$$B(t) = \sum_{m \notin Z_0} c_m e^{2\pi imt}, \tag{2.9}$$

$$c_m = h_m \Big/ \left(1 - \int_0^A \beta(a)\rho_0(a)e^{-2\pi ima}\,da\right), \quad m \notin Z_0.$$

By (2.8) there is a constant k for which

$$\left| 1 \Big/ \left(1 - \int_0^A \beta(a)\rho_0(a)e^{-2\pi ima}\,da\right) \right| \le k, \quad m \notin Z_0.$$

Thus $m^2|c_m|^2 \le k^2 m^2|h_m|^2$. Since $h \in P^1$ it follows that $\Sigma \, m^2|h_m|^2 < +\infty$ and hence $\Sigma \, m^2|c_m|^2 < +\infty$, which means $B(t)$ defined by (2.9) is absolutely continuous and hence differentiable almost everywhere. Since B satisfies (2.5) by construction, it easily follows that in fact $B \in P^1$.

From Eq. (2.5) (k is a generic constant not necessarily the same in different expressions)

$$|B(t)| \le \left(\int_0^A \beta^2 \rho_0^2 \, da \right)^{1/2} \left(\int_0^A B^2(t - a) \, da \right)^{1/2} + \|h\|_0$$

$$\le k \left(\int_0^1 B^2(t - a) \, da \right)^{1/2} + \|h\|_0 \le k \left(\sum_{m \notin Z_0} |c_m|^2 \right)^{1/2} + \|h\|_0,$$

so that

$$\|B\|_0 \le k \left(\sum_{m \notin Z_0} |h_m|^2 \right)^{1/2} + \|h\|_0 \le k \left(\int_0^1 h^2 dt \right)^{1/2} + \|h\|_0 \le k\|h\|_0.$$

Similar inequalities obtained from (2.5) when differentiated show that $\|B'\|_0 \le k\|h'\|_0$, and consequently $\|B\|_1 = \|Lh\|_1 \le k\|h\|_1$. ∎

The general solution of Eq. (2.1a) is

$$\rho(t, a) = \rho_0(a) \left(B(t - a) + \int_0^a h_2(t - a + s, s)/\rho_0(s) \, ds \right), \tag{2.10}$$

where B is an arbitrary differentiable function. This solution will also solve (2.1b) if and only if B solves the integral Eq. (2.5) with

$$h(t) = h_1(t) + \int_0^A \beta(a)\rho_0(a) \int_0^a h_2(t - a + s, s)/\rho_0(s) \, ds \, da. \tag{2.11}$$

Note that $(h_1, h_2) \in P^1 \times P^{1-}(\mu)$ implies that $h \in P^1$, and that if $B \in P^1$ solves (2.5) then ρ defined by (2.10) is a solution of (2.1) which lies in $P^1(\mu)$.

Define $N(\mu)$ to be the subspace of all solutions $\rho \in P^1(\mu)$ of (2.2) and let $N^\perp(\mu) = \{\rho \in P^1(\mu): \Omega[\rho, \tilde{\rho}] = 0 \ \forall \ \tilde{\rho} \in N(\mu)\}$, where

$$\Omega[x, \tilde{x}] := \int_0^A \beta(a) \, \mathrm{av}[x(t, a)\tilde{x}^*(t, a)]/\rho_0(a) \, da.$$

Here "*" denotes complex conjugation. Let $M^\perp(\mu) \in P^1 \times P^{1-}(\mu)$ be the closed subspace consisting of those (h_1, h_2) for which h defined by (2.11) lies in $N^\perp(\mu)$.

LEMMA 2.2

Assume $\mu \in \Delta$ and $\beta \in C^0$.

(a) dim $N(\mu)$ is finite.
(b) The system of Eqs. (2.1) has a solution $\rho \in P^1(\mu)$ if and only if $(h_1, h_2) \in M^\perp(\mu)$, in which case (2.1) has a unique solution $\rho \in N^\perp(\mu)$ and the operator $S: M^\perp(\mu) \to N^\perp(\mu)$ defined by $\rho = S(h_1, h_2)$ is linear and bounded.

Proof. The space $N(\mu)$ of homogeneous solutions of (2.2) is spanned by the real and imaginary parts of $\tilde{\rho}_m(t, a) = \rho_0(a) \exp(2\pi imt)$, $m \in Z_0$, and hence is finite dimensional. This proves (a).

Suppose $\rho \in P^1(\mu)$ solves (2.1). Then B in (2.10) solves (2.5) with h given by (2.11). By Lemma 2.1, $h \in N^\perp$, i.e. $(h_1, h_2) \in M^\perp(\mu)$.

Conversely, suppose $(h_1, h_2) \in M^\perp(\mu)$. Then h defined by (2.11) lies in $N^\perp(\mu)$ and by Lemma 2.1 Eq. (2.5) has a unique solution $Lh \in N^\perp$. With $B(t) = \sum_{m \in Z_0} c_m \exp(2\pi imt) + Lh$ for arbitrary coefficients c_m, $m \in Z_0$, $\rho(t, a)$ defined by (2.10) lies in $P^1(\mu)$ and solves (2.1). Now $h \in N^\perp(\mu)$ if and only if $h_m = 0$, $m \in Z_0$, i.e. if and only if

$$h_{1m} + \int_0^A \beta(a)\rho_0(a) \int_0^a \mathrm{av}[h_2(t - a + s, s)e^{-2\pi imt}]/\rho_0(s) \, ds \, da = 0, \quad m \in Z_0, \tag{2.12}$$

where h_{1m} are the Fourier coefficients of $h_1(t)$:

$$h_{1m} = \int_0^1 h_1(t)e^{-2\pi imt}\,dt. \tag{2.13}$$

The unique solution ρ of (2.1) which lies in $N^\perp(\mu)$, i.e. which satisfies $\Omega[\rho, \rho_0 \exp(2\pi im(t - a))] = 0$ for all $m \in Z_0$, is obtained by choosing the constants c_m to be equal to h_{1m}. Thus (2.1) has a unique solution $\rho \in N^\perp(\mu)$ given by $\rho = S(h_1, h_2)$, where

$$S := \rho_0(a)\left(\sum_{m\in Z_0} h_{1m}e^{2\pi im(t-a)} + (Lh)(t - a) + \int_0^a h_2(t - a + s, s)/\rho_0(s)\,ds\right), \tag{2.14}$$

with h_{1m} given by (2.13).

Simple inequalities show, using (2.14) and Lemma 2.1, that

$$|S/\rho_0| \le k\|h_1\|_0 + k\|h\|_1 + A\|h_2\|_\mu$$

$$\left|\frac{\partial}{\partial a}(S/\rho_0)\right| \le k\|h_1\|_0 + k\|h\|_1 + \|h_2\|_\mu + A\left\|\frac{\partial}{\partial t}h_2\right\|_\mu,$$

$$\left|\frac{\partial}{\partial t}(S/\rho_0)\right| \le k\|h_1\|_0 + k\|h\|_0 + k\|h\|_1 + A\left\|\frac{\partial}{\partial t}h_2\right\|_\mu,$$

and consequently $\|S(h_1, h_2)\|_{\mu,1} \le k(\|h_1\|_1 + \|h_2\|_{\mu,1-})$. ∎

We will be interested below in the case when dim $N(\mu) = 1$, that is, when

$$\int_0^A \beta(a)\rho_0(a)\,da = 1, \tag{2.15a}$$

$$\int_0^A \beta(a)\rho_0(a)e^{-2\pi ima}\,da \ne 1, \quad 0 \ne m \in Z. \tag{2.15b}$$

In this case the homogeneous Eqs. (2.2) have exactly one independent solution $\rho \in P^1(\mu)$, namely the time-independent or equilibrium solution $\rho = \rho_0(a)$ of (2.4). Then $N(\mu)$ is spanned by $\rho_0(a)$ and

$$N^\perp(\mu) = \left\{\rho \in P^1(\mu): \int_0^A \beta(a)\,\mathrm{av}[\rho(t, a)]\,da = 0\right\}.$$

By (2.12), $(h_1, h_2) \in M^\perp(\mu)$ and the nonhomogeneous Eq. (2.1) have solutions if and only if the single constraint

$$\mathrm{av}[h_1] + \int_0^A \beta(a)\rho_0(a)\int_0^a \mathrm{av}[h_2(t - a + s, s)]/\rho_0(s)\,ds\,da = 0 \tag{2.16}$$

holds. This "orthogonality condition" on (h_1, h_2) is an averaged version of that derived in [15] for $(h_1, h_2) \in R \times B_\mu$ and for equilibrium solutions $\rho \in B_{\mu,1}$ of (2.3).

The condition (2.15a) implies that $n = 1$ is an "eigenvalue" of the problem (2.4) with $\beta(a)$ replaced by $n\beta(a)$. In fact, $n = 1$ is then the only value of n for which this homogeneous problem has a nontrivial solution (namely $\rho = \rho_0$). The next lemma deals with a generalization of this result when μ and β suffer small-amplitude α time-periodic perturbations. Consider the equations

$$\rho_t(t, a) + \rho_a(t, a) + (\mu(a) + r_2(\alpha)(t, a))\rho(t, a) = 0, \tag{2.17a}$$

$$\rho(t, 0) - n_0\int_0^A (\beta(a) + r_1(\alpha)(t, a))\rho(t, a)\,da = 0, \tag{2.17b}$$

where $n_0 \in R$ and where r_i for each $\alpha \in (-\alpha_0, \alpha_0)$ is a real-valued function of t and a such that

$$r_i(0)(t, a) \equiv 0,$$

and for which the operators defined by

$$l_1(\alpha, \rho) := \int_0^A r_1(\alpha)\rho \, da, \quad l_2(\alpha, \rho) := r_2(\alpha)\rho \tag{2.18}$$

map

$$l_1 : (-\alpha_0, \alpha_0) \times P^1(\mu) \to P^1 \quad \text{and} \quad l_2 : (-\alpha_0, \alpha_0) \times P^1(\mu) \to P^{1^-}(\mu) \tag{2.19}$$

and are $q \geq 1$ times Fréchet differentiable in (α, ρ).

Lemma 2.3

Suppose $\mu \in \Delta$ and $\beta \in C^0$ satisfy (2.15) and the r_i are, as described above, real-valued functions for which the operators l_i in (2.18) satisfy (2.19) and are $q \geq 1$ times Fréchet differentiable. Then for sufficiently small $|\alpha| < \tilde{\alpha}_0 \leq \alpha_0$ the homogeneous Eqs. (2.17) have a nontrivial solution

$$\rho(t, a) = \rho_0(a) + z(\alpha)(t, a) \tag{2.20}$$

for

$$n_0 = 1 + \lambda(\alpha), \tag{2.21}$$

where $z(\alpha) : (-\tilde{\alpha}_0, \tilde{\alpha}_0) \to N^{\perp}(\mu)$ and $\lambda : (-\tilde{\alpha}_0, \tilde{\alpha}_0) \to R$ are $q \geq 1$ times differentiable and where $z(0) = 0$, $\lambda(0) = 0$.

Proof. A substitution of (2.20) and (2.21) into (2.17) results in a system of the form (2.1) for z with

$$h_1 := \int_0^A (r_1(\alpha)(t, a) + \lambda(\beta(a) + r_1(\alpha)(t, a)))(\rho_0(a) + z) \, da,$$
$$h_2 := -r_2(\alpha)(t, a)(\rho_0(a) + z). \tag{2.22}$$

Because (2.15) holds it is necessary that the "orthogonality" condition (2.16) hold. Thus we reformulate (2.17) with (2.20)–(2.21) as follows. Given $z \in N^{\perp}(\mu)$ and $\alpha \in R$ we define λ so that (2.16) holds:

$$\left(1 + \int_0^A \beta(a) \, \text{av}[z(t, a)] \, da + \int_0^A \text{av}[r_1(\alpha)(t, a)(\rho_0(a) + z(t, a))] \, da\right)\lambda$$

$$= -\int_0^A \text{av}[r_1(\alpha)(t, a)(\rho_0(a) + z(t, a))] \, da + \int_0^A \beta(a)\rho_0(a) \int_0^a \text{av}[r_2(\alpha)(t - a + s, s)$$

$$\times (\rho_0(s) + z(t - a + s, s))]/\rho_0(s) \, ds \, da, \tag{2.23}$$

and then the equation for z is equivalent to the operator equation

$$z - S(h_1(\alpha, z), h_2(\alpha, z)) = 0, \tag{2.24}$$

where S is the operator of Lemma 2.2 and h_1, h_2 are defined by (2.22) with λ given by (2.23). Clearly, λ is well defined by (2.23) for $|\alpha|$ and $\|z\|_{\mu, 1}$ sufficiently small, say in an open neighborhood $\Gamma \subset R \times N^{\perp}(\mu)$ of $(\alpha, z) = (0, 0)$.

Under the stated hypotheses the operator $F: \Gamma \to N^{\perp}(\mu)$ defined by $F(\alpha, z) := z - S(h_1(\alpha, z), h_2(\alpha, z))$ is $q \geq 1$ times continuously differentiable and

$$F(0, 0) = 0, \quad F_z(0, 0) = I,$$

where $I : N^{\perp}(\mu) \to N^{\perp}(\mu)$ is the identity operator. The implicit-function theorem implies that (2.24) has a (local) $q \geq 1$ times continuously differentiable solution $z = z(\alpha) \in N^{\perp}(\mu)$, which yields (2.20) and, after a substitution into (2.23), λ in (2.21). $\blacksquare$

As pointed out above $z \in N^{\perp}(\mu)$ means under (2.15) that

$$\int_0^A \beta(a)\, \mathrm{av}[z(\alpha)(t, a)]\, \mathrm{d}a = 0$$

for each α.

With sufficient differentiability ($q \geq 1$ large enough) lower-order terms of any desired order in the α expansions of ρ and n can be computed by the familiar method of substituting expansions

$$\rho(t, a) = \rho_0(a) + \alpha z_1(t, a) + \alpha^2 z_2(t, a) + \cdots,$$

$$n_0 = 1 + \alpha\lambda_1 + \alpha^2\lambda_2 + \cdots$$

into (2.17), equating coefficients of like powers of α on both sides of both equations and solving the resulting recursive linear nonhomogeneous systems for z_i. These linear problems have the form (2.1) and are solvable by Lemma 2.2 when the orthogonality condition above is satisfied by an appropriate choice of λ_i.

From (2.23) it is easily seen that

$$\lambda_1 = -\int_0^A \mathrm{av}[r_1'(0)(t, a)]\rho_0(a)\, \mathrm{d}a + \int_0^A \beta(a)\rho_0(a) \int_0^a \mathrm{av}[r_2'(0)(t - a + s, s)]\, \mathrm{d}s\, \mathrm{d}a,$$

where $r_i'(0)$ denotes the partial derivative with respect to α at $\alpha = 0$. This gives a formula for the lowest-order correction to the "eigenvalue" n_0.

A natural way to study time-periodic oscillations in the death and fertility rates is to consider oscillations about (age-specific) averages μ and β, i.e. to assume that

$$\mathrm{av}[r_i(\alpha)(t, a)] \equiv 0, \quad \forall (\alpha, a) \in (-\alpha_0, \alpha_0) \times [0, A]. \tag{2.25}$$

In this case, $\lambda_1 = 0$ and $n_0 = 1 + \alpha^2\lambda_2 + \cdots$ so that there is a second-order correction to the "eigenvalue" n_0. Moreover, from (2.23) and (2.25),

$$\lambda_2 = -\int_0^A \mathrm{av}[r_1'(0)(t, a)z_1(t, a)]\, \mathrm{d}a$$

$$+ \int_0^A \beta(a)\rho_0(a) \int_0^a \mathrm{av}[r_2'(0)(t - a + s, s)z_1(t - a + s, s)]/\rho_0(s)\, \mathrm{d}s\, \mathrm{d}a,$$

where z_1 is the unique solution of (2.1) in $N^{\perp}(\mu)$ with

$$h_1 = \int_0^A r_1'(0)(t, a)\rho_0(a)\, \mathrm{d}a, \quad h_2 = -r_2'(0)(t, a)\rho_0(a),$$

as guaranteed by Lemma 2.2; that is, z_1 is given by (2.10):

$$z_1(t, a) = \rho_0(a)\left(B(t - a) - \int_0^a r_2'(0)(t - a + s, s)\, \mathrm{d}s\right),$$

where $B(t)$ is the unique solution of the integral Eq. (2.5) satisfying av$[B] = 0$ with [see (2.11)]

$$h(t) = \int_0^A r_1'(0)(t, a)\rho_0(a)\, da - \int_0^A \beta(a)\rho_0(a) \int_0^a r_2'(0)(t - a + s, s)\, ds\, da.$$

Under assumption (2.25) these formulas allow for a study of the effects, to lowest order, on the ''eigenvalue'' n_0 and ''eigensolution'' ρ (as determined by λ_2 and z_1) due to periodic oscillations about average death and fertility rates μ and β. An example is given in Sec. 4.

3. A NONLINEAR THEOREM

In [15,16] the autonomous version of the nonlinear Eqs. (1.3)–(1.4) in which the death and fertility rates D and f are general functionals of the population density ρ, but do not depend explicitly on time t, were studied using global and local bifurcation techniques with n as a bifurcation parameter. Under the normalization

$$\int_0^A f(0)(a) \exp\left(-\int_0^a D(0)(s)\, ds\right) da = 1$$

and certain minimal smoothness assumptions it was shown in [16] that a global continuum of equilibrium $\rho = \rho(a)$ solution pairs $(n, \rho) \in R \times B_\mu$ exists and bifurcates from the critical point $(1, 0)$. This continuum connects to the boundary of the domain of definition of the functionals D and f. In [15] this bifurcation branch was parameterized and studied locally near the bifurcation point $(1, 0)$.

Here we wish to allow D and f to depend explicitly on time t in a periodic manner as in (1.3)–(1.4). Specifically, let D and f suffer small-amplitude α time-periodic perturbations of the form

$$D = \mu + R_2(\alpha, \rho), \quad f = \beta + R_1(\alpha, \rho), \tag{3.1}$$

where $\mu \in \Delta$ and $\beta \in C^0$ satisfy (2.15) and the terms $R_i = R_i(\alpha, \rho)(t, a)$ are such that $R_i(0, 0)(t, a) \equiv 0$ and the operators defined by

$$n_1(\alpha, \rho) := \int_0^A R_1(\alpha, \rho)(t, a)\rho(t, a)\, da, \quad n_2(\alpha, \rho) := R_2(\alpha, \rho)(t, a)\rho(t, a) \tag{3.2}$$

are $q \geq 1$ times Fréchet differentiable as operators mapping

$$n_1 : (-\alpha_0, \alpha_0) \times P^1(\mu) \to P^1, \quad n_2 : (-\alpha_0, \alpha_0) \times P^1(\mu) \to P^1(\mu). \tag{3.3}$$

THEOREM 3.1

Assume that D and f have the form (3.1), where $\mu \in \Delta$ and $\beta \in C^0$ satisfy (2.15) and the perturbation terms R_i are such that $R_i(0, 0) = 0$ and (3.2) defines $q \geq 1$ times Fréchet-differentiable operators which satisfy (3.3). Then for $|\alpha| < \tilde{\alpha}_0 \leq \alpha_0$ and for $|\epsilon| < \epsilon_0$ sufficiently small, (1.3)–(1.4) has a (unique) solution of the form

$$\rho = \epsilon\rho_1(\alpha) + \epsilon w(\alpha, \epsilon) \quad \text{with} \quad n = n_0(\alpha) + n_1(\alpha, \epsilon), \tag{3.4}$$

where

$$\rho_1(\alpha) = \rho_0 + z(\alpha), \quad n_0 = 1 + \lambda(\alpha) \tag{3.5}$$

are as in Lemma 2.3 with $r_1 = R_1(\alpha, 0)$ and $r_2 = R_2(\alpha, 0)$ and where

$$w : (-\tilde{\alpha}_0, \tilde{\alpha}_0) \times (-\epsilon_0, \epsilon_0) \to N^\perp(\mu), \quad n_1 : (-\tilde{\alpha}_0, \tilde{\alpha}_0) \times (-\epsilon_0, \epsilon_0) \to R$$

are $q \geq 1$ times Fréchet differentiable and satisfy $w(\alpha, 0) \equiv 0$, $n_1(\alpha, 0) \equiv 0$.

Proof. If (3.1) and (3.4) are substituted into (1.3)–(1.4), then making use of (3.5) we obtain for each α equations of the form (2.1) for w where

$$h_1 = h_1(\alpha, \epsilon, w) := (R_2(\alpha, 0) - R_2(\alpha, \epsilon\rho_1 + \epsilon w))(\rho_1 + w) - R_2(\alpha, 0)w$$

$$h_2 = h_2(\alpha, \epsilon, w) := n_1 \int_0^A (\beta + R_1(\alpha, \epsilon\rho_1 + \epsilon w))(\rho_1 + w)\, da + n_0 \int_0^A R_1(\alpha, 0)w\, da$$

$$+ n_0 \int_0^A (R_1(\alpha, \epsilon\rho_1 + \epsilon w) - R_1(\alpha, 0))(\rho_1 + w)\, da.$$

Choosing n_1 so that $(h_1, h_2) \in M^\perp(\mu)$, i.e. choosing $n_1 = n_1(\alpha, \epsilon, w)$ so that (2.16) holds, these equations for $w \in N^\perp(\mu)$ can be equivalently reformulated using Lemma 2.2 as $H(\alpha, \epsilon, w) = 0$, where

$$H(\alpha, \epsilon, w) := w - S(h_1(\alpha, \epsilon, w), h_2(\alpha, \epsilon, w))$$

defines an operator $H: (-\alpha_0, \alpha_0) \times (-\epsilon_0, \epsilon_0) \times N^\perp(\mu) \to N^\perp(\mu)$. Note that n_1 so defined is easily seen to be well defined, since the coefficient of n_1 in (2.16) is 1 when $(\alpha, \epsilon, w) = (0, 0, 0)$, and $q \geq 1$ times differentiable for α, ϵ and $\|w\|_{\mu,1}$ sufficiently small. Moreover, a straightforward solution of this linear equation for n_1 shows that $n_1(\alpha, 0, 0) = 0$ for all small $|\alpha|$ and that $\partial n_1(0, 0, 0)/\partial w = 0$.

These facts yield

$$H(0, 0, 0) = 0, \quad \partial H(0, 0, 0)/\partial w = I,$$

and the implicit-function theorem in turn yields a unique solution $w = w(\alpha, \epsilon)$ of $H(\alpha, \epsilon, w) = 0$ satisfying $w(0, 0) = 0$. It is easy to see that $H(\alpha, 0, 0) \equiv 0$, so that $w(\alpha, 0) \equiv 0$, $n(\alpha, 0) \equiv 0$ for all small $|\alpha|$. ∎

For $q \geq 1$ sufficiently large lower-order ϵ coefficients in the expansions (3.4) can be found in the standard manner of deriving and successively solving linear problems of the form (2.1) (for small $|\alpha|$) by the methods in Sec. 2.

4. AN EXAMPLE

Suppose that the age-specific fertility rate oscillates cosinusoidally around an average density-dependent rate with a small-amplitude α (relative to the average):

$$f := \beta(a)(1 + \alpha \cos 2\pi t)g(\rho), \quad 0 \leq \beta \in C^0.$$

Here g is an arbitrary functional of density ρ satisfying $g(0) = 1$, which is sufficiently smooth for an application of the theorem in Sec. 3 ($q \geq 1$ times continuously differentiable). A specific example is the frequently used logistic-type model obtained by setting $g = [1 - \int_0^A k(a)\rho(t, a)\, da]_+$ [21].

With regard to the death rate D we assume in this simple example that it is density independent. Specifically, we choose

$$D = \mu := 1/(A - a) \in \Delta$$

so that the death rate is a monotonically increasing function of age $a \in [0, A)$. The system of Eqs. (1.3)–(1.4) then reduces to

$$\rho_t + \rho_a + \frac{1}{A - a}\rho = 0, \quad t > 0, \quad 0 < a < A < +\infty; \tag{4.1}$$

$$\rho(t, 0) = n \int_0^A \beta(a)(1 + \alpha \cos 2\pi t)g(\rho)\rho\, da, \quad t > 0. \tag{4.2}$$

For this example

$$\rho_0(a) = (A - a)/A, \quad 0 \le a \le A, \tag{4.3}$$

and the normalization of β is

$$\int_0^A \beta(a)\rho_0(a)\, da = 1. \tag{4.4}$$

According to the theorem of Sec. 3 (with $R_1 = \beta(a)(-1 + (1 + \alpha \cos 2\pi t)g(\rho))$ and $R_2 \equiv 0$), Eqs. (4.1)–(4.2) have nontrivial periodic solutions of the form (3.4). The lowest-order terms in the expansions (3.4) are given by the solution (3.5) of the linearized equations

$$\rho_t + \rho_a + \frac{1}{A - a}\rho = 0, \tag{4.5}$$

$$\rho(t, 0) = n \int_0^A \beta(a)(1 + \alpha \cos 2\pi t)\rho\, da, \tag{4.6}$$

which can be written

$$\rho_1 = \rho_0(a) + \alpha z_1(t, a) + \alpha^2 z_2(t, a) + \cdots, \quad n_0 = 1 + \lambda_1\alpha + \lambda_2\alpha^2 + \cdots. \tag{4.7}$$

From the remarks at the end of Sec. 2 it follows, in fact, that $\lambda_1 = 0$, as will also be seen from the analysis below.

To determine the lower-order coefficients in these α expansions, we substitute (4.7) into (4.5)–(4.6) and equate coefficients of like powers of α. This leads to the following three systems of equations:

$$\rho_{0a} + \frac{1}{A - a}\rho_0 = 0, \quad \rho_0(0) = \int_0^A \beta(a)\rho_0(a)\, da;$$

$$z_{1t} + z_{1a} + \frac{1}{A - a}z_1 = 0, \quad z_1(t, 0) = \int_0^A \beta(a)z_1(t, a)\, da + \lambda_1 + \cos 2\pi t; \tag{4.8}$$

$$z_{2t} + z_{2a} + \frac{1}{A - a}z_2 = 0, \quad z_2(t, 0) = \int_0^A \beta(a)z_2(t, a)\, da + \lambda_2$$

$$+ \int_0^A \beta(a)\rho_1(t, a)\, da \cos 2\pi t. \tag{4.9}$$

to be solved for the first three coefficients in (4.7). The first system is satisfied by the choice (4.3) for $\rho_0(a)$, in view of (4.4).

In order to solve the system (4.8) for z_1 it is necessary by Lemma 2.2 that

$$\text{av}[\lambda_1 + \cos 2\pi t] = 0,$$

i.e. $\lambda_1 = 0$, in which case

$$z_1(t, a) = B(t - a)\rho_0(a), \tag{4.10}$$

where $B(t)$ solves the integral equation

$$B(t) = \int_0^A \beta(a)\rho_0(a)B(t - a)\, da + \cos 2\pi t.$$

The solution of this integral equation is given by

$$B(t) = c_1 \cos 2\pi t + c_2 \sin 2\pi t; \tag{4.11}$$

$$c_1 = \frac{1 - C}{d}, \quad c_2 = \frac{S}{d}, \quad d = (1 - C)^2 + S^2; \tag{4.12}$$

$$C = \int_0^A \beta(a)\rho_0(a) \cos 2\pi a \, da, \quad S = \int_0^A \beta(a)\rho_0(a) \sin 2\pi a \, da. \tag{4.13}$$

Here it is necessary that

$$d > 0$$

in order that (2.15b) holds.

Equations (4.11)–(4.13) define the lowest-order oscillation in ρ_1 of the solution ρ given by (3.4):

$$\rho_1 = \frac{A - a}{A} (1 + \alpha B(t - a) + \cdots).$$

From (4.11) one can determine the age-specific phase and amplitude of this oscillation (relative to that in the fertility rate f) as they depend, through (4.12)–(4.13), on the age-specific inherent fertility rate $\beta(a)$ and the maximum lifespan A. Some general conclusions which can be drawn are as follows. Unless $c_2 = 0$, the "total birth rate" $B(t)$ is not in phase with the oscillation in the fertility rate f. Moreover, since $c_1 \geq 0$, the total birth rate B peaks later than f (with relative phase $\tan^{-1}(S/(1 - C))$ if and only if $S > 0$. The amplitude of B increases without bound as d approaches 0, where (2.15) fails to hold and a resonance occurs.

In order to determine (to lowest order) the effect that the oscillation in f has on the critical value n_0 of the average inherent net reproductive rate n we must (since $\lambda_1 = 0$) calculate λ_2. This is done by means of the required "orthogonality" condition (2.16) for the solution of (4.9). The result is

$$\lambda_2 = (C^2 + S^2 - C)/d.$$

Consequently, in this example there is a second-order adjustment in the critical bifurcation value of the averaged inherent net reproductive rate caused by the oscillation in the fertility rate f. There is an increase in this critical value if $\lambda_2 > 0$ and a decrease otherwise.

A simple example is given by the case when fertility is not age specific: $\beta(a) \equiv \beta_0 = $ const > 0 or, by the normalization (4.4), when $\beta(a) \equiv 2/A$. Then

$$C = (1 - \cos 2\pi A)/2\pi^2 A^2, \quad S = (2\pi A - \sin 2\pi A)/2\pi^2 A^2 \tag{4.14}$$

and $\lambda_2 \geq 0$, i.e. there is an increase in the critical value n_0, since

$$C^2 + S^2 - C = (1 + \pi^2 A^2)(1 + \cos(2\pi A + \phi))/2\pi^4 A^4 \geq 0, \tag{4.15}$$

where $\phi = \tan^{-1}(2\pi A/(\pi^2 A^2 - 1))$. In this example $S > 0$, so that as remarked above the total birth rate $B(t)$ given by (4.11) with (4.14) in the coefficients (4.12) is never exactly in phase with the oscillatory fertility rate f, and in fact peaks after f. Furthermore, in this example, the coefficient of the lowest-order oscillation in the total population size $P = \int_0^A \rho(t, a) \, da$ is easily computed to be

$$\int_0^A z_1(t, a) \, da = \int_0^A B(t - a)\rho_0(a) \, da = -2A((C^2 - S^2 - C) \cos 2\pi t - S \sin 2\pi t)/d.$$

Consequently, by (4.15) and $S > 0$, P oscillates out of phase with the fertility rate f with a phase difference $\tan^{-1}(S/(C^2 + S^2 - C))$ ranging from one-quarter to one-half cycle.

The purpose of this example is simply to illustrate the general results and techniques of Secs. 2 and 3. It is not intended here to study in depth the biological implications of this model or of small-amplitude oscillations in the vital rates in general. It is hoped to do this in a future paper.

5. SUMMARY

The theorem appearing in Sec. 2 establishes a local, parameterized branch of time-periodic solutions of the McKendrick Eqs. (1.3)–(1.4) when the age-specific vital rates D and f suffer small-amplitude α periodicities in time. The branch bifurcates from the trivial solution $\rho \equiv 0$ at a critical value n_0 of the averaged net reproductive rate n, as given by Lemma 2.3. This result is a generalization of the autonomous equilibrium results in [15], which correspond to $\alpha = 0$. This theorem and the lemmas of Sec. 2 permit the use of standard perturbation techniques to calculate lower-order approximations to the time-periodic solutions. A simple example is given in Sec. 4 in which fertility oscillates consinusoidally in time while the death rate is time and density independent. Besides illustrating the general results and techniques, this example shows some interesting biological results that such a model can imply, such as an increased critical bifurcation value for the average net reproductive rate due to the fertility-rate oscillation and certain phase relationships between fertility and the total birth rate and the total population size.

REFERENCES

1. E. R. Pianka, *Evolutionary Ecology*, 2nd Edn. Harper & Row, New York (1978).
2. M. Badii and A. Schiaffino, Asymptotic behaviour of positive solutions of periodic delay logistic equations. *J. Math. Biol.* **14**, 95–100 (1982).
3. M. Bardi, An equation of growth of a single species with realistic dependence on crowding and seasonal factors. *J. Math. Biol.* **17**, 33–43 (1983).
4. M. S. Boyce and D. J. Daley, Population tracking of fluctuating environments and natural selection for tracking ability. *Am. Naturalist* **115**(4), 480–491 (1980).
5. D. S. Cohen and S. Rosenblat, A delay logistic equation with variable growth rate. *SIAM J. Appl. Math.* **42**(3), 608–624 (1982).
6. B. D. Coleman, Nonautonomous logistic equations as models of the adjustment of populations to environmental change. *Math. Biosci.* **45**, 159–173 (1979).
7. B. D. Coleman, Y.-H. Hsieh and G. P. Knowles, On the optimal choice of r for a population in a periodic environment. *Math. Biosci.* **46**, 71–85 (1979).
8. J. M. Cushing, Periodic Kolmogorov systems. *SIAM J. Math. Anal.* **13**(5), 811–827 (1982).
9. J. M. Cushing, Stable positive periodic solutions of the time-dependent logistic equation under possible hereditary influences. *J. Math. Anal. Appl.* **60**(3), 747–754 (1977).
10. J. M. Cushing, Oscillatory population growth in periodic environments. To appear in *Theor. Pop. Biol.*
11. P. de Mottoni and A. Schiaffino, On logistic equations with time periodic coefficients. *I.A.C. "Mauro Picone", Rome, Series III*, **192** (1979).
12. R. M. Nisbet and W. S. C. Gurney, Population dynamics in a periodically varying environment. *J. Theor. Biol.* **56**, 459–475 (1976).
13. S. Rosenblat, Population models in a periodically fluctuating environment. *J. Math. Biol.* **9**, 23–36 (1980).
14. P. Sonneveld and J. van Kan, On a conjecture about the periodic solution of the logistic equation. *J. Math. Biol.* **8**, 285–289 (1979).
15. J. M. Cushing, Existence and stability of equilibria in age-structured population dynamics. *J. Math. Biol.* **20**, 259–276 (1984).
16. J. M. Cushing, Equilibria in structured populations, *J. Math. Biol.* **23**, 15–39 (1985).
17. J. Prüss, Equilibrium solutions of age-specific population dynamics of several species. *J. Math. Biol.* **11**, 65–84 (1981).
18. J. Prüss, On the qualitative behavior of populations with age-specific interactions. *Comput. Math. Appl.* **9**(3), 327–340 (1983).
19. G. F. Webb, *Theory of Nonlinear Age-Dependent Population Dynamics*. Monographs in Pure and Applied Mathematics, Vol. *89*, Marcel Dekker, New York, 1985.
20. K. E. Swick, Periodic solutions of a nonlinear age-dependent model of single species population dynamics. *SIAM J. Math. Anal.* **11**(5), 901–910 (1980).
21. F. Hoppensteadt, *Mathematical Theories of Populations: Demographics, Genetics and Epidemics*. Regional Conf. Series in Appl. Math. **20**. SIAM, Philadelphia, PA (1975).

Comp. & Maths. with Appls. Vol. 12A, Nos. 4/5, pp. 527–539, 1986
Printed in Great Britain.

0886–9553/86 $3.00 + .00
© 1986 Pergamon Press Ltd.

LOGISTIC MODELS OF STRUCTURED POPULATION GROWTH

G. F. Webb
Mathematics Department, Vanderbilt University, Nashville, TN 37235, U.S.A.

Abstract—A general model of structured population dynamics with logistic-type nonlinearity is considered. The model consists of a semilinear hyperbolic partial differential equation with a linear boundary condition and an initial value. The logistic nonlinearity corresponds to increased mortality as the population increases. The problem is treated as an abstract semilinear evolution equation. Sufficient conditions are given for the solutions to converge to an equilibrium solution. Applications are made to age-structured populations, size-structured populations, multispecies-structured populations, and cell populations.

INTRODUCTION

The purpose of this paper is to prove the convergence to equilibrium of solutions of structured population models with logistic nonlinearities. The prototype of such models is an age-structured population described as follows: Let $p(a, t)$ be the density of a population at time t with respect to an age variable a, so that the total population at time t between ages a_1 and a_2 is

$$\int_{a_1}^{a_2} p(a, t) \, \mathrm{d}a.$$

Let ω be a weight function and let

$$Q(t) = \int_0^\infty p(a, t)\omega(a) \, \mathrm{d}a$$

be the weighted integral of the density. Consider the logistic model of age-structured population growth given by

$$p_t(a, t) + p_a(a, t) = -[\mu(a) + \eta(Q(t))]p(a, t), \tag{1.1}$$

$$p(0, t) = \int_0^\infty \beta(a)p(a, t) \, \mathrm{d}a, \tag{1.2}$$

$$p(a, 0) = \phi(a), \tag{1.3}$$

where μ and β are the age-specific mortality and fertility moduli, respectively, $\eta(Q(t))$ is the density-dependent mortality, and ϕ is the known initial age distribution at time 0. The logistic term $\eta(Q(t))$ provides a mechanism for increased mortality as the portion $Q(t)$ of the population increases and the effects of crowding and resource limitation take hold.

The separated form of the right-hand side of (1.1) is essential to our development. This form for the mortality process has been described by M. Gurtin[1] and W. Streifer[2] as a separation into an endogenous part $\mu(a)$, which depends on age, but is not influenced by environmental limitations, and an exogenous part $\eta(Q(t))$, which is independent of age, but arises as resources become limited because of crowding. The form of the birth process (1.2) is also essential to our approach in that it is not effected by crowding. Nonlinear age-dependent population models having the form (1.1)–(1.3) have been studied by a number of researchers, including M. Gurtin and R. MacCamy[3,4], J. Prüss[5–7], M. Gurtin[1], and the author[8]. Recently, P. Marcati[9], S. Busenberg and M. Iannelli[10], and M. Iannelli[11] have treated such models and established the convergence of solutions to equilibrium age distributions. In this paper we will simplify their proofs and extend their results to the general context of a population structured by internal variables.

Our approach to this problem rests upon viewing it as a semilinear evolution equation in a Banach space. The linear part of this equation corresponds to a strongly continuous semigroup of bounded linear operators in this Banach space. The behavior of the solutions of the linear problem is determined by the dominant real eigenvalue of the infinitesimal generator of this semigroup. The nonlinear problem is subsequently analyzed as a simple nonlinear perturbation of the linear semigroup. In this paper we will first prove a general theorem concerning logistic nonlinear structured population dynamics and then apply it to a variety of biological populations.

2. AN ABSTRACT LOGISTIC EQUATION

In this section we investigate the stability of equilibrium solutions to an abstract logistic equation in a Banach space. The formulation of the abstract problem allows applications to diverse models of structured populations. Let X be a Banach lattice with norm $\| \ \|$ and let X_+ denote the cone of nonnegative elements of X (see [12], p. 369). We require the following hypotheses:

$T(t)$, $t \geq 0$ is a strongly continuous semigroup of bounded linear operators in X
with infinitesimal generator A and $T(t)x \in X_+$ for $x \in X_+$.   (2.1)

There exist a real number λ_0 and a direct sum decomposition $X = X_0 \oplus X_1$ with
associated projections P_i, $P_iX = X_i$, $i = 0, 1$, such that $P_iT(t) = T(t)P_i$,
$i = 0, 1$, $T(t)P_0 = e^{\lambda_0 t}P_0$, $t \geq 0$, and for some constants $M \geq 1$, $\omega < \lambda_0$,
$|T(t)P_1| \leq Me^{\omega t}|P_1|$, $t \geq 0$.   (2.2)

$\mathscr{F}$ is a bounded linear functional on X such that $\mathscr{F}x > 0$ for $x \in X_+ - \{0\}$.   (2.3)

η is a continuous increasing function from $[0, \infty)$ onto $[0, \infty)$.   (2.4)

Remark 2.1 The hypothesis (2.2) implies that $AP_0x = \lambda_0 P_0x$ for any $x \in X$ so that λ_0 is an eigenvalue of A. In a typical case the spectrum of A consists of a finite number of eigenvalues lying to the right of some line parallel to the imaginary axis and nonessential spectrum lying to the left of this line. The eigenvalue λ_0 is dominant in the sense that $\lambda_0 > \operatorname{Re} \lambda$ for any λ in the spectrum of A. Furthermore, $\lim_{t \to \infty}|e^{-\lambda_0 t}T(t) - P_0| = \lim_{t \to \infty}|e^{-\lambda_0 t}T(t)P_1| \leq \lim_{t \to \infty}Me^{-(\lambda_0 - \omega)t}|P_1| = 0$.

Consider the abstract logistic equation

$$w'(t) = Aw(t) - \eta(\mathscr{F}w(t))w(t), \quad t \geq 0, \quad w(0) = x. \tag{2.5}$$

Models such as (1.1)–(1.3) can be formulated as in (2.5), where the boundary condition (1.2) is incorporated into the domain in A.

THEOREM 2.1

Let (2.1)–(2.4) hold and let $x \in X_+ \cap D(A)$ such that $P_0x \in X_+ - \{0\}$. There exists a unique continuously differentiable function $w: [0, \infty) \to X_+ - \{0\}$ such that w satisfies (2.5). If $\lambda_0 < 0$, then $\lim_{t \to \infty}w(t) = 0$, and if $\lambda_0 \geq 0$, then $\lim_{t \to \infty}w(t) = \eta^{-1}(\lambda_0)P_0x/\mathscr{F}P_0x$.

Proof. We first claim that

$$T(t)x \in X_+ - \{0\} \quad \text{and} \quad \mathscr{F}(T(t)x) > 0 \quad \text{for } t \geq 0. \tag{2.6}$$

If on the contrary, $T(t_1x) = 0$ for some t_1; then $0 = P_0T(t_1)x = e^{\lambda_0 t_1}P_0x$, which contradicts $P_0x \in X_+ - \{0\}$. We also claim that

$$\lim_{t \to \infty} \mathscr{F}(T(t)Ax)/\mathscr{F}(T(t)x) = \lambda_0, \tag{2.7}$$

since

$$\mathscr{F}(T(t)Ax)/\mathscr{F}(T(t)x) = \mathscr{F}(\lambda_0 e^{\lambda_0 t}P_0x + T(t)P_1Ax)/\mathscr{F}(e^{\lambda_0 t}P_0x + T(t)P_1x)$$

$$= \mathscr{F}(\lambda_0 P_0x + e^{-\lambda_0 t}T(t)P_1Ax)/\mathscr{F}(P_0x + e^{-\lambda_0 t}T(t)P_1x).$$

We next show that for each $t_1 > 0$ there exists a continuous function $F: [0, \infty) \to [0, \infty)$ such that for $0 \le t \le t_1$,

$$F(t) = \exp\left[-\int_0^t \eta(F(s))\ ds\right]\mathscr{F}(T(t)x). \tag{2.8}$$

Let W be the closed convex subset of the Banach space $C([0, t_1], R)$ consisting of functions F such that $\|F\|_{C[0,t_1]} \le |\mathscr{F}|(|P_0| + M \max\{1, e^{\lambda_0 t_1}\}|P_1|)$. Define $K: W \to W$ by

$$(KF)(t) = \exp\left[-\int_0^t \eta(F(s))\ ds\right]\mathscr{F}(T(t)x), \quad F \in W, 0 \le t \le t_1.$$

By the hypothesis K is a continuous mapping of W into itself. Also, the image of K is compact by the Arzelá–Ascoli Theorem (see [13], p. 266). By the Schauder–Leray Theorem (see [14], p. 50) K has a fixed point F in W.

We claim that this fixed point of K is unique. Notice that F satisfies

$$F'(t) = [G(t) - \eta(F(t))]F(t), \tag{2.9}$$

where $G(t) = \mathscr{F}(T(t)Ax)/\mathscr{F}(T(t)x)$. Suppose that F and $\hat{F}$ are both fixed points of K. Since η is increasing,

$$\begin{aligned}
\tfrac{1}{2}d/dt[F(t) - \hat{F}(t)]^2 &= G(t)[F(t) - \hat{F}(t)]^2 \\
&\quad - [\eta(F(t))F(t) - \eta(\hat{F}(t))\hat{F}(t)][F(t) - \hat{F}(t)] \\
&\le \left(\sup_{0\le\tau\le t_1} G(\tau)\right)[F(t) - \hat{F}(t)]^2.
\end{aligned}$$

Since $F(0) = \hat{F}(0)$, this last differential inequality implies that $F \equiv \hat{F}$ on $[0, t_1]$. Since t_1 is arbitrary, we conclude that there exists a unique function $F: [0, \infty) \to [0, \infty)$ satisfying (2.8). Define $w: [0, \infty) \to X_+ - \{0\}$ by

$$w(t) = \exp\left[-\int_0^t \eta(F(s))\ ds\right]T(t)x, \quad t \ge 0.$$

Notice that $w(t) \in D(A)$, since $x \in D(A)$. Also, $\mathscr{F}w(t) = F(t)$ and so w is the unique continuously differentiable function satisfying (2.5).

If $\lambda_0 < 0$, then $T(t)x \to 0$ as $t \to \infty$, so that $w(t) \to 0$ as $t \to \infty$. If $\lambda_0 \ge 0$, then let $\in > 0$ and set $\delta = \eta(\eta^{-1}(\lambda_0) + \in) - \lambda_0$. By (2.7) there exists $t_1 > 0$ such that if $t \ge t_1$ then $|G(t) - \lambda_0| < \delta/2$. Suppose that $t \ge t_1$ and $F(t) \ge \eta^{-1}(\lambda_0) + \in$. Then, $\eta(F(t)) \ge \eta(\eta^{-1}(\lambda_0) + \in) = \lambda_0 + \delta > G(t) + \delta/2$. Thus, $F(t)$ is not $> \eta^{-1}(\lambda_0) + \in$ for all $t > t_1$, since (2.9) would then imply that $F'(t) < -(\delta/2)F(t)$, which in turn implies that $F(t) \le e^{-\delta t/2}F(t_1)$, a contradiction. Thus, there exists $t_2 > t_1$ such that $F(t_2) < \eta^{-1}(\lambda_0) + \in$. Assume that there exists $t > t_2$ such that $F(t) > \eta^{-1}(\lambda_0) + \in$ and let t_3 be the infimum of all such t_2. Then, $t_3 > t_2$, $F(t_3) \ge \eta^{-1}(\lambda_0) + \in > F(t)$ for $t_2 < t < t_3$ and as above, $F'(t_3) < 0$. But then $F(t)$ is decreasing at t_3 and this yields a contradiction. Hence, $F(t) < \eta^{-1}(\lambda_0) + \in$ for all $t > t_2$. A similar argument shows that $F(t) > \eta^{-1}(\lambda_0) - \in$ for all $t >$ some t_2' (unless $\lambda_0 = 0$). Therefore $\lim_{t\to\infty}F(t) = \eta^{-1}(\lambda_0)$. From (2.8) we have that

$$\begin{aligned}
\lim_{t\to\infty} \exp\left[\lambda_0 t - \int_0^t \eta(F(s))\ ds\right] &= \lim_{t\to\infty} F(t)/\mathscr{F}(e^{-\lambda_0 t}T(t)x) \\
&= \eta^{-1}(\lambda_0)/\mathscr{F}P_0 x,
\end{aligned} \tag{2.10}$$

which implies that $\lim_{t\to\infty}w(t) = \eta^{-1}(\lambda_0)P_0 x/\mathscr{F}P_0 x$.   ∎

Remark 2.2. Notice that if $w = \eta^{-1}(\lambda_0)P_0 x/\mathscr{F}P_0 x$, then $Aw = \eta(\mathscr{F}w)w$. Thus $w(t)$ converges to an equilibrium solution of (2.5). Furthermore, if $P_0x/\mathscr{F}P_0x$ is independent of x, then $w(t)$ converges to a unique equilibrium, as is the case if P_0 has rank one.

3. CHRONOLOGICAL AGE-STRUCTURED POPULATIONS

The classical linear model of age-dependent population dynamics has the formulation

$$l_t(a, t) + l_a(a, t) = -\mu(a)l(a, t), \tag{3.1}$$

$$l(0, t) = \int_0^\infty \beta(a)l(a, t)\, da, \tag{3.2}$$

$$l(a, 0) = \phi(a), \tag{3.3}$$

where μ and β are the age-specific mortality and fertility moduli, respectively (the units of μ and β are 1/time) and ϕ is the known initial age distribution. In [8] it is shown that the solutions of (3.1)–(3.3) correspond to a strongly continuous semigroup of bounded linear operators in the Banach space $L^1(0, \infty)$, satisfying (2.1) and (2.2). We collect these results ([8], Proposition 3.2, Proposition 3.7, and Theorem 4.10) into the following theorem.

THEOREM 3.1

Let $\mu \in L^\infty(0, \infty)$ and let $\beta \in L^\infty_+(0, \infty) - \{0\}$. Define $A: L^1(0, \infty) \to L^1(0, \infty)$ by

$$A\phi = -\phi' - \mu\phi, \quad D(A) = \left\{ \phi \in L^1(0, \infty): \phi \text{ is absolutely continuous,} \right.$$

$$\left. \phi' \in L^1(0, \infty), \text{ and } \phi(0) = \int_0^\infty \beta(a)\phi(a)\, da \right\}. \tag{3.4}$$

Then, A is the infinitesimal generator of a strongly continuous semigroup of bounded linear operators $T(t)$, $t \geq 0$ in $L^1(0, \infty)$ satisfying (2.1), and if $\phi \in D(A)$, then $l(a, t) = (T(t)\phi)(a)$ is the unique solution of (3.1)–(3.3). Let $\lambda = \lambda_0$ be the (necessarily) unique real solution of the equation

$$1 = \int_0^\infty e^{-\lambda a}\beta(a)\Pi(a, 0)\, da. \tag{3.5}$$

Let $\lambda_0 > -\underline{\mu}$ where $\underline{\mu} \equiv \sup_{a \to \infty} \text{ess-inf}_{b>a}\mu(b)$, and define $P_0: L^1(0, \infty) \to L^1(0, \infty)$ by

$$(P_0\phi)(a) = e^{-\lambda_0 a}\Pi(a, 0)V_{\lambda_0}(\phi)/M_{\lambda_0}, \quad \phi \in L^1(0, \infty), \quad a \geq 0, \tag{3.6}$$

where

$$\Pi(a, b) = \exp\left[-\int_b^a \mu(c)\, dc \right], \quad 0 \leq a \leq b; \tag{3.7}$$

$$V_{\lambda_0}(\phi) = \int_0^\infty \beta(a)e^{-\lambda_0 a}\left[\int_0^a e^{\lambda_0 b}\Pi(a, b)\phi(b)\, db \right] da, \quad \phi \in L^1(0, \infty); \tag{3.8}$$

$$M_{\lambda_0} = \int_0^\infty \beta(a)ae^{-\lambda_0 a}\Pi(a, 0)\, da. \tag{3.9}$$

Then, (2.2) is satisfied for this λ_0, P_0, $P_1 = I - P_0$, and any $\omega < \lambda_0$ sufficiently close to λ_0. Consequently,

$$\lim_{t \to \infty} |e^{-\lambda_0 t}T(t) - P_0| = 0.$$

Remark 3.1. The Equation (3.5) is known as the *characteristic equation* and its unique real solution λ_0 is known as the *intrinsic growth constant*. The quantity $\Pi(a, b)$ is the *survival probability* of an individual from age b to age a. The quantity $V_{\lambda_0}(\phi)$ is called the *natural reproductive value* of the initial age distribution ϕ, and the quantity M_{λ_0} is called the *mean age of childbirth*. The connection of the characteristic equation to the ultimate behavior of the solutions of (3.1)–(3.2) was discovered by F. Sharpe and A. Lotka[15] in 1911, and the first rigorous treatment was given by W. Feller[16] in 1941. Recently, operator-theoretic methods have been used to prove this result in [8,17–20]. In the classical theory of linear age-dependent population dynamics the density $T(t)P_0\phi = e^{\lambda_0 t}P_0\phi$ is called a *stable age distribution,* since it has the property that the proportion of the population in the age range $[a_1, a_2]$

$$\int_{a_1}^{a_2} (T(t)P_0\phi)(a) \, da \bigg/ \int_0^{\infty} (T(t)P_0\phi)(a) \, da$$

is constant for all time t.

Remark 3.2. In [8] stronger hypotheses are placed on μ and β than stated in Theorem 3.1. In particular μ and β are required to be continuous in [8], but this hypothesis may be weakened to measurable and essentially bounded using the methods in [20]. Also, μ is require to be bounded below above 0 in [8], but this hypothesis may be replaced by the condition $\lambda_0 > -\mu$, as is easily seen from the proof given in [8], Theorem 4.6.

Theorem 2.1 now applies immediately to the logistic model of chronological age-structured population dynamics (1.1)–(1.3).

THEOREM 3.2

Let $\mu \in L^\infty(0, \infty)$, let $\beta, \omega \in L^\infty_+(0, \infty) - \{0\}$, let $\lambda_0 > -\mu$, let η satisfy (2.4), and let $\phi \in L^1_+(0, \infty) \cap D(A)$ such that $P_0\phi \in L^1_+(0, \infty) - \{0\}$. The unique solution of (1.1)–(1.3) satisfies $\lim_{t\to\infty}\|p(\cdot, t)\|_{L^1(0,\infty)} = 0$ if $\lambda_0 < 0$, and $\lim_{t\to\infty}\|p(\cdot, t) - \eta^{-1}(\lambda_0)\phi_0/\int_0^\infty \phi_0(a)\omega(a) \, da\|_{L^1(0,\infty)} = 0$ if $\lambda_0 \geq 0$, where $\phi_0(a) = e^{-\lambda_0 a}\Pi(a, 0)$, $a \geq 0$.

The ideas of Theorems 3.1 and 3.2 can be implemented for systems of hyperbolic partial differential equations with logistic type nonlinearities. As an example of such a system consider two interacting species with density functions $p_1(a, t)$ and $p_2(a, t)$. The densities satisfy the equations

$$p_{1t}(a, t) + p_{1a}(a, t) = -[\mu_1(a) + \mu_{11}(Q_1(t)) + \mu_{12}(Q_2(t))]p_1(a, t), \tag{3.10}$$

$$p_{2t}(a, t) + p_{2a}(a, t) = -[\mu_2(a) + \mu_{21}(Q_1(t)) + \mu_{22}(Q_2(t))]p_2(a, t), \tag{3.11}$$

$$p_1(0, t) = \int_0^{\infty} \beta_1(a)p_1(a, t) \, da, \tag{3.12}$$

$$p_2(0, t) = \int_0^{\infty} \beta_2(a)p_2(a, t) \, da, \tag{3.13}$$

$$p_1(a, t) = \phi_1(a), \tag{3.14}$$

$$p_2(a, t) = \phi_2(a). \tag{3.15}$$

Here μ_i, β_i are the age-specific mortality and fertility moduli for species $i = 1, 2$:

$$Q_i(t) = \int_0^{\infty} p_i(a, t)\omega_i(a) \, da,$$

with ω_i a weight function, μ_{ij}, $i, j = 1, 2$, correspond to intraspecies or interspecies contributions to the mortality processes, and ϕ_i is the known initial age distribution. If μ_{12} and μ_{21} are increasing functions, then this system corresponds to a model of two competing species, and if μ_{12} is decreasing and μ_{21} is increasing, then this system corresponds to a predator–prey model with p_1 the density of the predator and p_2 the density of the prey. Similar models of age-dependent interacting species have been treated in [8,21–25].

We require the following hypothesis:

$$\mu_1, \mu_2 \in L^\infty_+(0, \infty), \beta_1, \beta_2, \omega_1, \omega_2 \in L^\infty_+(0, \infty) - \{0\}, \text{ and } \mu_{11}, \mu_{12}, \mu_{21}, \mu_{22}$$

are continuously differentiable functions from $[0, \infty)$ to $[0, \infty)$.   (3.16)

For $i = 1, 2$ define

$$A_i\phi = -\phi' - \mu_i\phi, \quad D(A_i) = \Big\{\phi \in L^1(0, \infty): \phi \text{ is absolutely continuous,}$$

$$\phi' \in L^1(0, \infty), \text{ and } \phi_i(0) = \int_0^\infty \beta_i(a)\phi(a)\, da\Big\}. \tag{3.17}$$

By Theorem 3.1 A_i is the infinitesimal generator of a strongly continuous semigroup of bounded linear operators $T_i(t)$, $t \geq 0$ in $L^1(0, \infty)$ satisfying (2.1) and (2.2) with dominant real eigenvalue λ_{0i}, as in (3.5) and associated projection P_{0i} as in (3.6).

THEOREM 3.3

Let (3.16) hold, let $\lambda_{0i} > -\mu_i$, $i = 1, 2$, and let $\phi_i \in L^1_+(0, \infty) \cap D(A_i)$ such that $P_{0i}\phi_i \in L^1_+(0, \infty) - \{0\}$, $i = 1, 2$. There exists a unique solution to the problem (3.10)–(3.15). Suppose there exist $\hat{Q}_1, \hat{Q}_2 \geq 0$ such that

$$\mu_{11}(\hat{Q}_1) + \mu_{12}(\hat{Q}_2) = \lambda_{01} \quad \text{and} \quad \mu_{21}(\hat{Q}_1) + \mu_{22}(\hat{Q}_2) = \lambda_{02}. \tag{3.18}$$

If

$$\mu'_{11}(\hat{Q}_1)\hat{Q}_1 + \mu'_{22}(\hat{Q}_2)\hat{Q}_2 > 0 \quad \text{and} \quad \mu'_{11}(\hat{Q}_1)\mu'_{22}(\hat{Q}_2) > \mu'_{12}(\hat{Q}_1)\mu'_{21}(\hat{Q}_2), \tag{3.19}$$

then

there exists $t_1 > 0$ and $\delta > 0$ such that if $\|p_i(\cdot, t_1) - \hat{\phi}_i\|_{L^1(0,\infty)} < \delta$, $i = 1, 2$,

then $\lim\limits_{t\to\infty}\|p_i(\cdot, t) - \hat{Q}_i\hat{\phi}_i \big/ \int_0^\infty \hat{\phi}_i(a)\omega_i(a)\, da\|_{L^1(0,\infty)} = 0$, where $\hat{\phi}_i(a) =$

$$e^{-\lambda_{0i}a} \exp\Big[-\int_0^a \mu_i(c)\, dc\Big], \quad a \geq 0, i = 1, 2. \tag{3.20}$$

Proof. Under the hypothesis stated the existence of a unique solution $\{p_1(\cdot, t), p_2(\cdot, t)\}$ from $[0, \infty)$ to $L^1_+(0, \infty) \times L^1_+(0, \infty)$ for the problem (3.10)–(3.15) follows from Theorems 2.1, 2.3, 2.4, 2.5, 2.9, and Proposition 3.6 in [8]. In addition the following formulas hold for $i = 1, 2$:

$$p_i(\cdot, t) = \exp\Big[-\int_0^t \{\mu_{i1}(Q_1(s)) + \mu_{i2}(Q_2(s))\}\, ds\Big] T_i(t)\phi_i, \quad t \geq 0, \tag{3.21}$$

$$Q_i(t) = \exp\Big[-\int_0^t \{\mu_{i1}(Q_1(s)) + \mu_{i2}(Q_2(s))\}\, ds\Big] S_i(t), \quad t \geq 0, \tag{3.22}$$

$$Q'_i(t) = [S'_i(t)/S_i(t) - \mu_{i1}(Q_1(t)) - \mu_{i2}(Q_2(t))]Q_i(t), \quad t \geq 0, \tag{3.23}$$

where

$$S_i(t) = \int_0^\infty (T_i(t)\phi_i)(a)\omega_i(a)\, da, \quad t \geq 0,$$

and $S_i(t) > 0$ for all $t \geq 0$ as in (2.6).

As in (2.7) $\lim_{t\to\infty} S_i'(t)/S_i(t) = \lambda_{0i}$, $i = 1, 2$. Thus, the system of ordinary differential equations (3.23) can be written as

$$Q'(t) = LQ(t) + f(Q(t)) + g(t, Q(t)), \tag{3.24}$$

where

$$Q(t) = \{Q_1(t), Q_2(t)\}, \quad LQ(t) = \{\lambda_{01}Q_1(t), \lambda_{02}Q_2(t)\},$$

$$f(Q(t)) = -\{[\mu_{11}(Q_1(t)) + \mu_{12}(Q_2(t))]Q_1(t), [\mu_{21}(Q_1(t)) + \mu_{22}(Q_2(t))]Q_2(t)\},$$

$$g(t, Q(t)) = \{(S_1'(t)/S_1(t) - \lambda_{01})Q_1(t), (S_2'(t)/S_2(t) - \lambda_{02})Q_2(t)\}.$$

If (3.18) holds, then the linearization of the system (3.24) about $\{\hat{Q}_1, \hat{Q}_2\}$ yields

$$R'(t) = [L + f'(\{\hat{Q}_1, \hat{Q}_2\})]R(t) + o(R(t)) + g(t, R(t) + \{\hat{Q}_1, \hat{Q}_2\}), \tag{3.25}$$

where

$$R(t) = \{Q_1(t) - \hat{Q}_1, Q_2(t) - \hat{Q}_2\},$$

$$f(\{R_1, R_2\} + \{\hat{Q}_1, \hat{Q}_2\}) - f(\{\hat{Q}_1, \hat{Q}_2\}) = f'(\{\hat{Q}_1, \hat{Q}_2\})$$
$$\times \{R_1, R_2\} + o(\{R_1, R_2\}),$$

$$f'(\{\hat{Q}_1, \hat{Q}_2\})\{R_1, R_1\} = -\{[\mu_{11}'(\hat{Q}_1)R_1 + \mu_{12}'(\hat{Q}_2)R_2]\hat{Q}_1$$
$$+ [\mu_{11}(\hat{Q}_1) + \mu_{12}(\hat{Q}_2)]R_1,$$
$$[\mu_{21}'(\hat{Q}_1)R_1 + \mu_{22}'(\hat{Q}_2)R_2]\hat{Q}_2 + [\mu_{21}(\hat{Q}_1) + \mu_{22}(\hat{Q}_2)]R_2\}.$$

The hypothesis (3.18) and (3.19) means that the eigenvalues of $L + f'(\{\hat{Q}_1, \hat{Q}_2\})$ have negative real parts, so that $\lim_{t\to\infty} R(t) = 0$ if $R(t_1)$ is sufficiently small (see [26], p. 327 or [27], p. 89). The conclusion (3.20) now follows as in (2.10).   ∎

Remark 3.3. The local asymptotic stability of the nontrivial equilibrium $\{\hat{Q}_1, \hat{Q}_2\}$ [that is, the conclusion (3.20)] is a consequence of the domination of the interspecies mortality by the intraspecies mortality [that is, hypothesis (3.19)]. Results similar to Theorems 3.2 and 3.3 were obtained by J. Prüss in [6].

4. PHYSIOLOGICAL AGE-STRUCTURED POPULATIONS

For some biological populations the chronological age of individuals is not easily monitored. For these populations there may be some other physical characteristic such as size, volume, mass, or protein content which provides a suitable measure of individual differences. Models of populations structured by size or some other indicator of physiological age have been treated in [19,20,28–34].

Consider a population structured by physiological age s (which we will refer to as the size variable), but which is, in general, a physical characteristic of individuals different from chronological age. The total population at time t having sizes between s_1 and s_2 is

$$\int_{s_1}^{s_2} n(s, t)\, ds,$$

where $n(s, t)$ is the density with respect to size s at time t. The linear formulation of physiological age structured population dynamics is

$$n_t(s, t) + (g(s)n(s, t))_s = -\mu(s)n(s, t), \tag{4.1}$$

$$g(s_0)n(s_0, t) = \int_{s_0}^{\infty} \beta(s)n(s, t)\, ds, \tag{4.2}$$

$$n(s, 0) = \phi(s). \tag{4.3}$$

Here μ and β are the size-specific mortality and fertility moduli, respectively, s_0 is the size of all neonates, g is the growth rate of size in individuals (in the sense that

$$\int_{s_1}^{s_2} g(s)^{-1}\, ds$$

is the time required for an individual to grow from size s_1 to size s_2), and ϕ is the initial size distribution of the population at time 0. The units of μ and β are 1/time and the units of g are size/time.

In [19] it is shown that the solutions of (4.1)–(4.3) give rise to a strongly continuous semigroup of bounded linear operators $T(t)$, $t \geq 0$ in $L^1(s_0, \infty)$. We collect the results from [19] into the following.

Theorem 4.1

Let $\mu \in L^\infty(s_0, \infty)$, let $\beta \in L^\infty_+(s_0, \infty) - \{0\}$, and let $g: [s_0, \infty) \to [0, \infty)$ be continuously differentiable with g and g' bounded on $[s_0, \infty)$ and g bounded away from 0 on $[s_0, \infty)$. Define $A: L^1(s_0, \infty) \to L^1(s_0, \infty)$ by

$$A\phi = -(g\phi)' - \mu\phi, \quad D(A) = \Big\{ \phi \in L^1(s_0, \infty): \phi \text{ is absolutely continuous,}$$

$$(g\phi)' \in L^1(s_0, \infty), \text{ and } g(s_0)\phi(s_0) = \int_{s_0}^\infty \beta(s)\phi(s)\, ds \Big\}. \tag{4.4}$$

Then A is the infinitesimal generator of a strongly continuous semigroup of bounded linear operators $T(t)$, $t \geq 0$ in $L^1(s_0, \infty)$ satisfying (2.1), and if $\phi \in D(A)$, then $n(s, t) = (T(t)\phi)(s)$ is the unique solution of (4.1)–(4.3). Let $\lambda = \lambda_0$ be the (necessarily) unique real solution of the equation

$$1 = \int_{s_0}^\infty \exp[-\lambda\tau(s)]\, \frac{\beta(s)}{g(s)}\, \Lambda(s, s_0)\, ds, \tag{4.5}$$

let $\lambda_0 > -(\underline{\mu + g'})$, where $\underline{\mu + g'} \equiv \sup_{a \to \infty} \text{ess-inf}_{b>a}\mu(b) + g'(b)$, and define $P_0: L^1(s_0, \infty) \to L^1(s_0, \infty)$ by

$$(P_0\phi)(s) = \exp[-\lambda_0\tau(s)]\, \frac{g(s_0)}{g(s)}\, \Lambda(s, s_0) W_{\lambda_0}(\phi)/N_{\lambda_0}, \quad \phi \in L^1(s_0, \infty), \quad s \geq s_0, \tag{4.6}$$

where

$$\tau(s) = \int_{s_0}^s g(\sigma)^{-1}\, d\sigma, \ s \geq s_0, \tag{4.7}$$

$$\Lambda(s, r) = \exp\left[-\int_r^s \mu(\sigma)g(\sigma)^{-1}\, d\sigma \right], \quad s_0 \leq r \leq s, \tag{4.8}$$

$$W_{\lambda_0}(\phi) = \int_{s_0}^\infty \beta(s) \exp[-\lambda_{s_0}\tau(s)] \left\{ \int_{s_0}^s \exp[\lambda_{s_0}\tau(r)]\Lambda(s, r) \right.$$

$$\left. \times\ \phi(r)g(s)^{-1}\, dr \right\} ds, \quad \phi \in L^1(s_0, \infty), \tag{4.9}$$

$$N_{\lambda_0} = \int_{s_0}^\infty \beta(s)\tau(s) \exp[-\lambda_0\tau(s)]\Lambda(s, s_0)\, \frac{g(s_0)}{g(s)}\, ds. \tag{4.10}$$

Then (2.2) is satisfied for this λ_0, P_0, $P_1 = I - P_0$, and any $\omega < \lambda_0$ sufficiently close to λ_0. Consequently, $\lim_{t \to \infty} |e^{-\lambda_0 t} T(t) - P_0| = 0$.

Remark 4.1. The various quantities in Theorem 4.1 are analogous to those in Theorem 3.1. For example, $\Lambda(s, r)$ is the probability that an individual survives from size r to size s. Notice from (4.6), (4.7), and (4.8) that

$$(P_0\phi)(s) = \exp\left[-\int_{s_0}^{s} \{\lambda_0 + \mu(r) + g'(r)\}g(r)^{-1}\,dr \right] W_{\lambda_0}(\phi)/N_{\lambda_0}.$$

If $\lambda_0 + \mu(s) + g'(s) < 0$ for $s \in [s_1, s_2]$, then the stable size distribution $(T(t)P_0\phi)(s) = e^{\lambda_0 t}(P_0\phi)(s)$ will be increasing for $s \in [s_1, s_2]$. The biological interpretation of this phenomenon is that a "stacking up" occurs in the size range $[s_1, s_2]$ in the sense that more individuals grow into this range than grow or die out of it. In [34] Vansickle discusses such phenomena in fish populations with s corresponding to body size.

Remark 4.2. The proof of Theorem 4.1 given in [19] uses a transformation of the independent variable s to convert the problem to the case of Theorem 3.1. In [19] an additional hypothesis was required, namely that ess-inf$(\mu + g') > 0$. That this hypothesis may be omitted follows from the fact that ess-inf μ need not be positive in Theorem 3.1 (see Remark 3.2).

Theorem 2.1 now applies immediately to the logistic model of physiological age-structured population dynamics:

$$n_t(s, t) + (g(s)n(s, t))_s = -[\mu(s) + \eta(Q(t))]n(s, t), \tag{4.11}$$

$$g(s_0)n(s_0, t) = \int_{s_0}^{\infty} \beta(s)n(s, t)\,ds, \tag{4.12}$$

$$n(s, 0) = \phi(s), \tag{4.13}$$

where $Q(t) = \int_{s_0}^{\infty} n(s, t)\omega(s)\,ds$ for some weight function ω.

THEOREM 4.2

Let $\mu \in L^{\infty}(s_0, \infty)$, let $\beta, \omega \in L_+^{\infty}(s_0, \infty) - \{0\}$, let $g: [s_0, \infty) \to [0, \infty)$ be continuously differentiable with g and g' bounded on $[s_0, \infty)$ and g bounded away from 0 on $[s_0, \infty)$, let $\lambda_0 > -(\mu + g')$, let η satisfy (2.4), and let $\phi \in L_+^1(s_0, \infty) \cap D(A)$ such that $P_0\phi \in L_+^1(s_0, \infty) - \{0\}$. The unique solution of (4.11)–(4.13) satisfies $\lim_{t\to\infty}\|n(\cdot, t)\|_{L^1(s_0,\infty)} = 0$ if $\lambda_0 < 0$, and $\lim_{t\to\infty}\|n(\cdot, t) - \eta^{-1}(\lambda_0)\phi_0/\int_{s_0}^{\infty} \phi_0(s)\omega(s)\,ds\|_{L^1(s_0,\infty)} = 0$ if $\lambda_0 \geq 0$, where $\phi_0(s) = e^{-\lambda_0\tau(s)}\Lambda(s, s_0)$, $s \geq s_0$.

Remark 4.3. The ideas we have developed can be applied to another type of nonlinear structured population problem. This problem does not have logistic form, but rather models a renewable resource with constant yield harvesting. Similar models of harvested structured populations have been treated in [1,29,35–37]. Suppose the hypothesis and notation of Theorem 4.1 and consider a size-structured population with density $p(s, t)$ subject to a constant-yield size-dependent harvesting $h(s)$ with $h \in L_+^1(s_0, \infty)$. The equations of the model are

$$p_t(s, t) + (g(s)p(s, t))_s = -\mu(s)p(s, t) - h(s), \tag{4.14}$$

$$g(s_0)p(s_0, t) = \int_{s_0}^{\infty} \beta(s)p(s, t)\,ds, \tag{4.15}$$

$$p(s, 0) = \phi(s). \tag{4.16}$$

The solution of (4.14)–(4.16) is given by

$$p(\cdot, t) = T(t)\phi - \int_0^t T(\sigma)h\,d\sigma, \quad t \geq 0. \tag{4.17}$$

Let P_0 and P_1 act on (4.17) to yield

$$P_0 p(\cdot, t) = e^{\lambda_0 t} P_0 \phi - \int_0^t e^{\lambda_0 \sigma} P_0 h \, d\sigma, \quad t \geq 0, \tag{4.18}$$

$$P_1 p(\cdot, t) = T(t) P_1 \phi - \int_0^t T(\sigma) P_1 h \, d\sigma, \quad t \geq 0. \tag{4.19}$$

From (4.18) we obtain

$$P_0 p(\cdot, t) = P_0 \phi - t P_0 h, \; \lambda_0 = 0, \quad t \geq 0,$$
$$P_0 p(\cdot, t) = e^{\lambda_0 t}[P_0 \phi - P_0 h / \lambda_0] + P_0 h / \lambda_0, \; \lambda_0 \neq 0, \quad t \geq 0, \tag{4.20}$$

and from (4.19) we obtain

$$\|P_1 p(\cdot, t)\|_{L^1(s_0, \infty)} \leq M e^{\omega t} \|P_1 \phi\|_{L^1(s_0, \infty)} + (e^{\omega t} - 1) M \|P_1 h\|_{L^1(s_0, \infty)} / \omega, \quad t \geq 0. \tag{4.21}$$

Since $\omega < \lambda_0$ and $p(\cdot, t) = P_0 p(\cdot, t) + P_1 p(\cdot, t)$, we conclude from (4.20) and (4.21) the following behavior: If $\lambda_0 = 0$ (that is, the unharvested population is in equilibrium), then constant-yield harvesting extinguishes the population in finite time. If $\lambda_0 > 0$ (that is, the unharvested population has exponential growth), then constant-yield harvesting does not extinguish the population provided that $P_0 \phi > P_0 h / \lambda_0$ (in fact, $\lim_{t \to \infty} e^{-\lambda_0 t} p(\cdot, t) = P_0(\phi - h / \lambda_0)$ so that the population maintains exponential growth).

5. CELL POPULATIONS

Structured population models have been used to model the growth of cell populations by many authors. The earliest such models are found in [28,33,38,39]. Recently, the theory of linear operator semigroups has been applied to a linear model of physiological age-structured cell growth by O. Diekmann *et al.*[40]. The model in [40] derives from the earlier models of Bell and Anderson[28] and Sinko and Streifer[33], which also involved chronological age as well as physiological age. In this section we will describe the linear model in [40] and apply Theorem 2.1 to a nonlinear variation of it involving a logistic-type nonlinearity.

Consider a population of cells structured by a size variable s in which mother cells divide by fission into two equal-size daughter cells. There is a minimum size $s_1 > 0$ which cells must reach before fission can occur and there is a maximum size $s_2 > s_1$ at which fission must occur. Consequently, newly formed daughter cells must lie in the size range $[s_1/2, s_2/2]$ and all cells must lie in the size range $[s_1/2, s_2]$. It is required that $s_1 \geq s_2/2$, which means that the maximum size of a daughter cell is less than or equal to the minimum size of a mother cell. The following equations (which are derived in [40] and [33]) are satisfied by the density function $n(s, t)$ of the cell population:

$$n_t(s, t) + (g(s)n(s, t))_s = -[\mu(s) + b(s)]n(s, t) + 4b(2s)n(2s, t), \tag{5.1}$$

$$n(s_1/2, t) = 0, \tag{5.2}$$

$$n(s, 0) = \phi(s). \tag{5.3}$$

Here μ, b, and g are the size-specific rates at which cells die, divide, and grow, respectively, and ϕ is the initial size distribution of the population [the last term in (5.1) is interpreted to be 0 whenever $s \geq s_2/2$].

The following hypothesis is required.

μ is a nonnegative continuous function on $[s_1/2, s_2]$. $\tag{5.4}$

b is continuous on $[s_1/2, s_2)$, $b(s) = 0$ for $s \in [s_1/2, s_1]$, $b(s) > 0$ for $s \in (s_1, s_2)$, and $\lim_{s \to s_2} \int_{s_1}^s b(\sigma) \, d\sigma = \infty$.

$$\tag{5.5}$$

g is strictly positive and continuous on $[s_1/2, s_2]$ and $g(2s) < 2g(s)$ for $s \in [s_1/2, s_2/2]$ (which means that two daughter cells together gain greater size than the undivided mother would have).   (5.6)

Define Λ by

$$\Lambda(s, r) = \exp\left[-\int_r^s (\mu(\sigma) + b(\sigma))g(\sigma)^{-1}\, d\sigma\right], \quad s_1/2 \le r \le s \le s_2. \qquad (5.7)$$

Notice that $\Lambda(s, r)$ is the probability that a cell of size r reaches size s without dying or dividing. Define the Banach lattice $X = \{\phi\colon \phi$ is a continuous function from $[s_1/2, s_2]$ to R, $\phi(s_1/2) = 0$, and $\sup_{s_1/2 \le s < s_2}|\phi(s)|/\Lambda(s, s_1/2) < \infty\}$ with norm $\|\phi\| = \sup_{s_1/2 \le s < s_2}|\phi(s)|/\Lambda(s, s_1/2)$.

The following theorem is proved in [40].

THEOREM 5.1

Let (5.4)–(5.6) hold and define $A\colon X \to X$ by

$$(A\phi)(s) = -(g(s)\phi(s))' - [\mu(s) + b(s)]\phi(s) + 4b(2s)\phi(2s),$$

$D(A) = \{\phi \in X\colon g(s)\phi(s)$ is continuously differentiable on

$[s_1/2, s_2/2) \cup (s_2/2, s_2),\ \lim_{s \to s_2/2^-} \{-(g(s)\phi(s))' - [\mu(s) + b(s)]\phi(s)$

$+ 4b(2s)\phi(2s)\} = \lim_{s \to s_2/2^+} \{-(g(s)\phi(s))' - [\mu(s) + b(s)]\phi(s)\},$

$- (g(s_1/2)\phi(s_1/2))' - (\mu(s_1/2) + b(s_1/2))\phi(s_1/2) + 4b(s_1)\phi(s_1)$

$= 0,$ and $\sup_{s_1/2 \le s \le s_2, s \ne s_2/2} \{[-(g(s)\phi(s))' - [\mu(s) + b(s)]\phi(s)$

$+ 4b(2s)\phi(2s)]/\Lambda(s, s_1/2)\} < \infty\}.$ (5.8)

Then A is the infinitesimal generator of a strongly continuous semigroup of bounded linear operators $T(t)$, $t \ge 0$ in X satisfying (2.1), and if $\phi \in D(A)$, then $n(s, t) = (T(t)\phi)(s)$ is the unique solution of (5.1)–(5.3). Let $\lambda = \lambda_0$ be the (necessarily) unique real solution of the equation

$$1 = 4\int_{s_1/2}^{s_2/2} \exp[-\lambda(\tau(2\sigma) - \tau(\sigma))]b(2\sigma)\Lambda(2\sigma, \sigma)/g(2\sigma)\, d\sigma, \qquad (5.9)$$

and define $P_0\colon X \to X$ by

$$P_0\phi = \phi_0 U_{\lambda_0}(\phi)/Q_{\lambda_0}, \qquad (5.10)$$

where

$$\tau(s) = \int_{s_{1/2}}^s g(\sigma)^{-1}\, d\sigma, \quad s_1/2 \le s \le s_2, \qquad (5.11)$$

$$U_{\lambda_0}(\phi) = \int_{s_1/2}^{s_2/2} \exp[-\lambda_0(\tau(s_2/2) - \tau(\sigma))]\{\phi(\sigma)\Lambda(\sigma, s_1/2)^{-1}$$

$$+ 4b(2\sigma)g(2\sigma)^{-1}\Lambda(2\sigma, \sigma)\int_{s_2/2}^{2\sigma} \exp[-\lambda_0(\tau(2\sigma) - \tau(\xi))]$$

$$\times \phi(\xi)\Lambda(\xi, s_1/2)^{-1}\, d\xi\}\, d\sigma, \quad \phi \in X, \qquad (5.12)$$

$$Q_{\lambda_0} = \int_{s_1/2}^{s_2/2} (\tau(2\sigma) - \tau(\sigma)) \exp[-\lambda_0(\tau(2\sigma) - \tau(\sigma))]$$

$$\times\ 4b(2\sigma)\Lambda(2\sigma,\ \sigma)/g(2\sigma)\ d\tau, \tag{5.13}$$

$$\phi_0(s) = \Lambda(s,\ s_1/2)g(s)^{-1} \exp[-\lambda_0(\tau(s_1) - \tau(s_1/2))]$$

$$\times \int_{s_1/2}^{\min\{s,s_2/2\}} \exp[-\lambda_0(\tau(2\sigma) - \tau(\sigma))]$$

$$\times\ 4b(2\sigma)\Lambda(2\sigma,\ \sigma)/g(2\sigma)\ d\sigma. \tag{5.14}$$

Then (2.2) is satisfied for this λ_0, P_0, $P_1 = I - P_0$, and any $\omega < \lambda_0$ sufficiently close to λ_0. Consequently, $\lim_{t\to\infty}|e^{-\lambda_0 t}T(t) - P_0| = 0$.

If the mortality of cells is dependent upon the total population, or if cells enter a nonproliferating state as the total population increases, then Theorem 2.1 may be applied. The equations of such a nonlinear logistic version of this problem are

$$q_t(s, t) + (g(s)q(s, t))_s = -[\mu(s) + \eta(Q(t)) + b(s)]q(s, t) + 4b(2s)q(2s, t), \tag{5.15}$$

$$q(s_1/2, t) = 0, \tag{5.16}$$

$$q(s, 0) = \phi(s), \tag{5.17}$$

where $Q(t) = \int_{s_1/2}^{s_2} q(s, t)\omega(s)\ ds$ for some weight function ω and the term $\mu(s) + \eta(Q(t))$ accounts for either mortality or passage to a permanent state of nonproliferation. By virtue of Theorem 2.1 we have the following.

THEOREM 5.2

Let (5.4)–(5.6) hold, let $\Lambda(s, s_1/2)\omega(s) \in L_+^1(s_1/2, s_2) - \{0\}$, let η satisfy (2.4), and let $\phi \in X_+ \cap D(A)$ such that $P_0\phi \in X_+ - \{0\}$. The unique solution of (5.15)–(5.17) satisfies $\lim_{t\to\infty}\|q(\cdot, t)\|_X = 0$ if $\lambda_0 < 0$, and $\lim_{t\to\infty}\|q(\cdot, t) - \eta^{-1}(\lambda_0)\phi_0/\int_0^\infty \phi_0(s)\omega(s)\ ds\|_X = 0$ if $\lambda_0 \geq 0$.

REFERENCES

1. M. E. Gurtin, The mathematical theory of age-structured populations, (to appear).
2. W. Streifer, Realistic models in population ecology. *Adv. Eco. Res.* **8**, 211–166 (1974).
3. M. E. Gurtin and R. C. MacCamy, Nonlinear age-dependent population dynamics. *Arch. Rat. Mech. Anal.* **54**, 281–300 (1974).
4. M. E. Gurtin and R. C. MacCamy, Some simple models for nonlinear age-dependent population dynamics. *Math. Biosci.* **43**, 199–211 (1979).
5. J. Prüss, Equilibrium solutions of age-specific population dynamics of several species. *J. Math. Biol.* **11**, 65–84 (1981).
6. J. Prüss, On the qualitative behavior of populations with age-specific interactions. *Internat. J. Comput. Math. Appl.* **9**(3), 327–340 (1983).
7. J. Prüss, Stability analysis for equilibria in age-specific population dynamics (to appear).
8. G. F. Webb, *Theory of Nonlinear Age-Dependent Population Dynamics, Monographs and Textbooks in Pure and Applied Mathematics Series,* Vol. 89. Marcel Dekker, New York (1985).
9. P. Marcati, On the global stability of the logistic age dependent population growth. *J. Math. Biol.* **15**, 215–226 (1982).
10. S. Busenberg and M. Iannelli, A class of nonlinear diffusion problems in age-dependent population dynamics. *J. Nonlinear Analysis* **7**, 501–529 (1983).
11. M. Iannelli, Mathematical problems in the description of age structured populations (to appear).
12. K. Yosida, *Functional Analysis*, 2nd Edn. Springer-Verlag, (1968).
13. N. Dunford and J. Schwartz, *Linear Operators, Part I: General Theory.* Interscience (1958).
14. T. Saaty, *Modern Nonlinear Equations.* McGraw-Hill, New York (1967).
15. F. R. Sharpe and A. J. Lotka, A problem in age distributions. *Phil. Mag.* **21**, 435–438 (1911).
16. W. Feller, On the integral equation of renewal theory. *Ann. Math. Stat.* **12**, 243–267 (1941).
17. W. Desch and W. Schappacher, Spectral properties of finite-dimensional perturbed linear semigroups (to appear). G. Greiner, A typical Perron–Frobenius theorem with application to an age-dependent population equation, in *Infinite-Dimensional Systems, Proceedings, Retzhof* 1983 (Edited by F. Kappel and W. Schappacher), pp. 86–100. Springer-Verlag (1984).

18. Song Jian, Yu Jing-Yuan, Hu Shung-Ju and Zhu Guang-Tian, Spectral properties of population operators and asymptotic behavior of population semigroups. *Acta Math. Sci.* **2**(2), 139–148 (1982).
19. G. F. Webb, A Semigroup proof of the Sharpe–Lotka Theorem, in *Infinite-Dimensional Systems, Proceedings, Retezhof* 1983 (Edited by F. Kappel and W. Schappacher), pp. 254–268. Springer-Verlag, (1984).
20. G. F. Webb, Dynamics of population structured by internal variables. *Math. Zeit.* **189**, 319–335 (1985).
21. J. M. Cushing and M. Saleem, A predator–prey model with age structure. *J. Math. Biol.* **14**, 231–250 (1982).
22. K. Gopalsamy, Time lags and density dependence in age dependent two species competition. *Bull. Australian Math. Soc.* **25**(2), 271–292, (1982).
23. M. E. Gurtin and D. S. Levin, On predator–prey interaction with predation dependent on age of prey. *Math. Biosci.* **47**, 207–219 (1979).
24. A. Haimovici, On the age dependent growth of two interacting populations. *Boll. Un. Mat. Ital.* **15**, 405–429 (1979).
25. M. Rotenberg, Equilibrium and stability in populations whose interactions are age-specific. *J. theor. Biol.* **54**, 207–224 (1975).
26. E. A. Coddington and N. Levinson, *Theory of Ordinary Differential Equations.* McGraw-Hill, New York (1955).
27. J. K. Hale, *Ordinary Differential Equations, Interscience Series on Pure and Applied Mathematics,* Vol. 21. Wiley-Interscience, New York (1969).
28. G. I. Bell and E. C. Anderson, Cell growth and division. I. A mathematical model with applications to cell volume distributions in mammalian suspension cultures. *Biophys. J.* **7**, 329–351 (1967).
29. F. Brauer, Nonlinear age-dependent population growth under harvesting. *Internat. J. Comput. Math. Appl.* **9**(3), 345–352 (1983).
30. O. Diekmann, H. Lauwerier, T. Aldenberg and J. Metz, Growth, fission, and the stable size distribution. *J. Math. Biol.* **18**(2), 135–148 (1983).
31. L. Murphy, A nonlinear growth mechanism in age dependent population dynamics. *J. Theor. Biol.* **104**, 493–506 (1983).
32. L. Murphy, Density dependent cellular growth in an age structured colony. *Internat. J. Comput. Math. Appl.* **9**(3), 383–392 (1983).
33. J. W. Sinko and W. Streifer, A new model for age–size structure of a population. *Ecology* **48**, 910–918 (1967).
34. J. VanSickle, Analysis of a distributed parameter population model based on physiological age. *J. theor. Biol.* **64**, 571–586 (1977).
35. M. E. Gurtin and L. F. Murphy, On the optimal harvesting of age structured populations: some simple models. *Math. Biosci.* **55**, 115–136 (1981).
36. M. E. Gurtin and L. F. Murphy, On the optimal harvesting of persistent age-structured populations. *J. Math. Biol.* **13**, 131–148 (1981).
37. D. Sanchez, Linear age-dependent population growth with harvesting. *Bull. Math. Biol.* **40**, 377–385 (1978).
38. S. Rubinov, *Age-Structured Equations in the Theory of Cell Populations. Studies in Mathematical Biology, Part II.* Math. Assoc. Amer., Washington, D.C. (1978).
39. H. Von Foerster, Some remarks on changing populations, in *The Kinetics of Cellular Proliferation*, pp. 382–407. Grune & Stratton (1959).
40. O. Diekmann, H. Heijmans and H. Thieme, On the stability of the cell size distribution. *J. Math. Biol.* **19**, 227–248 (1984).

Comp. & Maths. with Appls. Vol. 12A, Nos. 4/5, pp. 541–550, 1986
Printed in Great Britain.

0886–9553/86 $3.00 + .00
© 1986 Pergamon Press Ltd.

ABSTRACT STABILITY THEORY AND APPLICATIONS TO HYPERBOLIC EQUATIONS WITH TIME DEPENDENT DISSIPATIVE FORCE FIELDS

PIERANGELO MARCATI[†]
Dipartimento di Matematica, Università di Trento, 38050 Povo (Tn), Italy

Abstract—In this paper an abstract theorem of asymptotic stability, using the theory of one parameter Liapunov functions, is provided. The advantage of this approach is to remove, in several important cases, the request of precompactness of the orbits. Applications are made to an abstract class of hyperbolic equations, with time dependent dissipations, including damped and strongly damped wave equations and the extensible beam equation. The final example provides the asymptotic stability of the periodic solution to the membrane equation with a time dependent damping term.

1. ABSTRACT RESULTS

The stability theory of second-order evolution equations has been investigated using several different points of views. Between the relevant amount of important literature on this subject, we wish to mention the results of Dafermos[1,2], Dafermos and Slemrod[3] and Haraux[4] which connected the invariance principle of La Salle[5] Hale[6] with the nonlinear semigroups theory. A direct applications of Hale's theorem in the infinite dimensional phase spaces is subject to the verification of a nontrivial assumption: the precompactness of the orbits.

Some theories have been developed to avoid this strong requirement, for example the use of the weak topologies is one of the main tool of the research of Baillon and Brezis[7] and Ball[8]; while in the author's paper[9] a new method, derived from the Matrasov theorem, has been proposed. This paper is based on the theory of one parameter Liapunov functions of Salvadori[10,11].

The advantage of this approach seems to be restricted to strongly dissipative systems (that is when the dissipation is related with the kinetic energy in an appropriate sense). However, in this case it provides the strong convergence of the solutions towards the equilibria, without any hypothesis of precompactness. On the contrary, this technique does not seem to be suitable to attack weakly dissipative problems like, for example, the stabilization of semilinear control systems (see Ball and Slemrod[12,13]). The purpose of the present paper is the extension of the results given in [9] to abstract evolution systems (processes, in the terminology of Dafermos and Hale) and to apply them to a wide class of hyperbolic equations having a time-dependent dissipative force field.

We shall give the basic definitions and theorems at level of general topological spaces (actually we shall use this terminology to denote a T_2 topological space).

In the sequel, let us denote by E a topological space. We shall define an evolution system $p(t, t_0, x_0)$ in the following way. Assume that $p : \mathbb{R} \times \mathbb{R} \times E \to E$ is separately continuous, we say that p is an evolution system if

$$\text{(i)} \quad p(t_0, t_0, x_0) = x_0,$$

$$\text{(ii)} \quad p(t_2, t_1, p(t_1, t_0, x_0)) = p(t_2, t_0, x_0)$$

for all $t_0, t_1, t_2 \in \mathbb{R}, x_0 \in E$.

Definition 1.1. We say that $S \subset E$ is a Liapunov stable subset of E, if and only if, for all open sets $U \supset S$ there exists V_U, $S \subset V_U \subset U$, such that $p(t, t_0, x_0) \in U$ provided that $x_0 \in V_U$, for all $t \geqq t_0$. A subset $H \subset E$ is called an invariant set if and only if $p(t, t_0, H) \subset H$, for all $t \geqq t_0$. An invariant subset of a single element is named an equilibrium point.

Definition 1.2. We say that $S \subset E$ is an attractor and an open set $U \supset S$ is said to be interior to the bacine of attraction of S, if and only if, for all open sets $\hat{U}, U \supset \hat{U} \supset S$, there

---

†Permanent address: Dipartimento di Matematica Pura e Appl.-Università, Via Roma 33, 67100 L'AQUILA (Italy).

exists $\hat{t} = \hat{t}(\hat{U}) \geqq t_0$ such that $p(t, t_0, x_0) \in \hat{U}$, for all $t \geqq \hat{t}$. Let us denote by

$$B(S) = l \cup \{V : V \text{ is interior to the bacine of attraction of } S\},$$

which is called the bacine of attraction of S. Finally, we say that S is a global attractor if $B(S) = E$.

The following result is concerned with the asymptotic properties of the evolution system p.

Theorem 1.1

Let E be a topological space and $K_\alpha \subset E$ a fundamental system of neighbourhoods of S. Assume that

(i) for all α there exists $\gamma(t, t_0)$ continuous in both variables and $W_\alpha : R \times E \to R$, continuous, such that $|W_\alpha(t, x)| \leqq \gamma(t, t_0)$, for all $t \geqq t_0$, $x \in E$;

(ii) there exists $c_\alpha : R \to R_+$ continuous function such that

$$\dot{W}_\alpha(t_0, x_0) = D^+ W_\alpha(t, p(t, t_0, x_0))\big|_{t=t_0} \leqq -c_\alpha(t_0) < 0$$

for all $t_0 \in R$, $x_0 \in E\backslash K_\alpha$.

$$\text{(iii)} \quad \lim_{t \to +\infty} (\gamma(t, t_0))^{-1} \int_{t_0}^t c_\alpha(S) \, dS = +\infty$$

Under the above hypotheses, if S is a stable subset of E, then it is a global attractor.

Proof. We want to prove the property stated in the definition (1.2). That is, for all open $W \supset S$ we want to find $t_1 > t_0$, such that

$$p(t_1, t_0, x_0) \in V_W.$$

Where V_W is taken in the sense of Definition 1.1. If we neglect the above fact, there exists $W_0 \supset S$, such that for all $t > t_0$

$$p(t, t_0, x_0) \in E\backslash W_0.$$

Hence, we can find K_α, $S \subset K_\alpha \subset W_0$, such that

$$p(t, t_0, x_0) \in E\backslash K_\alpha.$$

Therefore, for all $t > t_0$, one has

$$\dot{W}_\alpha(t, p(t, t_0, x_0)) \leqq -c_\alpha(t).$$

The integration on (t_0, t), together with hypothesis (i), provides the following inequalities

$$-\gamma(t, t_0) \leqq W(t, p(t, t_0, x_0)) \leqq W(t_0, x_0) - \int_{t_0}^t c_\alpha(s) \, ds.$$

Using hypothesis (iii) we get a contradiction.

The analogy for discrete dynamical systems can be obtained in a similar way, hence, we will formulate the result omitting the proof.

Before standing the proposition we shall adapt the definition of evolution system to the discrete case.

A discrete evolution system (DES) is given by a two parameters family of nonlinear transformations on E verifying the following properties:

$$\text{(i)} \quad \text{for all } h, k \in N \quad P_{h,k} \in C(E),$$

$$\text{(ii)} \quad \text{for all } h \in N \quad P_{h,h} = I_E,$$

$$\text{(iii)} \quad \text{for all } h, k, l \in N \quad P_{h,k} \cdot P_{k,l} = P_{h,l}.$$

*Proposition*1.2 Given a (DES) on E, let S and $\{K_\alpha\}$ as in the above proposition. Moreover, assume that

(i) for all α, there exists

$$W_{\alpha,n} \in C(E, \mathbb{R}), \quad |W_{\alpha,n}(x)| \le \gamma_{n,k}, \quad \gamma_{n,k} > 0$$

for all $n \ge k$, $x \in E$.

(ii) There exists $\{c_{\alpha,n}\}$ such that

$$\Delta_+ W_{\alpha,n}(x) = W_{\alpha,n+1}(P_{n+1,n}(x)) - W_{\alpha,n}(x) \le -c_{\alpha,n} < 0,$$

for all $x \in E\backslash K_\alpha$.

(iii)

$$\lim_{n\to\infty} \gamma_{n,k}^{-1} \sum_{j=k}^{n} c_{\alpha,j} = +\infty.$$

Then the Liapunov stability of S implies the global attractivity.

2. APPLICATIONS

This section contains the applications of the above results to quite general hyperbolic equations including some models arising in continuum mechanics.

Consider the following abstract problem

$$\text{(H)} \qquad \frac{\mathrm{d}}{\mathrm{d}t} A\left(\frac{\mathrm{d}u}{\mathrm{d}t}\right) + B\left(t, u, \frac{\mathrm{d}u}{\mathrm{d}t}\right) + C(u) = 0,$$

where the following hypotheses hold.

I. There exist three Banach spaces K, D, P such that the following continuous embedding are fulfilled

$$P \subsetneq D \subsetneq K. \tag{2.1}$$

II. Let $A : K \to K^*$ a continuous bounded operator and $f \in C^1(K, \mathbb{R})$ a continuous functional such that $A = \mathrm{Grad}\, f$. Moreover there exist $a_0 \in C(\mathbb{R}_+, \mathbb{R}_+)$ increasing function and $a \in C([0, 1] \times \mathbb{R}_+, \mathbb{R}_+)$ such that

$$f(w) \ge a_0(|w|_K), \quad A(0) = 0,$$
$$\langle A(w) - A(\lambda w), w \rangle \ge a(\lambda, |w|_K), \quad w \in K, \tag{2.2}$$

where

$$\int_0^1 a(\lambda, r)\, \mathrm{d}\lambda \ge a_0(r), \quad a_0(0) = 0. \tag{2.3}$$

III. Let $C : P \to P^*$ a continuous bounded operator such that $C = \mathrm{Grad}\, g$, where $g \in C^1(P, \mathbb{R})$. In addition one has

$$g(u) \ge c_0(|u|_P), \quad g(0) = 0,$$
$$\langle C(u), u \rangle \ge c_0(|u|_P),$$

where $c_0 \in C(\mathbb{R}_+, \mathbb{R}_+)$ is an increasing function, such that $c_0(0) = 0$ and $c(r) \to +\infty$ as $r \to +\infty$.

IV. Let $B : \mathbb{R} \times P \times D \to D^*$ be a bounded continuous nonlinear operator such that for all $u \in P$, $v \in D$ and $t \in R$

$$|B(t, u, v)|_{D^*} \leq b_1(t)\{|v|_D + |v|_D^{p-1}\}$$
$$\langle B(t, u, v), v \rangle \geq b_0(t)|v|_D^p, \tag{2.5}$$

where $2 \leq p < +\infty$ and $b_0(t) \geq b_0 > 0$.

Moreover there exists $b^* > 0$ such that $b_1(t) \leq b^* b_0(t)$ for a sufficiently large t. In addition when $p > 2$, the following growth condition is required:

$$\lim_{t \to +\infty} (t - t_0)^{-p/(p-2)} \int_{t_0}^{t} b_0(s)\, ds = 0. \tag{2.6}$$

Proposition 2.1. Let us denote by $\hat{E}$ the subset of $P \times K$ such that for all initial data in $\hat{E}$ the Cauchy problem (H) admits a globally defined solution verifying

$$u \in C([0, T]; P)$$
$$u' \in L^p(0, T; D) \cap C([0, T]; K)$$
$$A(u') \in AC([0, T], P^*) \tag{2.7}$$

for all $T > 0$. Therefore, S is stable and globally attractive in $\hat{E}$ (in other words, it is asymptotically stable).

Proof. If we set $V : P \times K \to R$

$$V(u, v) = \langle A(v), u \rangle - f(v) + g(u). \tag{2.8}$$

It is not difficult to see that $\dot{V}(u, v) = -\langle B(t, u, v), v \rangle \leq 0$, hence, the stability of S easily holds.

In order to construct a family W_α to apply the abstract results, we observe that

$$\int_{t_0}^{t} \langle B(t, u(t), u'(t)), u'(t) \rangle\, dt \leq V(u(0), u'(0)) = V_0 \tag{2.9}$$

and in particular

$$\int_{t_0}^{t} b_0(t)|u'(t)|_D^p\, dt \leq V_0. \tag{2.10}$$

Denote by $E = \hat{E} \cap B_R(0)$, for a fixed $R > 0$, and

$$F_\epsilon(t, u_0, v_0) = V(u_0, v_0) + \epsilon\left[\langle A(v_0), u_0 \rangle \right.$$
$$\left. + \int_{t_0}^{t} \langle B(s, p(s, t_0, x_0)), u(s, t_0, x_0) \rangle\, ds\right], \tag{2.11}$$

where

$$x_0 = (u_0, v_0), \quad p(t, t_0, x_0) = (u(t, t_0, x_0), u'(t, t_0, x_0)). \tag{2.12}$$

Now, for all $\alpha \in (0, K)$, we want to find a positive $\epsilon(\alpha)$ such that

$$W_\alpha = F_{\epsilon(\alpha)}$$
$$\dot{F}_{\epsilon(\alpha)}(t, u_0, v_0) \leq -c_\alpha(t) \quad \text{in } E \backslash B_\alpha(0). \tag{2.13}$$

Hence, if we set $K_\alpha = B_\alpha(0) \cup E$, $S = \{0\}$, we will be able to apply the abstract Theorem 1.1, provided that the conditions (i) and (iii) be fulfilled.

Indeed, since along the motion $V \leqq V_0$, one has

$$\langle u'(t), A(u'(t)) \rangle - f(u'(t)) \leqq V_0, \tag{2.14}$$

hence

$$a_0(|u'(t)|_K) \leqq \int_0^1 \langle u'(t), A(u'(t)) - A(\lambda u'(t)) \rangle \, d\lambda$$

$$\leqq \langle u'(t), A(u'(t)) \rangle - f(u'(t)) \leqq V_0. \tag{2.15}$$

Then it follows

$$\int_{t_0}^t \langle u'(s), B(s, u(s), u'(s)) \rangle \, ds \leqq \int_{t_0}^t |u(s)|_P |B(s, u(s), u'(s))|_{P^*} \, ds$$

$$\leqq \mu(D^*, P^*) \left(\int_{t_0}^t |u(s)|_P^p \, ds \right)^{1/p} \left(\int_{t_0}^t |B(s, u(s), u'(s))|_D^{p'} \, ds \right)^{1/p'}, \tag{2.16}$$

where $\mu(D^*, P^*) = \sup \{|x|_{P^*} : |x|_{D^*} = 1\}$.

Since, for all t, $g(u(t)) \leqq V_0$, we get $|u(t)|_P \leqq c_0^{-1}(V_0)$ and

$$\int_{t_0}^t |u(s)|_P^p \, ds \leq (t - t_0)^{1/p} c_0^{-1}(V_0). \tag{2.17}$$

Moreover,

$$\int_{t_0}^t |B(s, u(s), u'(s))|_{D^*}^{p'} \, ds \leqq c(p) \left[\int_{t_0}^t (b_1^{p'}(s)|u'(s)|_D^p + b_1^{p'}(s)|u'(s)|_D^p) \, ds \right]. \tag{2.18}$$

Using the hypothesis that $b_1(t) \leqq b^* b_0(t)$, for t sufficiently large we get

$$\int_{t_0}^t b_1^{p'}(s)|u'(s)|^p \, ds \leqq b^* V_0 \tag{2.19}$$

and if $p > 2$

$$\int_{t_0}^t b_1^{p'}(s)|u'(s)|^{p'} \, ds \leqq \left(\int_{t_0}^t (b_1^{p'}(s)/b_0^{p'/p}(s))^{p-1/(p-2)} \, ds \right)^{p-2/p-1}$$

$$\left(\int_{t_0}^t b_0(s)|u'(s)|^{p'} \, ds \right) \leqq (b^*)^{p'} V_0 \left(\int_{t_0}^t b_0(s) \, ds \right)^{p-2/(p-1)}, \tag{2.20}$$

otherwise if $p = 2$, (2.20) reduces to (2.19).

In this way we obtain

$$\gamma_0(t, t_0) = (t - t_0)^{1/p} c_0^{-1}(V_0) c(p) \left[b^* V_0 + (b^*)^{p'} V_0 \left(\int_{t_0}^t b_0(s) \, ds \right)^{p-2/p-1} \right]. \tag{2.21}$$

By the boundedness of A, there exists a positive d_1 depending only on R, such that

$$F_\epsilon(t, u_0, v_0) \leqq d_1 + \gamma_0(t, t_0) = \gamma(t, t_0), \tag{2.22}$$

and hence, condition (i) of Theorem 1.1 is fulfilled. To prove condition (ii), assume that for all $t \geq t_0$, one has $p(t, t_0, x_0) \in E \backslash K_\alpha$, that is

$$\alpha \leq \sup \left(|u(t)|_P, |u'(t)|_K \right) \leq R,$$

therefore,

$$\dot{F}_\epsilon(t, u_0, v_0) = -\langle B(t, u_0, v_0), v_0 \rangle + \epsilon[\langle A(v_0), v_0 \rangle - \langle C(u_0), u_0 \rangle].$$

Hence,

$$\dot{F}_\epsilon(t, u_0, v_0) \leq -b_0 |v_0|_D^p + \epsilon[K_1 |v_0|_K - c_0(|u_0|_P)]$$
$$\leq -b_2 |v_0|_K^p + \epsilon[K_1 |v_0|_K - c_0(|u_0|_P)], \quad (2.23)$$

where

$$K_1 = \sup \left\{ |A(v_0)|_{K^*} : |v_0|_K \leq R \right\} \text{ and } b_2 = b_0 \mu(D, K)^{-1/p}.$$

To complete the proof it must be shown that (2.13) holds. By putting $\xi = |v_0|_K \; \eta = c_0(|u_0|_P)$ the proof is reduced to investigate the real-valued function

$$\phi_\epsilon(\xi, \eta) = -b_2 \xi^p + \epsilon[K_1 \xi - \eta]. \quad (2.24)$$

Indeed, for all $\alpha \in (0, R)$, it is possible to find $\epsilon(\alpha) > 0$ such that

$$\phi_{\epsilon(\alpha)}(\xi, \eta) < 0 \quad (2.25)$$

on the compact sets

$$\alpha \leq \sup \left(\xi, c_0^{-1}(\eta) \right) \leq R.$$

If we set

$$c_\alpha = -\min \left\{ \phi_{\epsilon(\alpha)}(\xi, \eta) : \alpha \leq \sup \left(\xi, c_0^{-1}(\eta) \right) \leq R \right\} \quad (2.26)$$

condition (iii) of Theorem 1.1 is reduced in this case to

$$\lim_{t \to +\infty} \gamma(t, t_0) c_\alpha(t - t_0) = +\infty. \quad (2.27)$$

By (2.21) there exist $K_2, K_3 > 0$ so that

$$\gamma(t, t_0) = (t - t_0)^{1/p} \left[K_2 + K_3 \left(\int_{t_0}^t b_0(s) \, ds \right)^{p-2/p-1} \right]. \quad (2.28)$$

Then (2.27) is achieved if

$$\lim_{t \to +\infty} \left[c_\alpha(t - t_0)^{p/p-1} \right] \Big/ \left[K_2 + K_3 \left(\int_{t_0}^t b_0(s) \, ds \right)^{p-2/p-1} \right] = +\infty \quad (2.29)$$

which is a consequence of (2.6).

    Example 1. Let Ω be an open-bounded subset of $\mathbf{R}^3$ with smooth boundary. We wish to study

(E1) $$\Box u + (1 + t^{1/2}) u_t^3 + u^3 = 0$$

under the boundary condition $u|_{\partial \Omega} = 0$.

One has

$$P = H_0^1(\Omega) \quad D = L^4(\Omega) \quad K = L^2(\Omega).$$

Moreover, $b_1(t) = b_0(t) = 1 + t^{1/2}$ and $p = 4$. Since

$$\lim_{t \to +\infty} (t - t_0)^{-2} \int_{t_0}^t (1 + s^{1/2}) \, ds = 0$$

by the above proposition it follows

$$\lim_{t \to +\infty} |u(t)|_{H_0^1} = \lim_{t \to +\infty} |u'(t)|_{L^2} = 0.$$

Example 2 (*Extensible beam*). Let $x \in (0, 1)$ and $t \geq 0$, the transverse motion of an extensible beam is described by means of the following equation (see Ball[14]):

$$\text{(E2)} \quad \rho \partial_t^2 u + \alpha \partial_x^4 u - \left[\beta + \kappa \int_0^1 |u_x|^2 \, d\xi \right] \partial_x^2 u + \lambda(t) \partial_t \partial_x^4 u$$

$$- \sigma \left[\int_0^1 u_x u_{xt} \, d\xi \right] \partial_x^2 u + \delta u_t = 0,$$

where ρ, α, β, κ, σ, $\delta > 0$, $x \in (0, 1)$, $t \geq 0$. Under the boundary conditions

$$u(0, t) = u(1, t) = u_x(0, t) = u_x(1, t) = 0.$$

Assume that

$$\lambda(t) \geq \lambda_0 > 0.$$

Hence, we can apply Proposition 2.1, by considering

$$P = H^2(0, 1) \cap H_0^1(0, 1) \quad D = P \quad K = L^2(0, 1)$$

and

$$g(u) = (\alpha/2) \int_0^1 |u_{xx}|^2 \, d\xi + (\beta/2) \int_0^1 |u_x|^2 \, d\xi + (\kappa/4) \left(\int_0^1 |u_x|^2 \, d\xi \right)^2$$

$$C(u) = \text{Grad } g(u) = \alpha \partial_x^4 u - \left[\beta + \kappa \int_0^1 |u_x|^2 \, d\xi \right] \partial_x^2 u$$

$$A(v) = v \quad f(v) = \left(\frac{1}{2} \right) |v|_K^2$$

$$B(t, u, v) = \lambda(t) \partial_x^4 v - \sigma \left[\int_0^1 u_x v_x \, d\xi \right] u_{xx} + \delta v,$$

therefore, one has

$$\langle B(t, u, v), v \rangle \geq \lambda(t) |v_{xx}|_{L^2}^2 + \delta |v|_{L^2}^2$$

and if $u \in B_R(0)$ in P, there exists $q(R) > 0$ such that

$$|B(t, u, v)|_{D^*} \leq \lambda(t) |v_{xx}|_{L^2}^2 + q(R) |v|_{L^2}^2.$$

Therefore, it follows $b_0(t) = \lambda(t)$, $b_1(t) = \lambda(t) + q(R)$. From Proposition 2.1 we established

$$\lim_{t \to +\infty} |u(t)|_{H^2 \cap H^1_0} = \lim_{t \to +\infty} |u'(t)|_{H^2 \cap H^1_0} = 0.$$

Example 3. The following example shows that when $p = 2$ the range of the applications for Proposition 2.1 is sufficiently large. Let Ω be an open-bounded subset in $\mathbb{R}^3$, we shall be concerned with the following damped wave equation

$$u_{tt} - \Delta u + j(t)u_t = 0$$

(E3)
$$u|_{\partial\Omega} = 0.$$

The general theorem can be applied, by putting

$$P = H^1_0(\Omega) \quad D = K = L^2(\Omega)$$

$$A(v) = v, \; f(v) = \left(\frac{1}{2}\right)|v|^2_{L^2}, \quad C(u) = -\Delta u, \quad g(u) = \left(\frac{1}{2}\right)|\nabla u|^2_{L^2}$$

$$B(t, u, v) = j(t)v.$$

If we assume

$$\lim_{t \to +\infty} j(t) > 0$$

then there exists $t_0 > 0$ such that the conditions of Proposition 2.1 are fulfilled.
Hence, we obtain

$$\lim_{t \to +\infty} |u(t)|_{H^1_0} = \lim_{t \to +\infty} |u'(t)|_{L^2} = 0.$$

Example 4. Consider the following example of strongly damped wave equation

$$u_{tt} - \Delta u - j(t)\Delta u_t = 0,$$

(E4)
$$u|_{\partial\Omega} = 0,$$

Ω as in the above example.
Assume that j fulfills the same conditions as stated in the foregoing case, then to apply Proposition 2.1 we require

$$P = H^1_0(\Omega) = D \quad K = L^2(\Omega)$$

$$A(v) = v, \quad f(v) = \left(\frac{1}{2}\right)|v|^2_{L^2}, \quad g(u) = \left(\frac{1}{2}\right)|\nabla u|^2_{L^2}$$

$$B(t, u, v) = -j(t)\Delta v.$$

Also in this case Proposition 2.1 holds, with $p = 2$. Hence,

$$\lim_{t \to +\infty} |u(t)|_{H^1_0} = \lim_{t \to +\infty} |u_t(t)|_{L^2} = 0.$$

Example 5. Let Ω be as in the above example. Our purpose in this case is to provide an example having a nonquadratic kinetic energy. Consider the following equation:

(E5)
$$\frac{\partial}{\partial t}\left(\frac{\partial u}{\partial t}\right)^3 - \Delta u - j(t)\Delta \frac{\partial u}{\partial t} = 0$$

under the boundary condition $y|_{\partial\Omega} = 0$.

Denote by

$$P = H_0^1(\Omega), \quad D = P, \quad K = L^4(\Omega).$$

Moreover, assume that j fulfills the foregoing condition and

$$A(v) = v^3, \quad A : L^4(\Omega) \to L^{4/3}(\Omega)$$

$$f(v) = \left(\frac{1}{4}\right)|v|_{L^4(\Omega)}^4$$

$$C(u) = -\Delta u, \quad g(u) = \left(\frac{1}{2}\right)|\nabla u|_{L^2(\Omega)}^2$$

$$B(t, u, v) = -j(t)\Delta v.$$

We can apply Proposition 2.1 with $p = 2$, in this way it follows

$$\lim_{t \to +\infty} |u(t)|_{H_0^1} = \lim_{t \to +\infty} |u_t(t)|_{L^4} = 0.$$

Example 6. (Periodic solution to damped wave equation). Let Ω be an open-bounded subset of $\mathbf{R}^n$ with smooth boundary $\partial\Omega$. The motion of vibrating membrane under the action of a periodic forcing term $f(t)$, is described by the following equation

$$\Box y + j(t)\beta(y_t) = f(t),$$

(E6)
$$y|_{\partial\Omega} = 0.$$

We assume that the time dependence of the friction term (via the function j) is periodic. However, under the same hypotheses on j stated above, the assumption of strong monotonicity on β, the general theory developed in Haraux[15] provides the existence of a periodic solution $\mathbf{u}$ to our equation. We shall prove here the stability and the global attractivity of $\mathbf{u}$. Let y be a solution to (E6), denote by $z = y - \mathbf{u}$. Hence, one has

$$\Box z + j(t)(\beta(\mathbf{u}_t(t) + z_t(t)) - \beta(\mathbf{u}_t(t))) = 0.$$

We shall prove that, for this equation, $z = 0$ is a stable globally attracting solution. Denote by

$$P = H_0^1(\Omega), \quad D = K = L^2(\Omega)$$

and by

$$A(v) = v, \quad f(v) = \left(\frac{1}{2}\right)|v|_{L^2}^2, \quad C(z) = -\Delta z, \quad g(z) = \left(\frac{1}{2}\right)|\nabla z|_{L^2}^2$$

$$B(t, z, v) = j(t)(\beta(\mathbf{u}_t(t) + v) - \beta(\mathbf{u}_t(t))),$$

$$p = 2.$$

In this way a standard application of Proposition 2.1 implies the desired properties on $z = 0$.

REFERENCES

1. C. M. Dafermos, An invariance principle for compact processes. *J. Diff. Eqns* **9**, 239–252 (1971).
2. C. M. Dafermos, Uniform processes and semicontinuous Liapunov functionals. *J. Diff. Eqns* **11**, 401–415 (1972).
3. C. M. Dafermos and M. Slemrod, Asymptotic behavior of nonlinear contraction semigroups. *J. Funct. Anal.* **13**, 97–106 (1973).
4. A. Haraux, Comportment a l'infini pour certains systemes dissipatifs nonlineaires. *Proc. R. Soc. Edinburgh* **84A**, 213–234 (1979).
5. J. La Salle, The stability of dynamical systems. *SIAM Reg. Conf. Ser. Appl. Math.* **25** (1976).

6. J. K. Hale, Dynamical systems and stability. *J. Math. Anal. Appls.* **26**, 39–59 (1969).

7. J. B. Baillon and H. Brezis, Une remarque sur le comportament asymptotique des semigroups non lineaires. *Houston J. Math.* **2**, 5–7 (1976).

8. J. M. Ball, On the asymptotic behavior of generalized processes with applications to nonlinear evolution equations. *J. Diff. Eqns* **27**, 224–265 (1978).

9. P. Marcati, Stability for second order abstract evolution equations. *Nonlinear Analysis TMA* **8**, 237–252 (1984).

10. L. Salvadori, Uso di due indici nel problema della stabilità. *Conf. Semin. Mat. Univ. Bari.* **146** (1977).

11. L. Salvadori, Famiglie ad un parametro di funzioni di Liapunov nello studio della Stabilità. *Symp. Math.*, Vol IV. Academic Press, New York (1971).

12. J. M. Ball and M. Slemrod, Feedback stabilization of distributed semilinear control system. *Appl. Math. Opm.* **5**, 169–179 (1979).

13. J. M. Ball and M. Slemrod, Nonharmonic Fourier series in the stabilization of distributed semilinear control systems. *Comm. Pure Appl. Math.* **32**, 555–587 (1979).

14. J. M. Ball, Stability theory for an extensible beam. *J. Diff. Eqns* **14**, 399–418 (1975).

15. A. Haraux, *Nonlinear Evolution Equations-Global Behavior of Solutions*, Lecture Notes in Math. **841** Springer-Verlag, Berlin (1981).

Comp. & Maths. with Appls. Vol. 12A, Nos. 4/5, pp. 551–556, 1986
Printed in Great Britain.

0886–9553/86 $3.00 + .00
1986 Pergamon Press Ltd.

SINGULARITIES OF HYPERBOLIC PDEs IN TWO COMPLEX VARIABLES

PETER A. McCoy

Mathematics Department, United States Naval Academy, Annapolis, MD 21402, U.S.A.

Abstract—Symmetric Poisson processes (SPP) arise as analytic solutions of a hyperbolic partial differential equation in C^2 and expand as series of symmetric products of ultraspherical polynomials. A pair of locally valid reciprocal integral equations is constructed that associates each SPP with a unique analytic function of a single complex variable. The singularities of the SPP are determined from the known singularities of the associate by analytically continuing the transform pair via the envelope method. Thus, a classical theorem of Nehari, concerning the singularities of solutions of an ODE that expand as Legendre series, is extended.

INTRODUCTION

In a neighborhood of a point of analyticity, a function of a single complex variable admits a Taylor's series expansion that may be analytically continued by Hadamard's argument[1–2] to determine global information about the location and nature of its singularities. This notion proved fruitful in the analysis of singularities of analytic functions that expand as Legendre series (Nehari[3]) by associating them with those of Taylor's series. And, it generalizes as the envelope method (Gilbert[4–6]), where an extensive theory exists for elliptic PDEs.

Typical elliptic PDEs with analytic continuation to C^2 become formally hyperbolic under an isotropic transformation. Such an example is the Laplacian; $\delta_{xx} + \delta_{yy} = 4i\delta_{zw}$ with $z = x + iy$, $w = x - iy$ for $(z, w) \in C^2$, where the flow of information about the solution of the hyperbolic equation from the elliptic problem is cursory and no independent theory results. It is of interest then to consider the singularity problem for a hyperbolic PDE on C^2 that, unlike the usual problems in function theory, is not equivalent to an elliptic PDE in E^2 under such a transformation.

A natural problem to consider arises from the equation (Bochner[7], McCoy[8])

$$\{\delta_x\{(1 - x^2)^{\lambda + 1/2}(1 - y^2)^{\lambda + 1/2}\delta_x\} - \delta_y\{(1 - x^2)^{\lambda + 1/2}(1 - y^2)^{\lambda + 1/2}\delta_y\}\}F(x, y) = 0, \quad (1a)$$

$\lambda \geq 0$ that describes symmetric Poisson processes on a square in E^2. The analytic solutions (SPP) expand as series in products of ultraspherical polynomials and will provide a generalization of Nehari's work on Legendre series when they admit analytic continuation as functions of two complex variables. This is accomplished through a locally valid integral transform and its inverse that uniquely associate a SPP with an analytic function of one complex variable. The envelope method is used in the process of analytically continuing the transforms to generate the manifolds of possible singularities for each associated function pair. The true singularities of a SPP are precisely the possible singularities that under the inverse transform generate true singularities of the associate.

THE RECIPROCAL INTEGRAL TRANSFORMS

Ultraspherical polynomials $\{P_n^\lambda(x)\}_{n=0}^\infty$ form an orthogonal system

$$\int_{-1}^{+1} P_n^\lambda(s)P_m^\lambda(s)(1 - s^2)^{\lambda - 1/2} \, ds = (\omega_n)^{-1}\delta_{nm},$$

$$\omega_n = 2^{1-2\lambda}\Gamma(n + 2\lambda)/\Gamma(\lambda)^2(n + \lambda)\Gamma(n + 1) \sim n^{2\lambda + 1}$$

on $-1 < x < 1$ where they are particular solutions of the differential equation

$$L_x P_n(x) := \{\delta_x((1 - x^2)^{\lambda + 1/2}\delta_x) + n(n + 2\lambda)(1 - x^2)^{\lambda + 1/2}\}P_n^\lambda(x) = 0, \quad \lambda > 0.$$

(Note: $\lambda = 0$ is recovered as a limit[9]). The $P_n^{1/2}(x) = P_n(x)$, Legendre's polynomial of degree n, and $P_n^1(x) = U_n(x)$ which is the nth degree Tchebyshev polynomial of the second kind. The symmetric products

$$R_n(x, y): = P_n^\lambda(x)P_n^\lambda(y), \quad n = 0, 1, 2, \ldots$$

are particular solutions of the Poisson process equation

$$\{(1 - y^2)^{\lambda + 1/2}L_x - (1 - x^2)^{\lambda + 1/2}L_y\}R_n(x, y) = 0, \tag{1b}$$

on the square $-1 < x, y < +1$ that form a basis for the analytic solutions.

To begin, let $\{a_n\}_{n=0}^\infty$ be a sequence of real constants such that

$$\limsup_{n \to \infty} |a_n|^{1/n} = \zeta < 1$$

and let σ be the sum of the semi major and minor axes of the ellipse

$$E_\sigma: = \{t \in C : |t - 1| + |t + 1| < \sigma\}.$$

Furthermore, let the real analytic functions f and F be defined by

$$f(t) = \sum_{n=0}^\infty a_n t^n, \quad |t| < \zeta^{-1}, \tag{2a}$$

$$F(z, w) = \sum_{n=0}^\infty a_n \omega_n R_n(z, w), \quad (z, w) \in E_\sigma \times E_\sigma. \tag{2b}$$

To see that (2b) is valid for sufficiently small σ, note from Darboux's extension of the LaPlace–Heine formula[9], that the asymptotic estimate

$$R_n(z, w) \sim \sigma^{2n}$$

is uniform on compacta of $E_\sigma \times E_\sigma$. This insures the existence of a uniformly and absolutely convergent dominant series. Thus, on the referenced product domain, $F(z, w)$ is the analytic continuation of $F(x, y)$ to C^2 (see [4]).

The first step is to establish that the functions (2a–b) are associated. The analytic kernel

$$K(s, z, w): = \sum_{n=0}^\infty s^n \omega_n R_n(z, w)$$

is constructed in a sufficiently small neighborhood of the origin in C^3. One finds from the orthogonality of the ultraspherical polynomials that

$$F(z, w) = \frac{1}{2\pi i} \int_\gamma f(t)K(t^{-1}, z, w) \frac{dt}{t} \tag{3a}$$

for $\gamma: |t| = (\zeta + \epsilon)^{-1}$; and, that

$$f(t) = \int_{-1}^{+1} \int_{-1}^{+1} F(x, y)K(t, x, y)(1 - x^2)^{\lambda - 1/2}(1 - y^2)^{\lambda - 1/2} \, dx \, dy, \tag{3b}$$

where, here and throughout, the principal branches of the radicals are defined in the usual way.

Following basic principles[4], the domains of (3a–b) are taken as initial domains of definition. Such a domain is the maximal open set with the property that any two of its points may be connected by a path consisting of a finite union of arcs none of which includes a singularity of the integrand of the defining transform. It may be possible to continue (3a–b) as function elements by extending the initial domain of definition. Now the contours are deformed to avoid singularities of the integrand that are attempting to cross them during the continuation process. The maximal extension of the initial domain is referred to as the domain of association. Here it includes the origin by construction.

On their domains of association, (3a–b) generate the integral operators

$$F(z, w) = \frac{1}{2\pi i} \int_{L_t} f(t) K(t^{-1}, z, w) \frac{dt}{t}$$

$$f(t) = \int_{L_w} \int_{L_z} F(z, w) K(t, z, w)(1 - z^2)^{\lambda - 1/2}(1 - w^2)^{\lambda - 1/2} \, dz \, dw, \quad \lambda \geq \frac{1}{2}.$$

The simple closed contour L_t is homologous to γ. And, L_w and L_z are simple open contours joining -1 to $+1$ that homologous to the interval $[-1, +1]$. Each homology is necessarily modulo the singularities.

The next step is to put these operators into a more usable form. The kernel is summed (Watson[10]) and written here as an integral

$$K(t, z, w) = \frac{1}{2\pi i} \int_{L_\tau} B(t, \tau, \Omega) \frac{d\tau}{\tau},$$

$$B(t, \tau, \Omega) = (2\lambda/(2i)^{2\lambda - 1})(1 - t^2)(\tau - \tau^{-1})^{2\lambda - 1}/(1 - 2t\Omega + t^2)^{\lambda + 1}$$

with the generating variable

$$\Omega := zw + (1 - z^2)^{1/2}(1 - w^2)^{1/2}(\tau + \tau^{-1})/2.$$

The simple closed contour L_τ is homologous to the unit circle $|\tau| = 1$. This brings us to summarize the discussion.

THEOREM 1

The dual integral equations

$$F(z, w) := H(f(t)) = -\left(\frac{1}{2\pi i}\right)^2 \int_{L_\tau} \int_{L_t} f(t) t^{2\lambda - 1} B(t, \tau, \Omega) \, dt \, \frac{d\tau}{\tau}, \tag{4a}$$

$$f(t) := H^{-1}(F(z, w))$$

$$= \frac{1}{2\pi i} \int_{L_\tau} \int_{L_w} \int_{L_z} F(z, w) B(t, \tau, \Omega)(1 - z^2)^{\lambda - 1/2}(1 - w^2)^{\lambda - 1/2} \, dz \, dw \, \frac{d\tau}{\tau} \tag{4b}$$

$(\lambda \geq 1/2)$ form a reciprocal transform pair that uniquely associates the analytic function elements f and the SPP F on their domains of association.

With this regularity theorem in hand, what remains is to examine the singularities. This is considered in the next section.

THE SINGULAR MANIFOLDS

During the construction of the domain of association, certain points may be reached for which no contour deformation will avoid a singularity. These points are possible singularities of the function element and are located on the analytic manifold of singularities of the integrand and on the derived manifold constructed by the envelope method. If the contour is open, certain end pinch singularities arise on account of the fact that the terminal points of the contour are fixed; and, therefore cannot move to avoid singularities that may approach during the continuation.

Let us begin with the SPP F and suppose that its associate has a finite singularity at $t = \alpha$. The points we are seeking are located on the manifold

$$M(f; \alpha) = \{(z, w) \in C^2 : \tau\Omega = 0; \tau \in L_\tau, t = \alpha\}, \tag{5a}$$

drawn from the (integrand of) the operator $H(f)$ and on its derived manifold

$$M_*(f; \alpha) = \{(z, w) \in C^2 : \delta_\tau(\tau\Omega) = 0; \tau \in L_\tau, t = \alpha\}. \tag{5b}$$

Eliminating τ from the resulting system

$$\tau\Omega\big|_{t=\alpha} = (\sigma - zw)\tau - (1 - z^2)^{1/2}(1 - w^2)^{1/2}(\tau^2 + 1)/2 = 0, \tag{6a}$$

$$\delta_\tau(\tau\Omega)\big|_{t=\alpha} = (\sigma - zw) - (1 - z^2)^{1/2}(1 - w^2)^{1/2}\tau = 0, \tag{6b}$$

with $\sigma := (\alpha + \alpha^{-1})/2$ gives the manifold of possible singularities of the SPP as

$$P(F; \alpha): = M(f; \alpha) \cap M_*(f; \alpha)$$
$$= \{(z, w) \in C^2 : (\sigma - zw)^2 = (1 - z^2)(1 - w^2), \alpha^2 - 2\alpha\sigma + 1 = 0\}.$$

It is not necessary to include $t = 0$ on the manifold because it is an interior point of the domain association of the principal branch of F.

The reasoning is reversed for the associate. Suppose that F has a finite singularity at $(z, w) = (z_0, w_0)$. The manifolds corresponding to (5a–b) are

$$M^{-1}(F; (z_0, w_0)): = \{t \in C : \tau\Omega = 0; \tau \in \mathcal{L}_\tau, (z, w) = (z_0, w_0)\} \tag{7a}$$

and the derived manifold

$$M_*^{-1}(F; (z_0, w_0)): = \{t \in C : \delta_\tau(\tau\Omega) = 0; \tau \in \mathcal{L}_\tau, (z, w) = (z_0, w_0)\}. \tag{7b}$$

Continuing, τ is eliminated from the system

$$\tau\Omega\big|_{(z,w)=(z_0,w_0)} = 0, \quad \delta_\tau(\tau\Omega)\big|_{(z,w)=(z_0,w_0)} = 0, \tag{8a–b}$$

where $\alpha = (t + t^{-1})/2$ to generate the manifold of possible singularities of f,

$$P^{-1}(F; (z_0, w_0)): = M^{-1}(F; (z_0, w_0)) \cap M_*^{-1}(F; (z_0, w_0))$$
$$= \{t \in C : (\alpha - z_0 w_0)^2 = (1 - z_0^2)(1 - w_0^2), t^2 - 2\alpha t + 1 = 0\}.$$

The symmetry displayed between the manifolds P and P^{-1} is expected because of the symmetry of the key variable

$$\Omega/2t = (t + t^{-1})/2 - (zw + (1 - z^2)^{1/2}(1 - w^2)^{1/2}(\tau + \tau^{-1})/2).$$

Following the envelope method we infer then that, with the exception of possible end pinch singularities, the point (z_0, w_0) is a true singularity of F if, and only if, it is on a manifold

$$\{(z, w) \in C^2 : (\alpha - zw)^2 = (1 - z^2)(1 - w^2), \alpha^2 - 2\alpha\sigma + 1 = 0\},$$

where $t = \alpha$ is a true singularity of the associate f.

What remains is to check for end pinch singularities. These are located at the branch manifolds

$$E: = \{(z, w) \in C^2 : 1 - z^2 = 0\} \cup \{(z, w) \in C^2 : 1 - w^2 = 0\}$$

on the manifold $P^{-1}(F; (z_0, w_0))$ and produce the singularities

$$P^{-1}(F; (z_0, w_0)) \cap E = \{(\pm 1, \pm(1 + \alpha^2)/2\alpha), (\pm(1 + \alpha^2)/2\alpha, \pm 1)\}.$$

Thus, the end pinch singularities appear in a natural way as shadows cast on the C plane that correspond exactly to the Nehari singularities of Legendre series when $\lambda = 1/2$ and the positive sign is selected. We have proved the following result.

THEOREM 2

Let the real analytic expansions

$$F(z, w) = \sum_{n=0}^{\infty} a_n \omega_n P_n^{\lambda}(z) P_n^{\lambda}(w), \quad \lambda \geq \frac{1}{2},$$

$$f(t) = \sum_{n=0}^{\infty} a_n t^n$$

define functions on their initial domains of definition. Then the point $(z_0, w_0) \in C^2$ reached by analytically continuing the function element F and avoiding the Nehari points

$$(\pm 1, \pm(1 + \alpha^2)/2\alpha) \quad \text{and} \quad (\pm(1 + \alpha^2)/2\alpha, \pm 1)$$

is a singularity of F if, and only if,

$$(\sigma - z_0 w_0)^2 = (1 - z_0^2)(1 - w_0^2), \quad \sigma = (1 + \alpha^2)/2\alpha,$$

where $t = \alpha \in C$ is a singularity reached by analytically continuing the function element f.

Remarks. Let us examine a particular Nehari point of F, say at $(z, 1)$, where the associate of f is singular at $t = e^{-i\theta_0}$. In accordance with Theorem 2, $z = \cos(\theta_0) = (e^{i\theta_0} + e^{-i\theta_0})/2$ is a singular point of $F(\cos\theta, 1)$. Conversely, the singularity $(\cos(\theta_0), 1)$ produces the singularity $t = e^{i\theta_0}$ of the associate. This demonstrates an old theorem of Szego[11] generalized in [3]. And it provides an application in the E^2 plane.

The applications in the E^2 plane can be taken further. The SPP under discussion are real analytic (*viz.* real coefficients) and have associates whose singularities necessarily occur in conjugate pairs. Moreover, the singular points of F are symmetric; $(z_0, w_0) = (w_0, z_0)$. Let us put these observations to use for singularities of $F(x, y)$ that are non-Nehari. These must be symmetrically located relative to the lines bisecting the I–III and II–IV quadrants. And, for each singularity of the associate at $t = \alpha$, there is a correspondent reflection at $t = \bar{\alpha}$. Thus, for each singularity of the associate not on an axis or quadrant bisector, there are 8 singular points of F. This is illustrated by application of Theorem 2 to such a singularity at $\alpha = \xi + i\eta$. The result defines the singularities of $F(x, y)$ at

$$x^2 = \xi^2/y^2 = [(\xi^2 + \eta^2 + 1) \pm ((\xi^2 - 1)^2 + \eta^2)^{1/2}((\xi^2 + 1)^2 + \eta^2)^{1/2}]/2,$$

$$y^2 = \xi^2/x^2 = [(\xi^2 + \eta^2 + 1) \pm ((\xi^2 - 1)^2 + \eta^2)^{1/2}((\xi^2 + 1)^2 + \eta^2)^{1/2}]/2.$$

With a little reflection, the terms arising in the products are simply distances from the singularity of the associate to (i) the origin, and (ii) the endpoints of the segment $[-1, +1]$. Because of the symmetries, these can be replaced by the distances from the conjugate singularity to (i) the origin and, (ii) the endpoints of the segment with endpoints $\{-i, +i\}$. This is, of course, because the initial domains of definition correspond to the Cartesian product of two ellipses with foci at $\{-1, +1\}$ and $\{-i, +i\}$ with semi axes whose lengths are calculated from the coefficients as in [3,9].

REFERENCES

1. P. Dienes, *The Taylors Series,* 1st Ed. Dover Publications, New York (1957).
2. E. Hille, *Analytic Function Theory,* Vols. 1 and 2. Blaisdell Publishing Company, Waltham (1962).
3. Z. Nehari, On the singularities of Legendre series. *J. Rational Mech. Anal.* **5**, 987–992 (1956).
4. R. P. Gilbert, *Function Theoretic Methods in Partial Differential Equations, Math. in Science and Engineering,* Vol. 54. Academic Press, New York (1969).
5. R. P. Gilbert, *Constructive Methods for Elliptic Equations, Lecture Notes in Math.,* Vol. 365. Springer-Verlag, New York (1974).
6. R. P. Gilbert, Integral operator methods in bi-axially symmetric potential theory. *Contrib. Diff. Eqns.* **2**, 441–456 (1963).
7. S. Bochner, Sturm-Liouville and heat equations whose eigenfunctions are ultraspherical or associated Bessel functions, Proc. Conf. on Differential Equations, University of Maryland, pp. 23–48 (1955).
8. P. A. McCoy, Hyperbolic boundary value problems arising from symmetric Poisson processes, *Comp. & Maths. with Appls.,* **11**, 307–313 (1985).
9. G. Szego, *Orthogonal Polynomials,* Pub. 23, 3rd ed. Am. Math. Soc., Providence (1967).
10. G. N. Watson, Notes on generating functions of polynomials: (3) polynomials of Legendre and Gegenbauer. *J. London Math. Soc.* **8**, 289–292 (1933).
11. G. Szego, On the singularities of zonal harmonic expansions. *J. Rational Mech. Anal.* **3**, 561–564 (1954).

Comp. & Maths. with Appls. Vol. 12A, Nos. 4/5, pp. 557–563, 1986
Printed in Great Britain.

0886–9553/86 $3.00 + .00
© 1986 Pergamon Press Ltd.

ON THE TRICOMI PROBLEM

G. Adomian
Center for Applied Mathematics, The University of Georgia, Athens, GA, U.S.A.

and

N. Bellomo
Department of Mathematics, Politecnico, Torino, Italy

Abstract—The Tricomi equation is solved by the decomposition method[6–8] on the half-space $x, y > 0$ and also for suitable boundary conditions given on the Boundary of the hyperbolic–elliptic domain.

1. INTRODUCTION

The Tricomi equation[1,2] is well known to be a fundamental mathematical model to solve the problem of defining the flow field around wing shapes at transonic speeds. This equation can be classified as a "mixed type" equation[3], which is elliptic in the positive half-space $(y > 0)$ of a cartesian plane with axes x, y and hyperbolic in the negative half-space $(y < 0)$, the boundary $y = 0$ being the parabolic separation line.

The Tricomi problem[4], as detailed in the next section, consists in assigning, as boundary conditions, the unknown function over a given profile in the elliptic plane intersecting the parabolic line in two distinguishable points and on a characteristic line departing from one of the two points. Such a problem is then well posed, in the terms specified in the next section, and a uniqueness theorem was already supplied by Tricomi in his celebrated monograph[4]. Analytical solutions of the Tricomi problem are useful in those applications such that the boundary conditions are assigned for transonic wing profiles[3,5].

This paper supplies an analytical solution for the above-mentioned boundary-value problem using Adomian's decomposition method for partial differential equations[6–8]. The second section provides a detailed description of the Tricomi problem. The decomposition method is used, in the third section, to obtain a class of solution for such a problem. An application and a discussion follow.

2. THE TRICOMI PROBLEM

Consider the Tricomi equation

$$y\phi_{xx} + \phi_{yy} = 0, \tag{1}$$

defined, with reference to Fig. 1, in the domain $D = E \cup I$, where the boundary of E is defined by a given convex profile:

$$x \in [0, 1] : y_e = y_e(x), \qquad x < x^* : \mathrm{d}y_e/\mathrm{d}x \geq 0, \qquad x > x^* : \mathrm{d}y_e/\mathrm{d}x \leq 0, \tag{2}$$

and by the segment $y = 0$, $x \in [0, 1]$.

On the other hand, the boundary of I is defined by the above-mentioned segment and by the two characteristic lines, which in the figure are indicated by the dash lines departing, respectively, from O and A:

$$1° : x^2 = -(4/9)y^3, \qquad 2° : (x - 1)^2 = -(4/9)y^3, \tag{3}$$

which satisfy the differential equation

$$\mathrm{d}y/\mathrm{d}x = \pm 1/y. \tag{4}$$

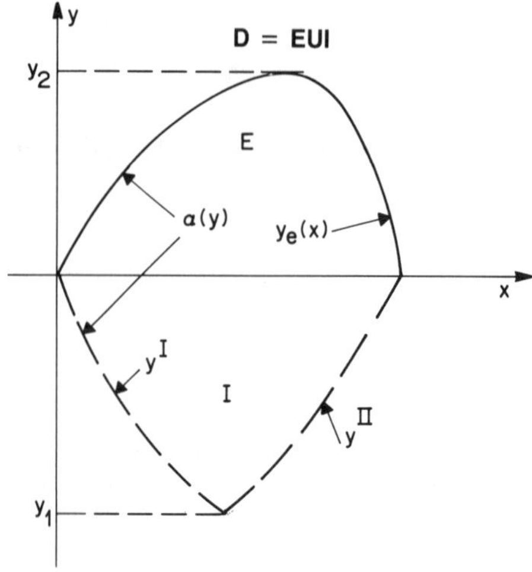

Fig. 1. Representation of the domain D.

Let us now denote by B the space of all functions $\phi = \phi(x, y)$ such that

(a) ϕ is a C^2 function for every $(x, y) \in D$.

(b) ϕ is continuous on the boundary of D.

(c) ϕ_x and ϕ_y are continuous on the boundary of D besides that in O and A, where they can tend to infinity with an order which is less than 1 with respect to the space coordinates.

After these preliminaries, the Tricomi problem can be defined as follows:

Finding a function $\phi = \phi(x, y)$ such that $\phi \in$ B and Eq. (1) is satisfied with the boundary conditions

$$y = y_e : \phi(x, y = y_e(x)) = \phi_e(x); \qquad y = y^I : \phi(x, y = y^I(x)) = \phi^I(x). \tag{4}$$

In the analysis of the above problem we shall denote by $\tau(x)$ and $v(x)$ two functions such that

$$\tau(x) = \phi(x, y = 0), \qquad v(x) = \phi_y(x, y = 0), \tag{5}$$

and by $u(y)$ and $v(y)$ the functions such that

$$u(y) = \phi(x = 0, y), \qquad v(y) = \phi_x(x = 0, y) \tag{6}$$

3. ANALYSIS

The analysis will be carried out in two steps. As the first step, the relatively easier problem will be considered of determining the function $\phi(x, y)$ in the half-space $x \geq 0$, given the values of the functions u and v, and a class of quasi-analytical solutions will be determined. Such a problem will be defined the "half-space problem." Then afterwards, as the second step, the direct Tricomi problem will be considered according to the description of the second section.

Step 1: *The half-space problem*

Following Adomian[6–8] let

$$L_x = \partial^2/\partial x^2, \qquad L_y = \partial^2/\partial y^2, \tag{7}$$

whose inverse L_x^{-1} and L_y^{-1} exists for every $q = q(x, y) \in$ B in the form

$$L_x^{-1}q = \int_0^x dr \int_0^r q(s, y)ds, \qquad L_y^{-1}q = \int_0^y dr \int_0^r q(x, s)ds. \tag{8}$$

Equation (1) can now be rewritten in one of the two forms:

$$L_y \phi = -y L_x \phi, \qquad L_x \phi = -(1/y) L_y \phi. \tag{9}$$

Each of them can be formally integrated applying the inverse operators L_y^{-1} to the first one and L_x^{-1} to the second one, then the addition, divided by 2, provides the following operator-equivalent formulation of Eq. (1):

$$\lambda = 1 : \phi(x, y) = \phi_0(x, y) - \frac{1}{2} \lambda \mathscr{B} (x, y)\phi(x, y), \tag{10}$$

where

$$\phi_0(x, y) = \frac{1}{2} (\tau(x) + y\nu(x) + u(y) + xv(y)) \tag{11}$$

and the operator $\mathscr{B}$ is defined as follows:

$$\mathscr{B} \cdot = (L_y^{-1} y L_x + L_x^{-1}(1/y)L_y)\cdot, \tag{12}$$

which is a mixed-type integral–differential operator. The decomposition method can now be applied to writing the solution of Eq. (10) in the form

$$\phi(x, y) \cong \phi_0(x, y) + \sum_{j=1}^{n} \lambda^i \phi_j(x, y). \tag{13}$$

Then Eq. (13) can be substituted into (10) and if the terms with the same power of λ are equated, the following recursive formula is found:

$$\phi_{j+1} = -\frac{1}{2} \mathscr{B} \phi_j(x, y), \tag{14}$$

which supplies all terms appearing in (13) in a sequence of easily computable quadratures, each defined by the preceding one.

Remark 1. The solution of the "half-space problem" already contains the indirect solution of the Tricomi problem as the actual expression of ϕ_0 supplies suitable boundary conditions on the lines y_e and y^l consistent with the data on $x = 0$ and $y = 0$ $(x > 0)$ with ϕ_0, which can be chosen in a way that $\phi \in B$.

Remark 2. The structure of Eq. (10) defines a fixed-point formulation of the problem; therefore the classical Cacciopoli–Banach fixed-point theorem[9] can be applied in order to prove the existence and uniqueness of the solutions of Eq. (10).

A proof, as specified in *Remark* 2, is supplied in the final section.

Step 2: *The Tricomi Problem*

Before approaching the problem defined in Sec. 2, it is useful to define (with obvious meanings of the adopted symbols and with reference to Fig. 1) the following curves and boundary conditions:

$$y \in [y_1, y_2] : \alpha = \alpha(y), \qquad c = c(y) = \phi(x = \alpha(y), y), \qquad d = d(y) = \phi_x(x = \alpha(y), y);$$
$$x \in [0, 1] : a = a(x) = \phi(x, y = y_e(x)), \qquad b = b(x) = \phi_y(x, y = y_e(x)).$$

After these preliminaries we can consider the boundary-value problem with the boundary conditions defined in Eq. (4), which is such that the functions $\tau = \tau(x)$ and $\nu = \nu(x)$ as well as $u(y)$ and $v(y)$ are not arbitrary, being defined by conditions (4).

In particular, the boundary conditions define, following the analysis of Tricomi himself, (Ref. [3], Chap. 3) the functions $\tau(x)$ and $v(x)$ when $y_e(x)$ is a normal line. Consequently we can formulate the following.

ALTERNATIVE I

Finding, at given $\tau(x)$ and $v(x)$, with $x \in [0, 1]$, two functions $u(y)$ and $v(y)$ defined for $y \in [y_1, y_2]$ and a solution $\phi(x, y) \in B$ of Eq. (10) such that the boundary conditions (4) are satisfied. In this line u and v can be decomposed as follows:

$$u = u(y) \cong \sum_{j=1}^{n} \lambda^j u_j(y), \qquad v = v(y) \cong \sum_{j=1}^{n} \lambda^j v_j(y), \tag{15}$$

whose expressions can be substituted in Eq. (10) together with expression (13), with the result

$$\sum_{j=0}^{n} \lambda^j \phi_j(x, y) = \frac{1}{2}(\tau(x) + yv(x)) + \frac{1}{2}\sum_{j=0}^{n} \lambda^j(u_j(y) + xv_j(y)) - \frac{1}{2}\lambda\,\mathcal{B}\sum_{j=0}^{n} \lambda^j \phi_j(x, y). \tag{16}$$

Then, equating the terms with the same power of λ yields

$$\phi_0(x, y) = \frac{1}{2}(\tau(x) + yv(x) + u_0(y) + xv_0(y)), \tag{17}$$

$$\phi_{j+1}(x, y) = \frac{1}{2}(u_{j+1}(y) + xv_{j+1}(y)) - \frac{1}{2}\mathcal{B}\phi_j(x, y), \tag{18}$$

where we have by definition

$$\phi_0(0, y) = u_0(y) = \frac{1}{2}(\tau(0) + yv(0) + u_0(y)), \tag{19a}$$

$$\phi_{x0}(0, y) = v_0(y) = \frac{1}{2}(\tau_x(0) + yv_x(0) + v_0(y)), \tag{19b}$$

$$\phi_{j+1}(0, y) = u_{j+1}(y) = \frac{1}{2}u_{j+1}(y) - \frac{1}{2}\psi_{j+1}(0, y), \tag{19c}$$

$$\phi_{x(j+1)}(0, y) = v_{j+1}(y) = \frac{1}{2}v_{j+1}(y) = \frac{1}{2}\psi_{x(j+1)}(0, y), \tag{19d}$$

where

$$\psi_{j+1}(x, y) = \mathcal{B}\phi_j(x, y). \tag{20}$$

Consequently

$$u_0(y) = \tau(0) + yv(0), \qquad v_0(y) = \tau_x(0) + yv_x(0), \tag{21}$$

$$u_{j+1}(y) = \psi_{j+1}(0, y), \qquad v_{j+1} = \psi_{x(j+1)}(0, y); \tag{22}$$

and as a final result,

$$\phi(x, y) \cong \frac{1}{2}(\tau(x) + yv(x) + \tau(0) + \tau_x(0) + y(v(0)$$

$$+ v_x(0))) + \frac{1}{2}\sum_{j=1}^{n} (\psi_j(0, y) + y\psi_{xj}(0, y) + \psi_j(x, y). \tag{23}$$

where the terms ψ_j are supplied by the recurrent formula (20).

ALTERNATIVE II

Finding, given boundary conditions in the form of Eq. (4), a function $\phi(x, y) \in B$ such that Eq. (10) and the boundary conditions are satisfied.

As a consequence of the above formulation of the problem, the assumption that $\tau(x)$ and $v(x)$ are determined *a priori* is removed and the solution of the Tricomi problem has to be found directly considering the conditions on the boundary of D.

The convexity of y_e with respect to the coordinate axes enables us to define the following inverse operators, which exists for $q \in B$:

$$A_x^{-1} = \int_{\alpha(y)}^x dr \int_{\alpha(y)}^r q(s, y)ds, \qquad A_y^{-1} = \int_{y_e(x)}^y dr \int_{y_e(x)}^r q(x, s)ds. \tag{24}$$

The same analysis which provided a solution for the half-space problem results in the operator equation

$$\phi(x, y) = \frac{1}{2}(a(x) + yb(x) + c(y) + xd(y)) - \frac{1}{2}\mathscr{B}*\phi(x, y), \tag{25}$$

where

$$\mathscr{B}*\cdot = (A_x^{-1}yL_y - A_y^{-1}(1/y)L_x)\cdot \tag{26}$$

where $a = a(x)$ and $b = b(x)$ are given, whereas the functions $c = c(y)$ and $d = d(y)$ have to be determined from the boundary conditions.

After these preliminaries, the analysis is completely analogous to the one of the Alternative I and the final result is

$$\phi(x, y) = \frac{1}{2}(a(x) + yb(x) + a(0) + a_x(0) + yb(0) + yb_x(0))$$

$$+ \frac{1}{2}\sum_{j=1}^n (\psi_j^*(0, y) + y\psi_{xj}^*(0, y) + \psi_j^*(x, y)), \tag{27}$$

where

$$\psi_j^*(x, y) = \mathscr{B}*\phi_j(x, y). \tag{28}$$

Remark 3. The analysis of the various problems considered in this section, as well as of the solutions which have been proposed in Eqs. (13, 14) for the half-space problem and by Eqs. (23) and (27) for the "Tricomi problem" in Alternatives I and II, respectively, should, in all cases, prove that the proposed solutions are the unique solutions of Eq. (10).

4. APPLICATION AND DISCUSSION

The various alternatives considered in the preceding section have been analyzed on the basis of the Adomian decomposition method and a solution procedure has been proposed for each case. Since all problems, however, have been formally solved in the same fashion, we shall restrict our attention in the application and analysis of this last section to the "relatively easier" half-space problem.

Now, the class of problems characterized by boundary conditions ϕ_0 such that the operator "$-\mathscr{B}$" acting upon ϕ_0 reproduces ϕ_0,

$$\phi_0 = -\mathscr{B}\phi_0, \tag{29}$$

will be considered. Keeping this in mind, consider now the closed convex subset $\mathscr{A}$ of a complete

G. ADOMIAN and N. BELLOMO

Banach space:

$$\{\mathscr{A} : \phi \in \Phi \subseteq B; \|\phi\| \le 2\|\phi_0\|\}, \tag{30}$$

where Φ is the class of functions fulfilling condition (29). Let $\mathscr{A}$ be equipped with the norm

$$\|\phi\| = \sup_{x,y \in D} |\phi|, \tag{31}$$

and rewrite Eq. (10) in the form

$$\phi(x, y) = U\phi(x, y) = \phi_0(x, y) - \frac{1}{2}\mathscr{B}\phi(x, y). \tag{32}$$

Condition (29) and the definition (30) imply the following:

    (i) $\phi \in \mathscr{A} \to U\phi \in \mathscr{A}$,

    (ii) $\phi_1, \phi_2 \in \mathscr{A} \to \left\| U\phi_1 - U\phi_2 \right\| \le \frac{1}{2}\left\| \phi_1 - \phi_2 \right\|$.

In fact,

$$U\phi = \phi_0 - \frac{1}{2}\mathscr{B}\phi = \phi_0 + \frac{1}{2}\phi \to \|U\phi\| = \left\| \phi_0 + \frac{1}{2}\phi \right\| \le \|\phi_0\| + \frac{1}{2}\|\phi\| \le 2\|\phi_0\|;$$

moreover,

$$\|U\phi_1 - U\phi_2\| = \left\| \phi_0 + \frac{1}{2}\phi_1 - \phi_0 - \frac{1}{2}\phi_2 \right\| = \frac{1}{2}\|\phi_1 - \phi_2\|.$$

The implications (i), (ii) contain the following result.

THEOREM

    Let $\phi_0 \in \Phi$; then Eq. (10) has a unique fixed point in $\mathscr{A}$ and consequently the solution of Eq. (10) exists and is unique in $\mathscr{A}$ and is given by Eqs. (13–14); i.e. the "decomposition method."

    Proof. Existence and uniqueness of the solutions follows from the already mentioned implications (i), (ii). This particular solution can be found also by Picard iterations:

$$\phi_1 = \phi_0 - \frac{1}{2}\mathscr{B}\phi_0 = \frac{3}{2}\phi_0; \ \phi_2 = \phi_0 - \frac{1}{2}\mathscr{B}\phi_1 = \frac{7}{4}\phi_0; \cdots; \ \phi_n = \phi_0(1 + (2^n - 1)/2^n).$$

Consequently

$$\phi(x, y) = \lim_{n \to \infty} \phi_n(x, y) = 2\phi_0(x, y).$$

If now the decomposition method is applied,

$$\phi_0 = \phi_0; \ \phi_1 = \frac{1}{2}\phi_0; \cdots; \ \phi_n = \left(\frac{1}{2}\right)^n \phi_0,$$

$$\phi(x, y) = \sum_{n=0}^{\infty} \left(\frac{1}{2}\right)^n \phi_0(x, y) = 2\phi_0(x, y).$$

The theorem is then proven.

Remark 4. An example of function fulfilling condition (29) is the following:

$$\phi_0(x, y) = 3x^2 - y^3;$$

thus Φ is a nonempty set. However, the theorem holds for other classes of functions. The exercise of finding suitable extensions is left to the reader. A further valuable analysis would be a study, analogous to the one of the above theorem, for both Alternatives I and II. In these cases the advantages of the application of the decomposition method is even more evident.

REFERENCES

1. C. Ferrari and F. Tricomi, *Transonic Aerodynamics*. Academic Press, New York (1968).
2. A. R. Manwell, *The Odograph Equations*. Oliver & Boyd, Edinburgh (1971).
3. C. Morawetz, Nonexistence of transonic flow past a profile. *Comm. Pure Appl. Math.* **14**, 357–367 (1964).
4. F. Tricomi, Sulle equazioni lineari alle derivate parziali di secondo ordine di tipo misto. *Memorie Accademia Lincei* **5**(14), 133–247 (1923).
5. S. Nocilla, Sulla determinazione del flusso transonico continuo attorno a profili alari simmetrici con curvatura regolare senze incidenza. *l'Aerotecnica, J. Italian Assoc. Aeron. Astron.* **4**, 245–260 (1973).
6. G. Adomian, *Stochastic Systems*. Academic Press, New York (1983).
7. R. Bellman and G. Adomian, *Partial Differential Equations*. Reidel, New York (1985).
8. L. V. Kantorovic and G. P. Akilov, *Funkzionalij Analiz*. Mir, Moskwa (1977).

Comp. & Maths. with Appls. Vol. 12A, Nos. 4/5, pp. 565–579, 1986
Printed in Great Britain.

0886-9553/86 $3.00 + .00
© 1986 Pergamon Press Ltd.

HIGH-ACCURACY FINITE-ELEMENT METHODS FOR POSITIVE SYMMETRIC SYSTEMS

WILLIAM LAYTON
School of Mathematics, Georgia Institute of Technology, Atlanta, Georgia 30332, U.S.A.

Abstract—A nonstandard-type "least'squares" finite-element method is proposed for the solution of first-order positive symmetric systems. This method gives optimal accuracy in a norm similar to the H^1 norm. When a regularity condition holds it is optimal in L^2 as well. Otherwise, it gives errors suboptimal by only $h^{1/2}$ (where h is the mesh diameter). Thus, it has greater accuracy than usual finite-element, finite-difference or least-squares methods for such problems. In addition, the spectral condition number of the associated linear system is only $O(h^{-1})$ vs. $O(h^{-2})$ for the usual least-squares methods.

Thus, the method promises to be an efficient, high-accuracy method for hyperbolic systems such as Maxwell's equations. It is also equally promising for mixed-type equations that have a formulation as a positive symmetric system.

1. INTRODUCTION

This paper considers a nonstandard finite-element method for the symmetric system

$$\mathscr{A}u \equiv \sum_{j=1}^{n} A_j(x) \frac{\partial u}{\partial x_j} + K(x)u = f(x), \quad x \in \Omega \subset \mathbb{R}^n. \tag{1.1}$$

The method proposed can be thought of as a nonstandard type of least-squares method. An operator $\mathscr{K}$ is constructed so that the Galerkin method applied to the regularized problem

$$\mathscr{K}^* \mathscr{A}u = \mathscr{K}^* f$$

is stable and convergent. Specifically, if $\mathscr{S}^h$ denotes a (vector) finite-element space, the Galerkin approximation $U \in \mathscr{S}^h$ to u is defined through the equations

$$(\mathscr{A}U, \chi + \delta \mathscr{A}^\circ \chi) = (f, \chi + \delta \mathscr{A}^\circ \chi), \quad \forall \chi \in \mathscr{S}^h, \tag{1.2}$$

where δ is a positive parameter and $\mathscr{A}^\circ$ is the principal part of $\mathscr{A}$.

Symmetric systems such as (1.1) arise in many important physical problems. Maxwell's equations, acoustic equations, hyperbolic systems, many elliptic equations and many equations of mixed type can be put into the positive symmetric form of Friedrichs[1]. Thus, a numerical method that works well for such problems would give a type-independent method. The method considered here gives an accurate scheme that is applicable to the physical variables in such problems. Suppose $\mathscr{S}^h$ consists of piecewise polynomials of degree $\leq k$ on a triangulation of diameter h. Optimal accuracy is achieved in a norm related to the H^1 norm

$$\|\mathscr{A}^\circ(u - U)\| + \|u - U\| = O(h^k). \tag{1.3}$$

In L^2 the qualitative properties of the method depend somewhat upon the type of the equation. For elliptic systems, and some hyperbolic systems, for example Maxwell's equations and some acoustic equations, optimal accuracy is achieved in L^2, $O(h^{k+1})$. For other problems that are not so regular, the errors are still nearly optimal in L^2:

$$\|u - U\| = O(h^{k + 1/2}).$$

Various other numerical methods have been tried for (1.1). Least-squares methods, Aziz, Fix and Leventhal[2], Aziz and Leventhal[3], Fix and Gunzburger[6], Stephan and Wendland[5], and Wendland[6], give optimal approximations in the norm in (1.3), and optimal approximations in L^2 for elliptic systems. For problems without the regularity of elliptic systems, standard least-squares methods lose more accuracy than the method proposed herein. Furthermore, the associated linear system is ill-conditioned compared to the one arising from the method (1.2). Mixed methods have also been studied extensively for first-order elliptic systems by Raviart and Thomas[7] and Fix, Gunzburger and Nicolaides[8]. The stability and accuracy of such schemes depend critically upon the types of elements and the geometry of the mesh via the Babuška–Brezzi condition. Patched variational principles[9,10] and least-squares methods (requiring $H^2(\Omega)$ elements)[11] have been tried for equations of mixed type, as well as first-order finite-difference[12] methods. The method studied here overcomes some of the difficulties of the previous approaches for each of these types of problems.

The work closest in spirit with the present paper seems to be the work of Katsanis[13,14] and Lesaint[11,15,16]. Katsanis[13] considers an $O(h^{1/2})$ finite-difference scheme for (1.1). Lesaint[16] considers a finite-element scheme based upon a weak formulation similar to the usual one. He proves $O(h^k)$ convergence in L^2 when piecewise polynomials of degree $\leq k$ are used.

The method (1.2) has appeared previously in a number of special cases of (1.1). Dendy[17] first proposed a method for hyperbolic equations similar to (1.2) and studied its convergence properties. Wahlbin[18] later noticed that if $\delta = O(h)$ and (1.2) is applied to the model equation $u_t + u_x = 0$, a method results that is stable in L^∞ as well as in L^2. In Layton[19], it was shown using Fourier analysis on a model problem that the spurious oscillations arising from a discontinuity in the true solution decay exponentially in h^{-1} and in the distance from the discontinuity. In Layton[20] it was noticed that (1.2) could be adapted to systems arising from, for example, control theory by interpreting it as a nonstandard least-squares method. Meanwhile, a method related to (1.2) (the streamline diffusion method) was studied for singularly perturbed convection/diffusion equations by Hughs and Brooks[21], Nävert[22], Raithby[23] and Axelsson[24].

Nävert, jointly with Johnson and Pitkäranta in [25], has also extended the results in [19] to the general case of variable coefficient system in a bounded domain. Namely, it was shown in [25] that the effects of discontinuities decay exponentially away from the discontinuity. A. H. Schatz (private communication) reports that similar results have been obtained with an accurate estimate of the smearing effects.

The method (1.2) can also arise from the "a-b-c" technique used by Friedrichs[1,26,27] to prove well posedness for the continuous equation.

The notation used is all standard. $\|\cdot\|$, $(\cdot,\cdot)$ denote the usual $(L^2(\mathbb{R}^n))^m$ norm and inner product respectively. $(H^k(\Omega))^m$ denotes the Sobolev space $(H^k(\Omega) = W_2^k(\Omega))^m$ defined in the usual manner[28].

1.1 *The continuous equation*

Let $\Omega \subset \mathbb{R}^n$ with $\Gamma = \partial\Omega$ smooth and noncharacteristic. Consider the BVP

$$\mathscr{A}u \equiv \sum_{j=1}^{u} A_j(x)\,\frac{\partial u}{\partial x_j} + K(x)u = f, \quad x \in \Omega, \quad (B - M)u = 0, \quad x \in \Gamma. \tag{1.4}$$

In the above, A_j are smooth, symmetric, $m \times m$ matrix functions of x, $K(x)$ is an $m \times m$, bounded, matrix function of x, $u: \Omega \subset \mathbb{R}^n \to \mathbb{R}^m$, and

$$B = \sum_{j=1}^{n} n_j A_j(x),$$

where $\mathbf{n} = (n_j)$ is the outward unit normal to Ω. Furthermore, $M(x)$ satisfies for $x \in \Gamma$ the following:

(i) $M(x)$ is a continuous function of $x \in \Gamma$.
(ii) $M_s(x) \equiv \frac{1}{2}(M(x) + M^*(x)) \geq 0$.

One important subclass of the problems we study herein is the class of *positive* symmetric systems, studied by Friedrichs[1,26].

Definition 1.1. $\mathscr{A}$ is positive if

$$D \equiv K + K^* - \sum_{j=1}^{n} \frac{\partial A_j}{\partial x_j} \geq d_0 I > 0 \text{ holds.}$$

In particular, Friedrichs has shown that if $\mathscr{A}$ is positive symmetric then a unique weak solution to (1.4) exists and that if $\text{Ker}(B - M) + \text{Ker}(B + M) = \mathbb{R}^m$, this solution is, in fact, a classical solution of (1.4).

The boundary-value problem that is adjoint to (1.4) is given by

$$\mathscr{A}^*v(x) \equiv - \sum_{j=1}^{n} \frac{\partial}{\partial x_j}(A_j(x)v) + K^*(x)v, \quad x \in \Omega, \quad (B + M^*)v = 0, \quad x \in \Gamma. \quad (1.5)$$

We let $\mathscr{A}^0$ denote the principle part of $\mathscr{A}$:

$$\mathscr{A}^{\circ}u = \sum_{j=1}^{n} A_j(x) \frac{\partial u}{\partial x_j}, \quad x \in \Omega.$$

Friedrichs[1] has shown that the following three identities hold true ($<\cdot, \cdot>$ denotes the $\mathbb{R}^n$ or $\mathbb{C}^n$ scalar product).

For all $u \in (H^1(\Omega))^m$,

$$2(\mathscr{A}u, u) = (Du, u) + \int_{\Gamma} \langle Bu, u \rangle \, d\Gamma. \quad (1.6)$$

For all $u, v \in (H^1(\Omega))^m$,

$$(\mathscr{A}u, v) = (u, \mathscr{A}^*v) + \int_{\Gamma} \langle Bu, v \rangle \, d\Gamma. \quad (1.7)$$

For all u satisfying $(B - M)u|_{\Gamma} = 0$ and for all v satisfying $(B + M^*)v|_{\Gamma} = 0$,

$$(\mathscr{A}u, v) = (u, \mathscr{A}^*v). \quad (1.8)$$

1.2 *An example of a symmetric system*

Positive symmetric systems arise directly and naturally in many problems of mathematical physics. Examples include Maxwell's equations and all derived systems from the Euler equations. Also, many second-order boundary-value problems of mathematical physics can be transformed directly into a positive symmetric system. Consider, for example, the Tricomi equation:

$$y\phi_{xx} - \phi_{yy} = f \quad \text{in } \Omega, \quad H\phi = 0 \quad \text{on } \Gamma = \partial\Omega, \quad (1.9)$$

where H represents an operator describing the usual boundary conditions for the Tricomi equation.

In physical problems leading to mixed-type equations (e.g., transonic flow), ϕ typically represents a potential function. We obtain a symmetric system from (1.9) by returning to the physical variables $\mathbf{u} = \nabla\phi$. Under this transformation (1.9) becomes the symmetric system

$$\begin{pmatrix} y & 0 \\ 0 & +1 \end{pmatrix}\begin{pmatrix} u \\ v \end{pmatrix}_x + \begin{pmatrix} 0 & -1 \\ -1 & 0 \end{pmatrix}\begin{pmatrix} u \\ v \end{pmatrix}_y = \begin{pmatrix} f \\ 0 \end{pmatrix}. \quad (1.10)$$

Then, the boundary conditions, $H\mathbf{u} = 0$, become Dirichlet boundary conditions on $\Gamma_{2,3}$, and nontangential, non-normal derivatives prescribed on Γ_4 and Γ_5 with no condition on Γ_1.

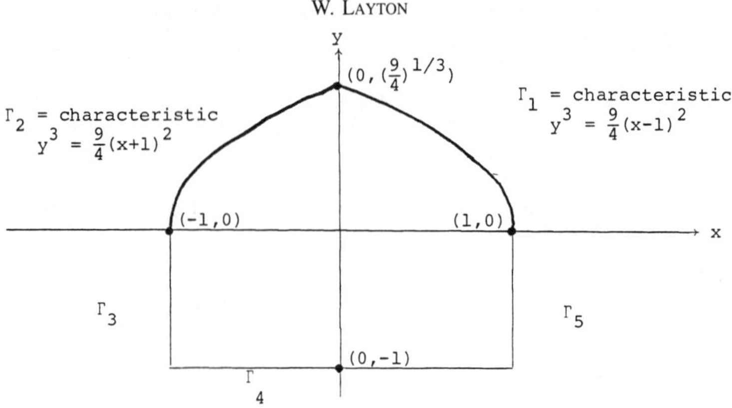

Fig. 1. The domain Ω in the Tricomi problem.

By a further transformation the above become a positive definite symmetric system (see, for example, [1,9,14,29]). These references contain all the details in the transformation and the precise formulation of the boundary conditions.

2. THE VARIATIONAL FORMULATION

In this section, we make the additional simplifying assumptions that $K(x)$ is a real symmetric matrice. Let δ denote a positive parameter, and define

$$\mathscr{B}_\delta(u, v) = \int \mathscr{A}u[v + \delta\mathscr{A}^\circ v]\, dx.$$

Let $\mathscr{H}^1$ denote the space of functions satisfying the boundary conditions of the original problem

$$\overset{\circ}{\mathscr{H}}{}^1 = \{v(x) \in (H^1(\Omega))^m | (B - M)v|_\Gamma = 0\}.$$

Equation (1.4) then has the variational formulation $u \in \overset{\circ}{\mathscr{H}}{}^1(\Omega)$ and satisfies:

$$\mathscr{B}_\delta(u, v) = (f, v + \delta\mathscr{A}^\circ v), \quad \forall v \in \overset{\circ}{\mathscr{H}}{}^1.$$

Let $\mathscr{S}^h$ denote a finite-element space, i.e., a finite-dimensional subspace of $\overset{\circ}{\mathscr{H}}{}^1$ consisting of typically continuous PW polynomials defined on some "triangulation" of Ω and satisfying the boundary conditions of (1.4). The modified Galerkin–FE approximation $U \in \mathscr{S}^h$ to u is calculated via the equations

$$\mathscr{B}_\delta(U, v) = (f, v + \delta\mathscr{A}^\circ v), \quad \forall v \in \mathscr{S}^h. \tag{2.1}$$

$\mathscr{S}^h$ will be assumed to satisfy the usual type of inverse estimates, typical of finite-element spaces used in practice.

Define the norm on $\mathscr{H}^1$ as

$$|||u|||^2 = \|u\|^2_{L^2(\Omega)} + \delta\|\mathscr{A}^\circ u\|^2_{L^2(\Omega)} + \int_\Gamma \langle M_s u, u \rangle\, d\Gamma,$$

where $M_s = \frac{1}{2}(M + M^*) \geq 0$. The basic error estimate follows closely the 1-D case (Layton[20]) when (1.4) is a *positive* symmetric system. When the system is not positive the argument is slightly more involved.

PROPOSITION 2.1
Assume $\mathscr{A}$ is positive, and $0 < \delta < d_0 \max|K|^{-2}$. Then, $\mathscr{B}_\delta$ is coercive in $|||\cdot|||$ on $\overset{\circ}{\mathscr{H}}{}^1$

uniformly in δ: for all $u \in \overset{\circ}{\mathscr{H}}{}^1$, with $C_1 = \min\{\tfrac{1}{2}, d_0/4\}$,

$$|\mathscr{B}_\delta(u, u)| \geq C_1 |||u|||^2.$$

Remark. If K is diagonal, the condition on δ is not needed.
Proof. By definition of $\mathscr{B}_\delta$:

$$\begin{aligned}
\mathscr{B}_\delta(u, u) &= (\mathscr{A}u, u) + \delta(\mathscr{A}^\circ u, \mathscr{A}^\circ u) + \delta(Ku, \mathscr{A}^\circ u) \\
&= \frac{1}{2}(Du, u) + \frac{1}{2}\int_\Gamma \langle Bu, u\rangle \, \mathrm{d}\Gamma + \delta\|\mathscr{A}^\circ u\|_{L^2(\Omega)} \\
&\quad + \delta(Ku, \mathscr{A}^\circ u).
\end{aligned}$$

Since $u \in \overset{\circ}{\mathscr{H}}{}^1$, $(B - M)u = 0$ on Γ. Thus, the above becomes

$$\begin{aligned}
|\mathscr{B}_\delta(U, u)| &\geq \frac{d_0}{2}\|u\|^2 + \delta\|\mathscr{A}^\circ u\|^2 + \frac{1}{2}\int_\Gamma \langle M_s u, u\rangle \, \mathrm{d}\Gamma \\
&\quad + \delta(Ku, \mathscr{A}^\circ u) \geq \left[\frac{d_0}{2} - \frac{\delta}{2}\max|K|^2\right]\|u\|^2 \\
&\quad + \frac{\delta}{2}\|\mathscr{A}^\circ u\|^2 + \frac{1}{2}\int_\Gamma \langle M_s u, u\rangle \, \mathrm{d}\Gamma.
\end{aligned}$$

Provided $\delta < d_0 \max|K|^{-2}$, the above is

$$\geq \frac{d_0}{4}\|u\|^2 + \frac{\delta}{2}\|\mathscr{A}^\circ u\|^2 + \frac{1}{2}\int_\Gamma \langle M_s u, u\rangle. \tag{2.2}$$

∎

Remark. The previous coercivity estimate also holds on the complex analog of $\mathscr{H}^1$. This fact will become important in Sec. 4.

If $\mathscr{A}$ is not positive then it is easy to see that $\mathscr{B}_\delta$ will still satisfy a type of Gårding inequality.

PROPOSITION 2.2

If $\mathscr{A}$ is a (possibly nonpositive) symmetric system, $0 < \delta < d_0 \max|K|^{-2}$, $\mathscr{B}_\delta$ satisfies

$$|\mathscr{B}_\delta(u, u)| \geq C_1 |||u|||^2 - \lambda\|u\|^2, \quad u \in \mathscr{H}^1,$$

where $C_1, \lambda > 0$ are independent of δ.

3. CONVERGENCE OF THE NUMERICAL METHOD

Throughout this section we assume that $\mathscr{S}^h$ satisfies the standard inverse and approximation assumptions typical of finite-element spaces used in practice. Specifically, for all $\phi \in \mathscr{S}^h \subset \mathscr{H}^1$,

$$h\|\phi\|_1 \leq C_{\mathrm{inv}}\|\phi\|,$$

and for all $u \in \overset{\circ}{\mathscr{H}}{}^1(\Omega) \cap H^s(\Omega)$,

$$\inf_{\chi \in \mathscr{S}^h} \{\|u - \chi\| + h\|u - \chi\|_1\} \leq Ch^s\|u\|_s, \quad 1 \leq s \leq k + 1.$$

First, we consider the case where the system is of positive type.

570 W. LAYTON

THEOREM 3.1

Assume that $\mathscr{A}$ is of positive type, and $\delta = O(h)$. Then

$$|||u - U||| \le C(\inf_{\chi \in \mathscr{S}^h} |||u - \chi||| + \delta^{-1/2}\|u - \chi\|)$$

holds with C ind. of δ. Thus

$$\|\mathscr{A}^{\circ}(u - U)\| \le Ch^k\|u\|_{k+1}, \quad \|u - U\| \le Ch^{k+1/2}\|u\|_{k+1}.$$

Proof. $U \in \mathscr{S}^h$ satisfies

$$\mathscr{B}_{\delta}(U - u, v) = 0, \quad \forall v \in \mathscr{S}^h.$$

Let $w \in \mathscr{S}^h$ be arbitrary, and set $\eta = u - w$, $\phi = U - w$. Then

$$\mathscr{B}_{\delta}(\phi, v) = \mathscr{B}_{\delta}(\eta, v), \quad \forall v \in \mathscr{S}^h.$$

Setting $v = \phi$ and using coercivity implies

$$\begin{aligned}
C|||\phi|||^2 &\le |\mathscr{B}_{\delta}(\phi, \phi)| = |\mathscr{B}_{\delta}(\eta, \phi)| \\
&\le |(\mathscr{A}^{\circ}\eta + K\eta, \phi + \delta\mathscr{A}^{\circ}\phi)| \\
&\le |(\mathscr{A}^{\circ}\eta, \phi) + (K\eta, \phi) + \delta(\mathscr{A}^{\circ}\eta, \mathscr{A}^{\circ}\phi) + \delta(K\eta, \mathscr{A}^{\circ}\phi)| \\
&\le C(\|\eta\|^2 + \|\phi\|^2) + \frac{\delta}{2\epsilon}\|\mathscr{A}^{\circ}\eta\|^2 + \frac{\delta\epsilon}{2}\|\mathscr{A}^{\circ}\phi\|^2 \\
&\quad + |(\mathscr{A}^{\circ}\eta, \phi)|.
\end{aligned}$$

Picking ϵ sufficiently small (independently of δ) gives

$$C|||\phi|||^2 \le C|||\eta|||^2 + |(\mathscr{A}^{\circ}\eta, \phi)|. \tag{3.1}$$

Consider now the last term. Since $\eta \in \mathscr{H}^1$, $\|\mathscr{A}^{\circ *}\phi\| \le C(\|\mathscr{A}^{\circ}\phi\| + \|\phi\|)$, and $B\eta = M\eta$ on Γ:

$$|(\mathscr{A}^{\circ}\eta, \phi)| = |(\eta, \mathscr{A}^{\circ *}\phi) + \int_{\Gamma} \langle B\eta, \phi \rangle \, d\Gamma| \le C_1\delta^{-1/2}\|\eta\|\delta^{1/2}(\|\mathscr{A}^{\circ}\phi\| + \|\phi\|)$$

$$+ C_2\left(\int_{\Gamma} \langle M_s\eta, \eta \rangle \, d\Gamma\right)^{1/2}\left(\int_{\Gamma} \langle M_s\phi, \phi \rangle \, d\Gamma\right)^{1/2}.$$

By the inverse estimate (since $\phi \in \mathscr{S}^h$)

$$\begin{aligned}
|(\mathscr{A}^{\circ}\eta, \phi)| &\le C\frac{\delta^{-1}}{2\epsilon}\|\eta\|^2 + C\frac{\delta\epsilon}{2}\|\phi\|_1^2 \\
&\quad + \frac{C}{2\epsilon}\|\eta\|_{L^2(\Gamma)}^2 + \frac{C\epsilon}{2}\|\phi\|_{L^2(\Gamma)}^2 \\
&\le C\frac{\delta^{-1}}{2\epsilon}\|\eta\|^2 + C(\delta h^{-1})\frac{\epsilon}{2}\|\phi\|^2 \\
&\quad + \frac{C}{2\epsilon}\left(\int_{\Gamma} \langle M_s\eta, \eta \rangle \, d\Gamma\right) + C\frac{\epsilon}{2}\left(\int_{\Gamma} \langle M_s\phi, \phi \rangle \, d\Gamma\right). \tag{3.2}
\end{aligned}$$

Using (3.2) in (3.1) gives, that if $\delta h^{-1} = O(1)$,

$$C|||\phi|||^2 \le C|||\eta|||^2 + C\delta^{-1}\|\eta\|^2.$$

In particular, we obtain, by choosing, for example, $w = \bar{u}$ = interpolant of u in $\mathscr{S}^h$, a quasioptimal estimate for $\|\mathscr{A}^\circ e\|$ and an almost optimal estimate for $\|e\|$:

$$\|\mathscr{A}^\circ(u - U)\| \le Ch^k\|u\|_{k+1}, \quad \|u - U\| \le Ch^{K+1/2}\|u\|_{k+1}. \quad\blacksquare$$

Remark. This implies optimality of the error in W_2^1 for systems of equations in one space dimension (see [20]) and optimality of the error in the "convected derivative" for scalar convection equations in two space dimensions (see [24]). Axelsson has also pointed out that no restriction on δ is needed for the estimate of $\|\mathscr{A}(u - U)\|$, and $\delta = O(h)$ is required to obtain the L^2 error estimate.

It is possible that $\|\mathscr{A}^\circ u\|$ is not equivalent to the H^1 norm. The simplest example where this occurs is

$$\mathscr{A}^\circ u \equiv \frac{\partial}{\partial y}\, u(x, y),$$

where $\|\mathscr{A}^\circ u\|$ gives no information on $\|\partial u/\partial y\|$.

Next we consider the case of a general symmetric system.

THEOREM 3.2

Suppose that the hypotheses of Proposition 2.2 holds, $\delta = O(h)$ and that if e satisfies $\mathscr{B}_\delta(e, \chi) = 0$ for all $\chi \in \mathscr{S}^h$,

$$\|e\| \le Ch(\|e\| + \|\mathscr{A}^\circ e\|). \tag{3.3}$$

Then there is an $h_0 > 0$ such that U exists uniquely for $h \le h_0$ and the error in the method satisfies

$$\||u - U\|| \le C \inf_{\chi \in \mathscr{S}^h} (\||u - \chi\|| + C\delta^{-1/2}\|u - \chi\|),$$

where C is independent of δ. Thus

$$\|u - U\| \le Ch^{k+1/2}\|u\|_{k+1}, \quad \|\mathscr{A}^\circ(u - U)\| \le Ch^k\|u\|_{k+1}.$$

Proof. For the moment, assume that U exists and is unique. As in the previous proof, let $\phi = U - \bar{u},\ \eta = u - \bar{u},\ e = u - U$. Proposition 2.2 then implies that

$$\||\phi\||^2 \le \frac{1}{C_1}\,|\mathscr{B}_\delta(\eta, \phi)| + \frac{\lambda}{C_2}\,\|\phi\|^2.$$

Note that $\|\phi\| \le \|e\| + \|\eta\|$. Next we bound above the $\mathscr{B}_\delta$ term precisely as in the previous proof. Thus,

$$\||e\||^2 \le 2\||\phi\||^2 + \||\eta\||^2 \le C\||\eta\||^2 + C\delta^{-1}\|\eta\|^2 + C\|e\|^2 + \|\eta\|^2.$$

Using (3.3) gives

$$\|e\|^2 \le Ch^2(\|e\|^2 + \|\mathscr{A}^\circ e\|^2) \le Ch\||e\||^2.$$

Thus, for h sufficiently small,

$$\||e\||^2 \le C\||\eta\||^2 + C\delta^{-1}\|\eta\|^2,$$

and the error estimate is proven.

Next, U is shown to exist. Since $\dim(\mathscr{S}^h) < \infty$, existence is implied by uniqueness. To prove uniqueness, let $B_\delta(U, \chi) = 0$ for all $\chi \in \mathscr{S}^h$. In particular, for $\chi = U$ we obtain

$$C_1\||U\||^2 \le |B_\delta(U, U)| + \lambda\|U\|^2 = \lambda\|U\|^2.$$

Using (3.3), $\|U\|^2 \le Ch^2(\|\mathscr{A}^\circ U\|^2 + \|U\|^2) \le Ch\|\|U\|\|^2$. Thus, there is an h_0 such that for $h \le h_0$ U exists and is unique. ∎

Assumption (3.3) will not hold for all symmetric systems. We will now consider (3.3) more carefully.

Following [20], let $\mathscr{K}_\delta \chi \equiv \chi + \delta \mathscr{A}^\circ \chi$, and consider the following problem adjoint to (2.1):

$$\mathscr{A}^* \mathscr{K}_\delta \psi = e, \quad \psi \in \mathscr{H}^1, \quad \mathscr{K}_\delta \psi \in \mathscr{H}^1. \tag{3.4}$$

In some cases, some regularity properties of the solution to (3.4) can be easily proven.

LEMMA 3.3

Suppose $\mathscr{A}$ is a positive symmetric operator and $A_j(x)$ are C^1; then, for δ sufficiently small, the solution to (3.4) satisfies

$$\|\psi\| + \|\mathscr{A}^\circ \psi\| + \delta\|\mathscr{A}^{\circ 2}\psi\| \le C\|e\|,$$

where C is independent of δ.

Proof. Let $\Phi = \mathscr{K}_\delta \psi$; then

$$(\Phi, \mathscr{A}\Phi) = (e, \Phi).$$

Since $\mathscr{A}$ is positive definite we obtain $\|\Phi\| \le C\|e\|$. The equation $\mathscr{A}^*\Phi = e$ then gives

$$\|\mathscr{A}^\circ \Phi\| + \|\Phi\| \le C\|e\|,$$

where the constant now depends upon the C^1 norm of $A_j(x)$. By a similar argument (since $\mathscr{K}_\delta$ is a positive symmetric operator for δ sufficiently small), we get that

$$\|\psi\| + \delta\|\mathscr{A}^\circ \psi\| \le C\|\mathscr{K}_\delta \psi\| = C\|\Phi\|.$$

Thus, $\|\psi\| + \|\mathscr{A}^\circ \psi\| \le C\|e\|$. Now consider $\mathscr{A}^\circ(\Phi)$; we have by the definition of Φ

$$([\mathscr{A}^\circ \psi] + \delta\mathscr{A}^\circ[\mathscr{A}^\circ \psi], \mathscr{A}^\circ \psi) = (\mathscr{A}^\circ \Phi, \mathscr{A}^\circ \psi).$$

For δ sufficiently small $I + \delta\mathscr{A}^\circ$ is a positive system. Thus, as before,

$$\|\mathscr{A}^\circ \psi\| + \delta\|\mathscr{A}^{\circ 2}\psi\| \le C\|\mathscr{A}^\circ \Phi\|.$$

Hence, the theorem follows. ∎

To verify (3.3) we will require a slightly stronger regularity assumption (RH) below. Another consequence of RH is that the $O(h^{k+1/2})$ estimate in the two previous theorems can be improved to $O(h^{k+1})$. The further regularity hypothesis we make to achieve this is:

$$\text{suppose that } \psi \text{ satisfies } \|\psi\|_1 \le C\|e\|. \tag{RH}$$

Remark. In the absence of (RH), optimal orders of convergence in L^2 cannot be expected for Galerkin-type methods. For example, Aziz and Leventhal[3] show that the L^2 error in the usual least-squares methods is $O(h^k)$ when a hypothesis akin to (RH) does not hold. (RH) is typical of elliptic problems and some nonelliptic problems[30]. Examples include Maxwell's equations and acoustic equations[30]. Note also that (RH) always holds in 1-D by Lemma 3.3.

PROPOSITION 3.4

In addition to the assumptions of Theorem 3.2, suppose (RH) holds. Then,

$$\|u - U\| \le C(h + \delta)(\|u - U\| + \|\mathscr{A}^\circ(u - U)\|).$$

Proof. Integration by parts reveals that ψ satisfies

$$\mathscr{B}_\delta(\phi, \psi) = (e, \phi), \quad \forall \phi \in \overset{\circ}{\mathscr{H}}{}^1. \tag{3.4}$$

Set $\phi = e = u - U$. Then,

$$\|e\|^2 \le B_\delta(e, \psi) = B_\delta(e, \psi - \chi), \quad \forall \chi \in \mathscr{S}^h,$$

$$\le C(\|e\| + \|\mathscr{A}^\circ e\|)(\|\psi - \chi\| + \delta\|\mathscr{A}^\circ(\psi - \chi)\|)$$

$$\le C(\|e\| + \|\mathscr{A}^\circ e\|) \inf_{\chi \in \mathscr{S}^h} \{\|\psi - \chi\| + \delta\|\mathscr{A}^\circ(\psi - \chi)\|\}$$

$$\le C(\|e\| + \|\mathscr{A}^\circ e\|)(h + \delta)\|\psi\|_1.$$

Thus, by (RH),

$$\|e\|^2 \le C(h + \delta)(\|e\| + \|\mathscr{A}^\circ e\|)\|e\|. \qquad \blacksquare$$

An interpolation argument then gives the intermediate rates of convergence.

COROLLARY 3.5

Suppose that $u \in (H^r(\Omega))^m \cap \overset{\circ}{\mathscr{H}}{}^1$, $1 \le r \le k + 1$, and the hypotheses of Theorem 3.2 and (RH) hold. Then, for $j = 0, 1$,

$$\|u - U\|_j \le Ch^{r-j}\|u\|_r, \quad 1 \le r \le k + 1. \tag{3.5}$$

Proof. By the previous proposition and Theorem 3.2, (3.5) holds with $r = k + 1$. Equation (3.5) holds with $r = 1$, $j = 0$, by Theorem 3.2, Proposition 3.1, and the approximation properties of $\mathscr{S}^h$. The result for $1 < r < k + 1$ follows by interpolation. $\qquad \blacksquare$

Remark. In the absence of (RH), one other case where optimal L^2 error estimates are possible is when a uniform tensor product mesh and tensor product basis functions are used. If each component of the basis functions are C^0, odd degree piecewise polynomials or odd degree smoothest splines, a cancellation argument can be employed as for the usual Galerkin–FE method (see Layton[31] for this case) to show that the error is optimal in L^2, $O(h^{k+1})$, provided $u \in H^{k+2}(\Omega) \cap \mathscr{H}^1(\Omega)$.

4. REMARKS ON CONDITIONING OF THE DISCRETE EQUATIONS AND IMPLEMENTATION OF THE METHOD

Since the method (1.2) can be interpreted as a nonstandard least-squares method, there is a legitimate concern that the condition number of the derived system might be significantly worse than (and perhaps even the square of) the condition number of the system arising from the usual finite-element method applied to original equations.

However, it is likely that this is not the case. We prove that the spectral condition number of the discrete system is $O(h^{-1})$ when $\delta = O(h)$.

The difficulty in estimating the (nonspectral) condition number is that the matrix $A = \mathscr{B}$ (ϕ_i, ϕ_j) is not symmetric, so it will have complex eigenvectors and eigenvalues.

Let $\{\phi_j\}$ denote a basis for $\mathscr{S}^h$, and let $\mathscr{S}^h(\mathbb{C})$ denote the complex valued analog of $\mathscr{S}^h$.

$$\phi \in \mathscr{S}^h(\mathbb{C}) \Longleftrightarrow \phi = \sum \alpha_j \phi_j, \quad \alpha_j \in \mathbb{C}.$$

$\mathscr{S}^h(\mathbb{C})$ is assumed to satisfy the inverse assumption and the following standard assumption:

$$\phi = \sum \alpha_j \phi_j \in \mathscr{S}^h(\mathbb{C}); \quad \text{then} \quad Ch^{-n}\|\phi\|^2 \le |\alpha| \le Ch^{-n}\|\phi\|^2. \tag{A1}$$

(A1) holds for most finite-element spaces on a locally quasiuniform mesh.

Proposition 4.1

In addition to the hypotheses of Theorem 3.1 assume that (A1), and the inverse hypothesis for $\mathcal{S}^h(C)$, holds $\delta = O(h)$, and $M = M^* > 0$. Then

$$|\lambda_{max}(A)| / |\lambda_{min}(A)| \le Ch^{-1}.$$

Proof. By the Remark 2.1 and Proposition 2.1,

$$C|||\phi|||^2 \le |\mathscr{B}_\delta(\phi, \phi)| \le C|||\phi|||^2$$

holds for all $\phi \in \mathcal{H}^1(C)$. (The upper bound on $|B_\delta|$ follows by an argument entirely analogous to the proof of Proposition 2.1). Let α denote the (complex) eigenvector of A corresponding to the eigenvalue λ. Then

$$\lambda = \frac{\alpha^T A \bar\alpha}{|\alpha|^2} = \frac{\mathscr{B}_\delta(\phi, \phi)}{|\alpha|^2}, \quad \phi = \sum \alpha_j \phi_j.$$

Let $\lambda = \lambda_{max}(A)$. Then

$$|\lambda_{max}| \le \frac{C|||\phi|||^2}{|\alpha|^2} \le \frac{Ch^{-1}\|\phi\|^2}{|\alpha|^2} \le Ch^n h^{-1}.$$

When $\lambda = \lambda_{min}(A)$,

$$|\lambda_{min}| \ge \frac{C|||\phi|||^2}{|\alpha|^2} \ge \frac{C\|\phi\|^2}{|\alpha|^2} \ge Ch^n,$$

so that the result follows.   ∎

Remark. Analogous results can be shown for the time-dependent problem. For example, if the time variable is discretized by the Crank–Nicholson (1-1 Padé) method and the spacial variables are discretized by the method analyzed herein, the spectral condition number of the linear system that must be inverted at each time step is $O(\Delta t/h)$.

Restrictions on meshwidth

In practice, δ should be chosen so that the condition on δ in Proposition 2.1 is automatically satisfied. Suppose Ω is divided into elements T_j, $\Omega = \cup_j T_j$. For example, we can choose for $x \in T_j$,

$$\delta(x) = \max\left\{ \mathrm{diam}(T_j), \frac{d_0}{4} |K(x_j)|^{-1} \right\}, \quad (x_j \in T_j).$$

Boundary conditions

There are two natural methods of imposing the boundary conditions of the problem. In the method proposed here the boundary conditions are strongly imposed on $\mathcal{S}^h$ by simply adding them in as an extra set of linear equations to the discrete systems. This method is perhaps the simplest one.

Another method, proposed in Johnson, Nävert and Pitkäranta[21], is to impose the boundary conditions weakly by calculating U via

$$\mathscr{B}_\delta(U, v) - \frac{1}{2} \int_\Gamma \langle (B - M)U, v \rangle \, d\Gamma = (f, v + \delta\mathcal{A}^\circ v), \quad \forall v \in \mathcal{S}^h.$$

For solving boundary-control problems for Friedrichs' systems, this method is likely easier to implement.

Numerical integration

In practice, the integrals involved in the methods (2.1) must be discretized by an appropriate quadrature scheme. Thus, Eq. (2.1) is replaced by a perturbed set of equations

$$\tilde{\mathscr{B}}_\delta(\tilde{U}, v) = \tilde{F}_\delta(v), \quad \forall v \in \mathscr{S}^h, \tag{4.1}$$

where $\tilde{\mathscr{B}}_\delta$ is the quadrature approximation to $\mathscr{B}_\delta$ and $\tilde{F}_\delta(v)$ approximates $F_\delta(v) \equiv (f, v + \delta\mathscr{A}^\circ v)$.

If the quadrature scheme is chosen appropriately $\tilde{\mathscr{B}}_\delta$ will remain coercive in $\|\|\cdot\|\|$:

$$|\tilde{\mathscr{B}}_\delta(w, w)| \geq C_2\|\|w\|\|^2, \tag{4.2}$$

where $C_2 > 0$ is independent of δ. By adapting an analysis of Strang, one can show the following.

THEOREM 4.2

Let $\tilde{U}$ be defined by (4.1), and assume that (4.2) and the assumptions of Theorem 3.1 hold. Then

$$\|\|u - \tilde{U}\|\| \leq C \inf_{\chi \in \mathscr{S}^h} \left\{ \|\|u - \chi\|\| + \delta^{-1/2}\|u - \chi\| \right.$$

$$\left. + \sup_{w \in \mathscr{S}^h} \left[\frac{|\mathscr{B}_\delta(\chi, w) - \tilde{\mathscr{B}}_\delta(\chi, w)|}{\|\|w\|\|} + \frac{|\tilde{F}_\delta(w) - (f, w + \mathscr{A}^\circ w)|}{\|\|w\|\|} \right] \right\}. \tag{4.3}$$

Proof. Since (4.2) holds,

$$C_2\|\|\tilde{U} - \chi\|\|^2 \leq |\tilde{\mathscr{B}}_\delta(\tilde{U} - \chi, \tilde{U} - \chi)|$$

$$= |\mathscr{B}_\delta(u - \chi, U - \chi) + [\mathscr{B}_\delta - \tilde{\mathscr{B}}_\delta](\chi, U - \chi) + [\tilde{F}_\delta - F_\delta](U - \chi)|.$$

Following the proof of Theorem 3.1, with $\phi = \tilde{U} - \chi$, $\eta = u - \chi$ gives, for any $\epsilon > 0$,

$$|\mathscr{B}_\delta(\eta, \phi)| \leq \epsilon\|\|\phi\|\|^2 + C(\epsilon)\|\|\eta\|\|^2 + C\delta^{-1}\|\eta\|^2.$$

Thus

$$(1 - \epsilon)\|\|\phi\|\|^2 \leq C\|\|\eta\|\|^2 + C\delta^{-1}\|\eta\|^2 + C\frac{|\mathscr{B}_\delta(\chi, \phi) - \tilde{\mathscr{B}}_\delta(\chi, \phi)|^2}{\|\|\phi\|\|^2} + C\frac{|\tilde{F}_\delta(\phi) - F_\delta(\phi)|^2}{\|\|\phi\|\|^2}.$$

The result now follows from the triangle inequality $\|\|u - \tilde{U}\|\| \leq \|\|\phi\|\| + \|\|\eta\|\|$. ∎

Thus, if the quadrature scheme is sufficiently accurate the perturbation terms on the right-hand side of (4.3) are of the same order as the error in their basic method. Analogous results can also be shown for the quadrature error in time-dependent Friedrichs systems (considered in the next section), following, for example, Raviart[32], Baker and Dougalis[33], or Layton[34].

5. TIME-DEPENDENT PROBLEMS

Time-dependent problems for symmetric systems can be solved by ''space–time'' finite elements (if sufficient storage is available) by noting that time-dependent Friedrichs' systems are also Friedrichs' systems, so the basic method can be applied directly.

One efficient way of accomplishing this is to use finite elements that are discontinuous in the time variables. In this case the $\mathbf{x}$-t linear system then uncouples into ''time steps.'' For this approach see, for example, Johnson, Nävert and Pitkäranta[22,25], Lesaint[15] and Jamet[35].

In this section, a different method is used: the spacial variables are discretized via the nonstandard least-squares method. Then, the resulting system of ODEs is solved by a time-stepping method with the appropriate stability properties.

Consider the time-dependent problem

$$\frac{\partial u}{\partial t} + \mathcal{A}u = f(x, t, u), \quad x \in \Omega \subset \mathbb{R}^n, \quad 0 < t < T \leq \infty;$$

$$\mathcal{A}u \equiv \sum A_j \frac{\partial u}{\partial x_j} + Ku; \tag{5.1}$$

$$u(x, 0) = u_0(x), \quad x \in \Omega,$$

subject to the boundary conditions and assumptions on $\mathcal{A}$ given in Sec. 1.1. $f(\cdot, \cdot, u)$ is a semilinear term satisfying the Caratheodory conditions.

First, note that when $\mathcal{A}$ is a positive symmetric system we obtain an *a priori* bound on the solution of (5.1).

PROPOSITION 5.1

Assume $\mathcal{A}$ is a positive symmetric system and f satisfies the following monotonicity condition: for all $u, v \in L^2(\Omega)$,

$$(f(\cdot, t, u) - f(\cdot, t, v), u - v) \leq \alpha(t)\|u - v\|^2. \tag{5.2}$$

Then, with d_0 given in Definition 1.1 and

$$\beta(r) = \int_0^r \frac{1}{2}(\alpha(w) - d_0)dw, \quad \|u(t)\| \leq \|u_0\| + \int_0^t e^{\beta(s)}\|f(\cdot, s, 0)\| \, ds. \tag{5.3}$$

In particular, if $e^{\beta(s)} \in L^1(0, \infty)$ (as occurs if, for example, $\alpha(w) < d_0$), the equation is stable and well posed in $L^\infty(0, \infty; L^2)$:

$$\|u(t)\| \leq \|u_0\| + C \sup_{0 \leq t < \infty} \|f(\cdot, t, 0)\|. \tag{5.4}$$

Proof. Multiplying (5.1) by u, integrating over Ω, using the positivity of $\mathcal{A}$ and the monotonicity condition (5.2) gives

$$\frac{d}{dt}\|u\|^2 \leq (\alpha(t) - d_0)\|u\|^2 + \int_\Gamma \langle Bu, u \rangle \, ds + (f(\cdot, t, 0), u),$$

from which the result follows. ∎

The semidiscrete Galerkin approximation is a strongly differentiable map $U: (0, \infty) \to \mathcal{S}^h$ satisfying

$$U_t \mathcal{A}U - f(\cdot, t, U), v + \delta \mathcal{A}^\circ v) = 0, \quad \forall v \in \mathcal{S}^h,$$

$$U(0) \in \mathcal{S}^h \text{ approximates } u_0 \text{ well}, \quad (B - M)U = 0 \quad \text{on } \partial\Omega. \tag{5.5}$$

To compute with (5.5) the time variable must be discretized by some A-stable method. The trapezoidal method is one commonly used scheme for this. The analysis of the effects of such time discretizations is straightforward (and leads to expected results) so it will be omitted.

Henceforth, we focus attention on the linear case $f(x, t, u) \equiv f(x, t)$.

LEMMA 5.2 (*Stability of the method*)

Assume (5.2) holds, $\delta = O(h)$ is sufficiently small, $\mathcal{A}$ is a positive symmetric operator, $f(x, t, u) \equiv f(x, t)$, and $\mathcal{S}^h$ satisfies the inverse property. Then, there is C, independent of t, δ, and U, and there is a $\gamma > 0$ such that

$$\|U(t)\|^2 \leq e^{-\gamma t}\|U(0)\|^2 + C\gamma^{-1} \sup_{0 < s < t} \|f(\cdot, s)\|^2.$$

Proof. Setting $v = U$ in (5.5) and using the Cauchy–Schwarz–Young inequality gives, for any $\epsilon > 0$,

$$\frac{\mathrm{d}}{\mathrm{d}t}\|U\|^2 - \delta C(\epsilon)\|U_t\|^2 + \delta\epsilon\|\mathscr{A}^{\circ}U\|^2 + 2\mathscr{B}_{\delta}(U, U) \leq \epsilon\|U\|^2 + \epsilon\delta\|\mathscr{A}^{\circ}U\|^2 + C(\epsilon)\|f(\cdot, t)\|^2.$$

Collecting terms,

$$\frac{\mathrm{d}}{\mathrm{d}t}\|U\|^2 + \{2\mathscr{B}_{\delta}(U, U) - \epsilon\|\|u\|\|^2\} \leq C\|f(\cdot, t)\|^2 + C\delta\|U_t\|^2. \tag{5.6}$$

The quantity in braces is positive for ϵ sufficiently small. Thus, we need only bound $\delta\|U_t\|$. To this end, set $v = U_t$ in (5.5). This gives

$$\|U_t\|^2 + \delta(U_t, \mathscr{A}^{\circ}U_t) + \mathscr{B}_{\delta}(U, U_t) = (f(\cdot, t), U_t + \delta\mathscr{A}^{\circ}U_t). \tag{5.7}$$

Since $\mathscr{A}$ is a positive symmetric system,

$$(U_t, \mathscr{A}^{\circ}U_t) = (U_t, (K + D)U_t) + \int_{\Gamma} \langle BU_t, U_t \rangle \, \mathrm{d}\Gamma,$$

so, by using the inverse property of $\mathscr{S}^h$ on the right-hand side,

$$(U_t, (I + \delta(K + D))U_t) + \int_{\Gamma} \langle BU_t, U_t \rangle \leq -\mathscr{B}_{\delta}(U, U_t) + C(\epsilon)\|f\|^2 + \epsilon C_{\mathrm{inv}}\|U_t\|^2.$$

Thus, for ϵ, δ sufficiently small,

$$\|U_t\|^2 \leq C\|f(\cdot, t)\|^2 - C\mathscr{B}_{\delta}(U, U_t). \tag{5.8}$$

Expanding $\mathscr{B}_{\delta}(U, U_t)$ and using similar inequalities: for any $\epsilon_{1,2}$,

$$\delta\mathscr{B}_{\delta}(U, U_t) \leq C(\epsilon_1)\|U\|^2 + \epsilon_1\delta^2\|\mathscr{A}^*U_t\|^2 + C(\epsilon_1)\delta\|U\|^2$$
$$+ \delta\epsilon_1\|U_t\|^2 + \epsilon_2\delta\|\mathscr{A}U\|^2 + C(\epsilon_2)\delta^3\|\mathscr{A}U_t\|^2.$$

Using $\|\mathscr{A}^*w\| \leq C(\|\mathscr{A}w\| + \|w\|)$ in (5.8) and the inverse estimates in the above and substituting it into (5.8) gives

$$\delta\|U_t\|^2 \leq C(\delta\|f(\cdot, t)\|^2 + \delta\|U\|^2) + \epsilon_2\delta\|\mathscr{A}U\|^2.$$

Using this in (5.6) gives, for any $\epsilon > 0$,

$$\frac{\mathrm{d}}{\mathrm{d}t}\|U\|^2 + \{2\mathscr{B}_{\delta}(U, U) - \max\{\epsilon, \delta\}\|\|U\|\|^2\} \leq C(\epsilon)(1 + \delta)\|f(\cdot, t)\|^2. \tag{5.9}$$

The quantity in braces is positive definite in $\|\|\cdot\|\|^2$, and hence in L^2. Thus, for some $\gamma > 0$,

$$\|U(t)\|^2 \leq \mathrm{e}^{-\gamma t}\|U(0)\|^2 + C\gamma^{-1} \sup_{0 < s < t} \|f(\cdot, t)\|^2. \qquad \blacksquare$$

THEOREM 5.3 (*convergence*)

Under the assumptions of Lemma 5.2, there is a C independent of h, u, U and t such that

$$\|u(t) - U(t)\| \leq Ce^{-\gamma t}\|u_0 - U(0)\| + Ch^{k+1/2} \sup_{0 < s < T} \{\|u_t(s)\|_{k+1} + \|u(s)\|_{k+1}\}.$$

If, in addition, the regularity hypothesis (RH) holds for the steady-state equation, the same result holds with "$k + \frac{1}{2}$" replaced by "$k + 1$."

Proof. Let $u - U = (u - w) - (U - w) = \eta - \phi$, where w is projection of u into $\mathscr{S}^h$ wrt $\mathscr{B}_\delta$:

$$\mathscr{B}_\delta(u - w, v) = 0, \quad \forall v \in \mathscr{S}^h.$$

ϕ then satisfies the equation

$$(\phi_t, v - h\mathscr{A}^\circ v) + \mathscr{B}_\delta(\phi, v) = (\eta_t, v - h\mathscr{A}v)$$

for all $v \in \mathscr{S}^h$. Applying the stability result (Lemma 5.2) gives

$$\|\phi(t)\| \leq \|\phi(0)\| + C \sup_{0<s<t} \|\eta_t(s)\|.$$

The triangle inequality then yields

$$\|\phi(0)\| \leq \|(u - U)\| + \|(u - w)(0)\|, \quad \|u - U\| \leq \|\phi\| + \|\eta\|.$$

The estimate for the error in the steady-state problem now completes the proof.   ■

REFERENCES

1. K. O. Friedrichs, Symmetric positive linear differential equations. *C.P.A.M.* **11**, 333–348 (1958).
2. A. K. Aziz, G. Fix and S. Leventhal, Numerical solution of linear mixed problems, in *L. N. in Physics on Computing Methods in the Applied Sciences* (Edited by J. L. Lions), pp. 58–88. Springer-Verlag, New York (1976).
3. A. K. Aziz and S. Leventhal, Finite element approximation for first order systems. *SIAM J. N. A.* **15**, 1103–1111 (1978).
4. G. J. Fix and M. D. Gunzburger, On least squares approximations to indefinite problems of mixed type. NASA-ICASE, Report 76-26 (1976).
5. E. Stephan and W. L. Wendland, Remarks to Galerkin and least squares methods with finite elements for general elliptic problems. *Manuscripta geodaetica* **1**, 93–123 (1976), and *L.N.M.* Vol. 564, pp. 461–471. Springer-Verlag, New York (1976).
6. W. L. Wendland, *Elliptic Systems in the Plane*. Pitman, London (1979).
7. P. A. Raviart and J. M. Thomas, A mixed finite element method for second order elliptic problems. *L.N.M.* Vol. 666. Springer-Verlag, New York (1977).
8. G. J. Fix, M. D. Gunzburger and R. A. Nicolaides, On mixed finite element methods for first order elliptic systems. *Numer. Math.* **37** (1981).
9. G. J. Fix and M. Gurtin, On patched variational principles with applications to elliptic and mixed elliptic–hyperbolic problems. *Numer. Math.* **28**, 259–271 (1977).
10. G. J. Fix, A mixed finite element method for transonic flows. NASA-ICASE, Report 76-25 (1976).
11. P. Lesaint, Finite element methods for the transport equation. *R.A.I.R.O. Serie Math.*, 67–94 (August 1974).
12. A. Jameson, Numerical solution of nonlinear partial differential equations of mixed type, in *Num. Sol. of P.D.E.'s, Vol. III* (Edited by B. Hubbard). Academic Press, New York (1976).
13. T. Katsanis, Numerical solution of symmetric positive differential equations. *Math. Comp.* **22**, 763–783 (1968).
14. T. Katsanis, Numerical solution of Tricomi equation using the theory of symmetric positive differential equations. *SIAM J.N.A.* **6**, 236–253 (1969).
15. P. Lesaint, Continuous and discontinuous finite element methods for solving the transport equation, in *The Math. of F. Elts. and Appls. II* (Edited by J. R. Whiteman), pp. 151–161. Academic Press, New York (1976).
16. P. Lesaint, Finite element methods for symmetric hyperbolic systems. *Numer. Math.* **21**, 244–255 (1973).
17. J. E. Dendy, Two methods of Galerkin type achieving optimal L^2-accuracy for first order hyperbolics. *SIAM J.N.A.* **11**, 637–653 (1974).
18. L. Wahlbin, A dissipative Galerkin method for the numerical solution of first order hyperbolic equations, in *Math. Aspects of F. Elts. in Partial Diff. Eqns.* (Edited by C. deBoor). Academic Press (1974).
19. W. Layton, Estimates away from a discontinuity for dissipative Galerkin methods for hyperbolic equations. *Math. Comp* **36**, 87–92 (1981).
20. W. Layton, Galerkin methods for two-point boundary value problems for first order systems. *SIAM J.N.A.* **20**, 161–171 (1983).
21. T. J. R. Hughes and A. N. Brooks, A multidimensional scheme with no crosswind diffusion, in *F.E.M. for Convection Dominated Flows* (Edited by T. J. R. Hughes), AMD Vol. 34, pp. 19–35. ASME (1979).
22. U. Nävert, A finite element method for convection–diffusion problems. Ph.D. Thesis, Dept. of Computer Science, Chalmers Inst. of Tech., Göteborg, Sweden (1982).
23. G. D. Raithby, Skew upstream differencing schemes for problems involving fluid flow. *Comp. Math. Appl. Mech. Eng.* **9**, 153–164 (1976).

24. O. Axelsson, On the numerical solution of convection dominated, convection–diffusion equations. Report 83-34, Mathematics Inst., Univ. of Nijmegen, the Netherlands (1983).
25. C. Johnson, U. Nävert and J. Pitkaranta, Finite element methods for linear hyperbolic problems (preprint).
26. K. O. Friedrichs, Symmetric hyperbolic linear differential equations. *C.P.A.M.* **7**, 345–392 (1954).
27. K. O. Friedrichs and P. D. Lax, On symmetrizable differential operators, in *Proc. Symp. in Pure Math.*, Vol. 10, pp. 128–137. Am. Math. Soc. (1967).
28. R. A. Adams, *Sobolev Spaces*. Academic Press, New York (1975).
29. A. K. Aziz, On the numerical solution of equations of elliptic hyperbolic type. ICASE Report 74-18, NASA–ICASE (1974).
30. A. Majda, Coercive inequalities for nonelliptic symmetric systems. *C.P.A.M.* **28**, 49–89 (1975).
31. W. Layton, The accuracy of finite element approximnations to semilinear, first order, hyperbolic systems. Report 8308, Feb. 1983, Math. Dept., Catholic Univ., The Netherlands.
32. P. A. Raviart, The use of numerical integration in finite element methods for parabolic equations, in *Topics in Num. Anal. Proc. of Royal Irish Acad. Conf. in Numer. Anal.* (Edited by J. J. H. Miller), pp. 233–264. Academic Press, New York (1973).
33. G. A. Baker and V. A. Dougalis, The effect of quadrature error in finite element approximations for second order hyperbolic equations. *SIAM J.N.A.* **13**, 577–598 (1976).
34. W. Layton, Some effects of numerical integration in finite element approximations to degenerate evolution equations. *Calcolo* **21**, 45–60 (1984).
35. P. Jamet, Galerkin-type approximations which are discontinuous in time for parabolic equations in a variable domain. *SIAM J.N.A.* **15**, 912–928 (1978).
36. A. Brooks, A. Petrov–Galerkin, F.E. formulation for convection dominated flows. Thesis, Cal. Inst. of Tech. (1981).
37. A. G. Deacon and S. Osher, A finite element method for a boundary value problem of mixed type. *SIAM J.N.A.*, **16**, 756–778 (1979).
38. T. J. Hughs, T. E. Tezdugon and A. Brooks, Streamline upwind formulation for advection–diffusion, Navier–Stokes and first order hyperbolic equations. Presented at 4th Int. Conf. on F.E.M. in Fluids, Tokyo, Japan, July 1982.
39. P. Sermer and R. Mathon, Least squares methods for mixed type equations. *SIAM J.N.A.* **18**, 705–723 (1981).

Comp. & Maths. with Appls. Vol. 12A, Nos. 4/5, pp. 581–604, 1986
Printed in Great Britain.

0886–9553/86 $3.00 + .00
© 1986 Pergamon Press Ltd.

HIGHER-ORDER SINGLE-STEP FULLY DISCRETE APPROXIMATIONS FOR NONLINEAR SECOND-ORDER HYPERBOLIC EQUATIONS

LAURENCE A. BALES
Department of Mathematics, University of Tennessee, Knoxville, TN 37996-1300, U.S.A.

Abstract—Error estimates are proved for finite-element approximations to the solution of an initial boundary value problem for nonlinear second-order hyperbolic partial differential equations. Optimal order rates of convergence in L^2 are shown for single-step fully discrete schemes which are linearized by extrapolations. The order of accuracy v in time is 2, 3 or 4. A class of iterative methods for approximately solving the fully discrete equations is analyzed and optimal order rates of convergence are also obtained.

1. INTRODUCTION

This paper concerns approximations to the solution $u = u(x, t)$ of the following initial boundary value problem:

$$
\begin{aligned}
u_{tt} &= -L(t, u)u + f(t, u) \\
&\equiv \sum_{j,k=1}^{N} \frac{\partial}{\partial x_j}\left(a_{jk}(x, t, u) \frac{\partial u}{\partial x_k} \right) - a_0(x, t, u)u + f(x, t, u) \quad \text{in } \Omega \times [0, \tau] \\
u &= 0 \quad \text{in } \partial\Omega \times [0, \tau], \\
u(0) &= u^0, \; u_t(0) = u_t^0 \quad \text{in } \Omega.
\end{aligned}
\tag{1.1}
$$

Here Ω is a bounded domain in $\mathbb{R}^N$, $1 \le N \le 3$, with sufficiently smooth boundary $\partial\Omega$ and the functions u^0 and u_t^0 are given in Ω. It will be assumed that (1.1) has a unique solution $u(x, t) \in [m_1, m_2]$ for all $(x, t) \in \bar{\Omega} \times [0, \tau]$. Assumptions on the required smoothness of the solution will be made when needed. The coefficients and right-hand side in (1.1) will be assumed to satisfy a Lipshitz continuity condition on the interval $M_\delta = [m_1 - \delta, m_2 + \delta]$ for some $\delta > 0$; i.e., for some positive constant L,

$$
\begin{aligned}
|a_{jk}(x, t, v) - a_{jk}(x, t, w)| &\le L|v - w|, \\
|a_0(x, t, v) - a_0(x, t, w)| &\le L|v - w|, \\
|f(x, t, v) - f(x, t, w)| &\le L|v - w|,
\end{aligned}
\tag{1.2}
$$

for all $(x, t) \in \bar{\Omega} \times [0, \tau]$ and $v, w \in M_\delta$. In addition it will be assumed that $\{a_{jk}(x, t, v)\}_{j,k=1}^{N}$ is a family of symmetric, uniformly positive definite matrices with sufficiently smooth coefficients on $\bar{\Omega} \times [0, \tau] \times M_\delta$, that $a_0(x, t, v) \ge 0$ and that $a_0(x, t, v)$ and $f(x, t, v)$ are sufficiently smooth on $\bar{\Omega} \times [0, \tau] \times M_\delta$.

For $s \ge 0$, $1 \le p \le \infty$, $W^{s,p}(\Omega) = W^{s,p}$ will denote the usual Sobolev space (with norm $\|\cdot\|_{s,p}$) of real-valued functions on Ω. The norm on $W^{s,2} = H^s = H^s(\Omega)$ is denoted by $\|\cdot\|_s$. The inner product on $L^2(\Omega) = L^2 = H^0$ is denoted by $(\cdot, \cdot)$ and the associated norm by $\|\cdot\|$. H_0^1 denotes the subspace of functions in H^1 that vanish (in the sense of trace) on $\partial\Omega$. $\|\cdot\|_{L^\infty}$ denotes the norm on $L^\infty = L^\infty(\Omega)$. If $(X, \|\cdot\|_X)$ is a Banach space and $1 \le p \le \infty$, $L^p(0, t; X)$ will be the space of strongly measurable functions $v: (0, t) \to X$ such that

$$
\|v\|_{L^p(0,t;X)} = \left(\int_0^t \|v(s)\|_X^p \, ds \right)^{1/p} < \infty,
$$

with the usual modification if $p = \infty$.

Acknowledgment—The author would like to thank Professor Vassilios A. Dougalis for many improvements in the manuscript.

Define $Y = \{g \in W^{1,\infty}$ such that $g(x) \in M_\delta$ for all $x \in \Omega\}$. For $0 \le t \le \tau$ and $g \in Y$, $L(t, g)$ is a self-adjoint elliptic operator on L^2 with domain $D_L = H^2 \cap H_0^1$. Also, for $g \in Y$ the bilinear form $a(t, g)(\cdot, \cdot)$, defined by

$$a(t, g)(\phi, \psi) = \int_\Omega \left(\sum_{j,k=1}^{N} a_{jk}(t, g) \frac{\partial \phi}{\partial x_j} \frac{\partial \psi}{\partial x_k} + a_0(t, g)\phi\psi \right) dx \tag{1.3}$$

for $\phi, \psi \in H^1$, is coercive over $H_0^1 \times H_0^1$. For $f \in L^2$ and $g \in Y$, $T(t, g)f \equiv w \in D_L$ is defined as the solution of the problem

$$L(t, g)w = f \quad \text{in } \Omega, \quad w = 0 \quad \text{on } \partial\Omega. \tag{1.4}$$

With $L(t) \equiv L(t, u(t))$ and $T(t) \equiv T(t, u(t))$ for $t \in [0, \tau]$, $L(t)$ and $T(t)$ are a smooth family of bounded operators (for $l \ge 0$) from $H^{l+2} \cap D_L$ to H^l and from H^l to $H^{l+2} \cap D_L$, respectively.

In Sec. 2 a semidiscrete approximation u_h to the solution u of (1.1) is defined. In Sec. 3 the semidiscrete equation and a class of single-step methods (which have order $2 \le \nu \le 4$) for approximately solving stiff ordinary differential equations are used to define a fully discrete approximation to u. Extrapolations are used to linearize the fully discrete equations at each time step. In Sec. 4, an optimal-order error estimate is proved for the fully discrete approximation and, in Sec. 5, the same type of estimate is proved for schemes which use an iterative method to approximately solve the fully discrete equations at each time step.

This work is a generalization of the work in [1] for linear problems with time-dependent coefficients to the nonlinear problem (1.1). Techniques similar to the ones in this work were used by Douglas, Dupont and Ewing[2] and Bramble and Sammon[3], where approximation schemes for nonlinear parabolic equations were analyzed. In [2] Douglas, Dupont and Ewing analyzed schemes which are second order in time and in [3] Bramble and Sammon analyzed schemes which are third and fourth order in time. Also, Ewing[4] has considered schemes which are second order in time for approximately solving nonlinear second-order hyperbolic equations.

Semidiscrete approximations for linear parabolic equations with time-independent coefficients were analyzed by Bramble, Schatz, Thomée and Whalbin[5] and fully discrete approximations for linear parabolic problems with time-independent coefficients by Baker, Bramble and Thomée[6]. Fully discrete schemes (up to fourth order in time) for linear parabolic problems with time-dependent coefficients were analyzed by Bramble and Sammon[7]. Baker and Bramble[8] analyzed semidiscrete and single-step fully discrete approximations for second-order hyperbolic equations. An example of multistep Galerkin discretizations of parabolic and hyperbolic equations is the work of Baker, Dougalis, and Karakashian[9].

Throughout this work C (sometimes used with a subscript) is used to denote a general positive constant which is not necessarily the same in any two places.

2. SEMIDISCRETE APPROXIMATIONS

Let $0 < h < 1$ be a parameter, and $\{S_h\}_{0<h<1}$ a family of finite dimensional subspaces of $W^{1,\infty}$. A corresponding family of operators $T_h(t)$ which approximates $T(t)$ and has the following properties is assumed to be given.

(i) T_h is self-adjoint, positive semidefinite on L^2, and positive definite on S_h.

(ii) There is an integer $r \ge 2$, such that for integer $j \ge 0$ there exists constants $C(u, j)$ and $C(u)$ with

$$\text{(a)} \quad \|(T^{(j)} - T_h^{(j)})f\| \le C(u, j)h^s\|f\|_{s-2}, \tag{2.1}$$

$$\text{(b)} \quad \|(T - T_h)w\|_{L^\infty} \le C(u)h^r|\log h|^{\bar{r}}\|Tw\|_{r,\infty}$$

for all $f \in H^{s-2}$, $Tw \in W^{r,\infty}$, $2 \le s \le r$, where $T^{(j)}$ and $T_h^{(j)}$ denote the jth time derivative of T and T_h, respectively, and $\bar{r} = 0$ if $r > 2$ or $0 < \bar{r} < \infty$ if $r = 2$.

(iii) On S_h define $L_h(t) = (T_h(t))^{-1}$ and $L_h^{(k)}(t) = \mathrm{d}^k/\mathrm{d}t^k\, L_h(t)$. For integer $k \geq 0$, there exists a constant $C = C(u, k)$, which is independent of h, such that for $t, s \in [0, \tau]$,

$$|(L_h^{(k)}(t)\phi, \phi)| \leq C(u, k)(L_h(s)\phi, \phi) \quad \text{for all } \phi \in S_h. \tag{2.2}$$

(iv) The following inverse properties are satisfied:

$$\text{(a)} \quad (L_h\phi, \phi) \leq Ch^{-2}(\phi, \phi), \tag{2.3}$$

$$\text{(b)} \quad \|\phi\|_{L^\infty} \leq \quad Ch^{-N/2}\|\phi\|, \quad \text{for all } \phi \in S_h.$$

(v) For each $t \in [0, \tau]$ and $g \in Y$, there exists a symmetric bilinear form $a_h(t, g)(\cdot, \cdot)$ on $W^{1,\infty} \times W^{1,\infty}$ that is positive definite on S_h and an operator $L_h(t, g)\colon S_h \to S_h$ such that the following identities and estimates are satisfied:

$$L_h(t, u) = L_h(t), \tag{2.4}$$

$$a_h(t, g)(\phi, \psi) = (L_h(t, g)\phi, \psi), \tag{2.5}$$

$$\text{(a)} \quad ((L_h(t, g) - L_h(t))\psi, \phi) \leq C\|g - u(t)\|_{L^\infty}\|L_h^{1/2}(t)\psi\|\,\|L_h^{1/2}(t)\phi\|, \tag{2.6}$$

$$\text{(b)} \quad ((L_h(t, g) - L_h(t))\psi, \phi) \leq C\|g - u(t)\|\,\|\psi\|_{W^{1,\infty}}\|L_H^{1/2}(t)\phi\| \quad \text{for all } \phi, \psi \in S_h.$$

An example of a family of operators satisfying the above assumptions is given by the following. Suppose $S_h \subset H_0^1$ so that the elements of S_h vanish on $\partial\Omega$. Assume further that S_h has the approximation property

$$\inf_{\chi \in S_h}\{\|w - \chi\| + h\|w - \chi\|_1\} \leq Ch^s\|w\|_s \quad \text{for } 2 \leq s \leq r.$$

The operators $T_h\colon L^2 \to S_h$ are defined by $a(t, u)(T_h f, \chi) = (f, \chi)$ for all $\chi \in S_h$, and the bilinear forms $a_h(t, g)(\cdot, \cdot)$ are the same as $a(t, g)(\cdot, \cdot)$ in (1.3). Verification of the estimates (2.6a, b) follows from the Lipshitz conditions on the coefficients in (1.2). For more details and other examples see Baker and Bramble[8], Bales[10], Bramble and Sammon[7], Bramble, Schatz, Thomée and Wahlbin[5], Huang and Thomée[11] and Sammon[12].

From conditions (i), (ii) and (iii) above it follows that there exists a constant C which is independent of h such that for $j \geq 0$ and $s, t \in [0, \tau]$,

$$\|T_h^{1/2}(s)L_h^{(j)}(t)T_h^{1/2}(s)\phi\| \leq C\|\phi\| \tag{2.7}$$

and

$$\|L_h^{1/2}(s)T_h^{(j)}(t)L_h^{1/2}(s)\phi\| \leq C\|\phi\| \tag{2.8}$$

for all $\phi \in S_h$. These estimates are proved in Bales[10]. Also, from (i), (ii), (iii) and (iv) [(2.3a)] it follows that, for $j \geq 0$ and $s, t \in [0, \tau]$, there exists a constant C which is independent of h such that

$$\|L_h^{(j)}(t)T_h(s)\phi\| \leq C\|\phi\| \tag{2.9}$$

and

$$\|T_h(s)L_h^{(j)}(t)\phi\| \leq C\|\phi\| \tag{2.10}$$

for all $\phi \in S_h$. Proofs of the estimates (2.9) and (2.10) can be found in Bales[10] and Sammon[12].

$w(t) = T_h(t)L(t)u(t)$ will denote the "elliptic projection" of the solution $u(t)$ of (1.1) into S_h and $P\colon L^2 \to S_h$ will denote the orthogonal L^2-projection onto S_h. In Bramble and Sammon[7], the following estimate is proved. For $m \geq 0$, $2 \leq s \leq r$ and $t \in [0, \tau]$, there exists a constant

$C = C(u, m)$ such that

$$\|u^{(m)}(t) - w^{(m)}(t)\| \le C(u, m)h^s \sum_{j=0}^{m} \|u^{(j)}(t)\|_s. \tag{2.11}$$

Also, it follows from estimates in Bramble and Sammon[7] [using (2.9)] that

$$\|L_h(t)w^{(m)}(t)\| \le C(u, m) \sum_{j=2}^{m+2} \|u^{(j)}(t)\|. \tag{2.12}$$

From (2.1b) it follows that

$$\|u(t) - w(t)\|_{L^\infty} \le C(u)h^r |\log h|^{\bar{r}} \|u(t)\|_{r,\infty}. \tag{2.13}$$

In this work it will be assumed that for $t \in [0, \tau]$, $j = 0, 1$,

$$\|w^{(j)}(t)\|_{W^{1,\infty}} \le C. \tag{2.14}$$

The hyperbolic problem (1.1) can be written

$$Tu_{tt} + u = Tf, \quad u(0) = u^0, \quad u_t(0) = u_t^0.$$

A semidiscrete approximation for the solution u is defined as the mapping $u_h: [0, \tau] \to S_h$ satisfying the linear problem

$$T_h(u_h)_{tt} + u_h = T_h f \quad 0 \le t \le \tau,$$

$$u_h(0) = v^0, \tag{2.15}$$

$$(u_h)_t(0) = v_t^0,$$

where v^0 and v_t^0 are given elements of S_h. Equation (2.15) will be used to derive the fully discrete scheme in the next section. Note that if the coefficients or right-hand side in (1.1) depend on the solution u then the approximate solution operator T_h or the function f in (2.15) also depends on u.

3. FULLY DISCRETE APPROXIMATIONS

The interval $[0, \tau - k]$ is divided into M equal subintervals of length k and $t_n \equiv nk$ for $n = 0, 1, \ldots, M$. The methods are based on the following single-step two-derivative formula (see Lambert[13]):

$$y_{n+1} - y_n = k(-q_1 y'_{n+1} + p_1 y'_n) + k^2(-q_2 y''_{n+1} + p_2 y''_n). \tag{3.1}$$

For a smooth function $y(t)$, y_m, y'_m, and y''_m approximate $y(t_m)$,

$$\frac{dy}{dt}(t_m) \quad \text{and} \quad \frac{d^2y}{dt^2}(t_m) \quad \text{for } m = 0, 1, \ldots, M.$$

The constants p_1, p_2, q_1, and q_2 determine the properties of a given method.

The function $\mathscr{A}[y]$ defined by

$$\mathscr{A}[y] = y(t + k) - y(t) + k(q_1 y'(t + k) - p_1 y'(t)) + k^2(q_2 y''(t + k) - p_2 y''(t)) \tag{3.2}$$

is associated with (3.1). $\mathscr{A}[y]$ is the truncation error of the single-step method and is used to define the order of the method.

Definition 3.1

A method given by (3.1) is of order $v > 0$, if $\mathscr{A}[t^j] = 0$ for $j = 0, 1, \ldots, v$ and $\mathscr{A}[t^{v+1}] \neq 0$.

Definition 3.2

The stability region R associated with a method given by (3.1) is defined as $R = \{k\lambda$: where λ is any complex number and k any positive number such that when the method is applied to $y' = \lambda y$ with $y(t_0) = y_0$ given and with constant step size k, the sequence $\{y_n\}_{n=1}^M$ satisfies $|y_n| \leq |y_0|\}$.

For the methods given by (3.1), $1 \leq v \leq 4$. It will be assumed in this work that $2 \leq v \leq 4$ and that the stability region R contains the imaginary axis. Examples of methods satisfying the above assumptions are given in Bales[1] and Bramble and Sammon[7].

If the single-step method (3.1) has order v and $y(t)$ is a function with $v + 1$ time derivatives then it follows by Taylor expansions that

$$
\begin{aligned}
\mathscr{A}[y] = & \int_t^{t+k} \frac{(t + k - s)^v}{v!}\, y^{(v+1)}(s)\, ds \\
& + q_1 k \int_t^{t+k} \frac{(t + k - s)^{v-1}}{(v - 1)!}\, y^{(v+1)}(s)\, ds \\
& + q_2 k^2 \int_t^{t+k} \frac{(t + k - s)^{v-2}}{(v - 2)!}\, y^{(v+1)}(s)\, ds.
\end{aligned}
\tag{3.3}
$$

In order to apply the single-step formula (3.1) to descritize the semidiscrete Eq. (2.15) in the time variable, (2.15) will be written as a first-order system. With $\mathscr{U}_h \equiv \binom{u_h}{(u_h)_t}$ and $\mathscr{L}_h \equiv \binom{0 \quad I}{-L_h \quad 0}$, (2.15) can be written

$$
\begin{aligned}
(\mathscr{U}_h)_t &= \mathscr{L}_h \mathscr{U}_h + \binom{0}{Pf}, \\
\mathscr{U}_h(0) &= \binom{v^0}{v_t^0}.
\end{aligned}
\tag{3.4}
$$

From (3.4) it follows that

$$
(\mathscr{U}_h)_{tt} = (\mathscr{L}_h^2 + \mathscr{L}_h^{(1)})\mathscr{U}_h + \mathscr{L}_h \binom{0}{Pf} + \binom{0}{Pf^{(1)}} = (\mathscr{L}_h^2 + \mathscr{L}_h^{(1)})\mathscr{U}_h + \binom{Pf}{Pf^{(1)}}.
\tag{3.5}
$$

Equations (3.1), (3.4) and (3.5) can be used to derive the fully discrete approximation. This was done in Bales[10] for the linear homogeneous ($f = 0$) Eq. (1.1). For the nonlinear equation treated here the terms $\mathscr{L}_h^{(1)}$ and $Pf^{(1)}$ in (3.5) will be replaced by difference quotients and extrapolation will be used to linearize the fully discrete equation which is derived from (3.1), (3.4) and (3.5). For $0 \leq m \leq M$ and $\hat{g} \in Y$ the following notation will be used throughout the remainder of this work:

$$
q(x) \equiv 1 + q_1 x + q_2 x^2, \quad p(x) \equiv 1 + p_1 x + p_2 x^2;
$$

$$
L_m \equiv L_h(t_m), \quad L_m^{(1)} \equiv L_h^{(1)}(t_m), \quad L_m(\hat{g}) \equiv L_h(t_m, \hat{g});
$$

$$
\mathscr{L}_m \equiv \mathscr{L}_h(t_m), \quad \mathscr{L}_m^{(1)} \equiv \mathscr{L}_h^{(1)}(t_m), \quad \mathscr{L}_m(\hat{g}) \equiv \mathscr{L}_h(t_m, \hat{g});
$$

$$
Q_m \equiv q(k\mathscr{L}_m), \quad P_m \equiv p(k\mathscr{L}_m);
$$

$$
\tilde{Q}_m = Q_m + q_2 k^2 \mathscr{L}_m^{(1)}, \quad \tilde{P}_m = P_m + p_2 k^2 \mathscr{L}_m^{(1)};
$$

$$
f_m = Pf(t_m, u(t_m)), \quad f_m(\hat{g}) = Pf(t_m, \hat{g});
$$

and

$$f_m^{(1)} = P f^{(1)}(t_m, u(t_m)).$$

Given a sequence $U = \{U^j\}_{j=0}^n \subset (S_h \cap Y) \times S_h$ and approximations

$$\hat{U}^{n+1} = \begin{pmatrix} \hat{U}_1^{n+1} \\ \hat{U}_2^{n+1} \end{pmatrix} \in (S_h \cap Y) \times S_h \quad \text{and} \quad \hat{U}^{n+2} = \begin{pmatrix} \hat{U}_1^{n+2} \\ \hat{U}_2^{n+2} \end{pmatrix} \in (S_h \cap Y) \times S_h$$

to

$$\begin{pmatrix} u(t_{n+1}) \\ u_t(t_{n+1}) \end{pmatrix} \quad \text{and} \quad \begin{pmatrix} u(t_{n+2}) \\ u_t(t_{n+2}) \end{pmatrix},$$

respectively, define

$$A_{n+1} = q(k\mathcal{L}_{n+1}(\hat{U}_1^{n+1})) + q_2 k^2 \frac{\mathcal{L}_{n+2}(\hat{U}_1^{n+2}) - \mathcal{L}_n(U_1^n)}{2k},$$

$$B_n = p(k\mathcal{L}_n(U_1^n)) + p_2 k^2 \frac{\mathcal{L}_{n+1}(\hat{U}_1^{n+1}) - \mathcal{L}_{n-1}(U_1^{n-1})}{2k} \quad \text{for } n \geq 1, \tag{3.6}$$

and

$$B_0 = p(k\mathcal{L}_0(U_1^0)) + p_2 k^2 \frac{-\mathcal{L}_2(\hat{U}_1^2) + 4\mathcal{L}_1(\hat{U}_1^1) - 3\mathcal{L}_0(U_1^0)}{2k}.$$

Using the above notation and motivated by (3.1), (3.4) and (3.5), the fully discrete equations are

$$A_1 U^1 = B_0 U^0 + p_1 k \begin{pmatrix} 0 \\ f_0 \end{pmatrix} + p_2 k^2 \begin{pmatrix} f_0 \\ \dfrac{-f_2(\hat{U}_1^2) + 4f_1(\hat{U}_1^1) - 3f_0}{2k} \end{pmatrix}$$

$$- q_1 k \begin{pmatrix} 0 \\ f_1(\hat{U}_1^1) \end{pmatrix} - q_2 k^2 \begin{pmatrix} f_1(\hat{U}_1^1) \\ \dfrac{f_2(\hat{U}_1^2) - f_0}{2k} \end{pmatrix},$$

and for $n \geq 1,$ \hfill (3.7)

$$A_{n+1} U^{n+1} = B_n U^n + p_1 k \begin{pmatrix} 0 \\ f_n(U_1^n) \end{pmatrix} + p_2 k^2 \begin{pmatrix} f_n(U_1^n) \\ \dfrac{f_{n+1}(\hat{U}_1^{n+1}) - f_{n-1}(U_1^{n-1})}{2k} \end{pmatrix}$$

$$- q_1 k \begin{pmatrix} 0 \\ f_{n+1}(\hat{U}_1^{n+1}) \end{pmatrix} - q_2 k^2 \begin{pmatrix} f_{n+1}(\hat{U}_1^{n+1}) \\ \dfrac{f_{n+2}(\hat{U}_1^{n+2}) - f_n(U_1^n)}{2k} \end{pmatrix}.$$

The approximations $\hat{U}^{n+1}$ and $\hat{U}^{n+2}$ are provided by extrapolations of previous time levels. These extrapolations are described in the following algorithm. Here $P : L^2 \to S_h$ denotes the L^2 orthogonal projection onto S_h.

ALGORITHM 3.3

(1) Set

$$U^0 = \begin{pmatrix} w(0) \\ w^{(1)}(0) \end{pmatrix} \quad \text{or} \quad U^0 = (\tilde{Q}_0)^{-1}\left[\begin{pmatrix} P & 0 \\ 0 & P \end{pmatrix}\begin{pmatrix} u^0 \\ u_t^0 \end{pmatrix} + q_1 k \begin{pmatrix} u_t^0 \\ -L(0)u^0 \end{pmatrix} \right.$$

$$\left. + q_2 k^2 \begin{pmatrix} -L(0)u^0 \\ -\left(\dfrac{\mathrm{d}}{\mathrm{d}t}Lu\right)(0) \end{pmatrix} \right].$$

(2) Set

$$\hat{U}_1^1 = \begin{cases} U_1^0, & \nu = 2, \\ U_1^0 + kPu_t(0), & \nu = 3, \\ U_1^0 + kPu_t(0) + \dfrac{k^2}{2}Pu_{tt}(0), & \nu = 4, \end{cases}$$

and

$$\hat{U}_1^2 = \begin{cases} U_1^0, & \nu = 2, \\ U_1^0 + 2kPu_t(0), & \nu = 3, \\ U_1^0 + 2kPu_t(0) + 2k^2 Pu_{tt}(0), & \nu = 4, \end{cases}$$

in (3.6). Let U^1 be the corresponding solution to (3.7). If $\nu = 2$, go to step (5).

(3) Set

$$\hat{U}_1^2 = \begin{cases} 2U_1^1 - U_1^0, & \nu = 3, \\ 4U_1^1 - 3U_1^0 - 2kPu_t(0), & \nu = 4, \end{cases}$$

and

$$\hat{U}_1^3 = \begin{cases} 3U_1^1 - 2U_1^0, & \nu = 3, \\ 9U_1^1 - 8U_1^0 - 6kPu_t(0), & \nu = 4, \end{cases}$$

in (3.6). Let U^2 be the corresponding solution to (3.7). If $\nu = 3$, go to step (5).

(4) Set $\hat{U}_1^3 = 3U_1^2 - 3U_1^1 + U_1^0$ and $\hat{U}_1^4 = 6U_1^2 - 8U_1^1 + 3U_1^0$ in (3.6). Let U^3 be the corresponding solution to (3.7).

(5) For $\nu \leq n + 1 \leq M$, set

$$\hat{U}_1^{n+1} = \begin{cases} 2U_1^n - U_1^{n-1}, & \nu = 2, \\ 3U_1^n - 3U_1^{n-1} + U_1^{n-2}, & \nu = 3, \\ 4U_1^n - 6U_1^{n-1} + 4U_1^{n-2} - U_1^{n-3}, & \nu = 4, \end{cases}$$

and

$$\hat{U}_1^{n+2} = \begin{cases} 3U_1^n - 2U_1^{n-1}, & \nu = 2, \\ 6U_1^n - 8U_1^{n-1} + 3U_1^{n-2}, & \nu = 3, \\ 10U_1^n - 20U_1^{n-1} + 15U_1^{n-2} - 4U_1^{n-3}, & \nu = 4, \end{cases}$$

in (3.6). Let U^{n+1} be the corresponding solution to (3.7).

4. ERROR ESTIMATES

This section contains estimates for the difference between the fully discrete approximation U^{n+1} defined in Algorithm 3.3 and the function

$$W(t_{n+1}) = \begin{pmatrix} w(t_{n+1}) \\ w^{(1)}(t_{n+1}) \end{pmatrix},$$

where $w(t_{n+1}) = T_h(t_{n+1})L(t_{n+1})u(t_{n+1})$ is the elliptic projection of $u(t_{n+1})$ into S_h. The analysis involves the inner products on $S_h \times S_h$ denoted by

$$((\phi, \psi))_n = (\phi_1, \bar{\psi}_1) + (T_h(t_n)\phi_2, \bar{\psi}_2), \tag{4.1}$$

where

$$\Phi \equiv \begin{pmatrix} \phi_1 \\ \phi_2 \end{pmatrix} \quad \text{and} \quad \Psi = \begin{pmatrix} \psi_1 \\ \psi_2 \end{pmatrix}$$

can be complex-valued functions and $\bar{\psi}_1$ and $\bar{\psi}_2$ denote the complex conjugates of ψ_1 and ψ_2, respectively. The corresponding norm is denoted by

$$|||\Phi|||_n = ((\Phi, \Phi))_n^{1/2}, \tag{4.2}$$

for $n = 0, 1, \ldots, M$. From (2.8) it follows that the norms $|||\cdot|||_m$ and $|||\cdot|||_n$ are equivalent for any integers m and n between 0 and M.

From estimates in Bales[10], it follows that there exists a constant $C > 0$ which is independent of h and k such that for all $\Phi \in S_h \times S_h$

$$C|||\Phi|||_n \leq |||Q_n\Phi|||_n, \tag{4.3}$$

$$C|||k\mathcal{L}_n\Phi|||_n \leq |||Q_n\Phi|||_n, \tag{4.4}$$

and, if $q_2 \neq 0$,

$$C|||k^2\mathcal{L}_n^2\Phi|||_n \leq |||Q_n\Phi|||_n. \tag{4.5}$$

Also, if k is sufficiently small there exist positive constants C_1 and C_2 such that

$$C_1|||Q_n\Phi|||_n \leq |||\tilde{Q}_n\Phi|||_n \leq C_2|||Q_n\Phi|||_n \tag{4.6}$$

for all $\Phi \in S_h \times S_h$.

The error $E^n = U^n - W(t_n)$, for $n = 0, 1, \ldots, M$, will be estimated using the error equations

$$
\begin{aligned}
Q_1 E^1 = {} & \tilde{P}_0 E^0 + (Q_1 - \tilde{Q}_1)E^1 \\
& + \left[-\tilde{Q}_1 W(t_1) + \tilde{P}_0 W(t_0) + p_1 k \begin{pmatrix} 0 \\ f_0 \end{pmatrix} + p_2 k^2 \begin{pmatrix} f_0 \\ f_0^{(1)} \end{pmatrix} - q_1 k \begin{pmatrix} 0 \\ f_1 \end{pmatrix} - q_2 k^2 \begin{pmatrix} f_1 \\ f_1^{(1)} \end{pmatrix} \right] \\
& + p_2 k^2 \begin{pmatrix} 0 \\ \dfrac{-f_2(\hat{U}_1^2) + 4f_1(\hat{U}_1^1) - 3f_0}{2k} - f_0^{(1)} \end{pmatrix} \\
& - q_1 k \begin{pmatrix} 0 \\ f_1(\hat{U}_1^1) - f_1 \end{pmatrix} - q_2 k^2 \begin{pmatrix} f_1(\hat{U}_1^1) - f_1 \\ \dfrac{f_2(\hat{U}_1^2) - f_0}{2k} - f_1^{(1)} \end{pmatrix} \\
& + (\tilde{Q}_1 - A_1)E^1 + (B_0 - \tilde{P}_0)E^0 \\
& + (\tilde{Q}_1 - A_1)W(t_1) + (B_0 - \tilde{P}_0)W(t_0)
\end{aligned}
$$

and, for $n \geq 1$,   (4.7)

$$
\begin{aligned}
Q_{n+1}E^{n+1} = {} & \tilde{P}_n E^n + (Q_{n+1} - \tilde{Q}_{n+1})E^{n+1} \\
& + \left[-\tilde{Q}_{n+1}W(t_{n+1}) + \tilde{P}_n W(t_n) + p_1 k \binom{0}{f_n} + p_2 k^2 \binom{f_n}{f_n^{(1)}} \right. \\
& \left. - q_1 k \binom{0}{f_{n+1}} - q_2 k^2 \binom{f_{n+1}}{f_{n+1}^{(1)}} \right] \\
& + p_1 k \binom{0}{f_n(U_1^n) - f_n} + p_2 k^2 \left(\frac{f_n(U_1^n) - f_n}{\dfrac{f_{n+1}(\hat{U}_1^{n+1}) - f_{n-1}(U_1^{n-1})}{2k} - f_n^{(1)}} \right) \\
& - q_1 k \binom{0}{f_{n+1}(\hat{U}_1^{n+1}) - f_{n+1}} - q_2 k^2 \left(\frac{f_{n+1}(\hat{U}_1^{n+1}) - f_{n+1}}{\dfrac{f_{n+2}(\hat{U}_1^{n+2}) - f_n(U_1^n)}{2k} - f_{n+1}^{(1)}} \right) \\
& + (\tilde{Q}_{n+1} - A_{n+1})E^{n+1} + (B_n - \tilde{P}_n)E^n \\
& + (\tilde{Q}_{n+1} - A_{n+1})W(t_{n+1}) + (B_n - \tilde{P}_n)W(t_n),
\end{aligned}
$$

which are derived from (3.7). The following technical lemma will be used to prove estimates for the last four terms in (4.7).

LEMMA 4.1

If $g \in Y$ and $\psi \in S_h$, then there is a constant C which is independent of h, k, g, and ψ such that

$$
\|T_{n+1}^{1/2}(L_{n+1}(g) - L_{n+1})\psi\| \leq
\begin{cases}
C\|g - u(t_{n+1})\|_{L^\infty}\|L_{n+1}^{1/2}\psi\| \\
C\|g - u(t_{n+1})\| \, \|\psi\|_{W^{1,\infty}}
\end{cases}
\tag{4.8}
$$

and

$$
\|(L_{n+1}(g) - L_{n+1})\psi\| \leq
\begin{cases}
\dfrac{C}{h} \, \|g - u(t_{n+1})\|_{L^\infty}\|L_{n+1}^{1/2}\psi\|, \\[2ex]
\dfrac{C}{h} \, \|g - u(t_{n+1})\| \, \|\psi\|_{W^{1,\infty}}.
\end{cases}
\tag{4.9}
$$

Proof. From (2.6) it follows that

$$
(T_{n+1}^{1/2}(L_{n+1}(g) - L_{n+1})\psi, \, \phi) \leq
\begin{cases}
C\|g - u(t_{n+1})\|_{L^\infty}\|L_{n+1}^{1/2}\psi\| \, \|L_{n+1}^{1/2}(T_{n+1}^{1/2}\phi)\|, \\
C\|g - u(t_{n+1})\| \, \|\psi\|_{W^{1,\infty}}\|L_{n+1}^{1/2}(T_{n+1}^{1/2}\phi)\|.
\end{cases}
$$

Therefore,

$$
\frac{(T_{n+1}^{1/2}(L_{n+1}(g) - L_{n+1})\psi, \, \phi)}{\|\phi\|} \leq
\begin{cases}
C\|g - u(t_{n+1})\|_{L^\infty}\|L_{n+1}^{1/2}\psi\|, \\
C\|g - u(t_{n+1})\| \, \|\psi\|_{W^{1,\infty}}.
\end{cases}
$$

Equation (4.8) follows from these estimates. Similar estimates are used with (2.3a) to prove (4.9). ∎

The techniques used to prove the estimates in the next lemma will be used to estimate several of the terms in (4.7). It follows from this lemma that if h, k, $\|\hat{U}_1^{n+2} - u(t_{n+2})\|_{L^\infty}$, $\|\hat{U}_1^{n+1} - u(t_{n+1})\|_{L^\infty}$ and $\|U_1^n - u(t_n)\|_{L^\infty}$ are sufficiently small and $k \leq C^*h$ for any constant C^*, then A_{n+1} is invertible so that U^{n+1} is well defined in (3.7).

590   L. A. Bales

LEMMA 4.2

For all $\Phi \in S_h \times S_h$, there is a constant C which is independent of h and k such that for $\hat{U}_1^{n+1}$, $\hat{U}_1^{n+2}$ and $U_1^n \in Y$,

$$\||(A_{n+1} - Q_{n+1})\Phi\||_{n+1} \le C\left(\frac{k}{h}\left(\|\hat{U}_1^{n+1} - u(t_{n+1})\|_{L^\infty} + \|\hat{U}_1^{n+2} - u(t_{n+2})\|_{L^\infty}\right.\right.$$

$$\left.\left. + \|U_1^n - u(t_n)\|_{L^\infty}\right) + k\right)\||Q_{n+1}\Phi\||_{n+1}. \quad (4.10)$$

Proof. Note that

$$A_{n+1} - Q_{n+1} = q_1 k\begin{pmatrix} 0 & 0 \\ -L_{n+1}(\hat{U}_1^{n+1}) + L_{n+1} & 0 \end{pmatrix}$$

$$+ q_2 k^2\begin{pmatrix} -L_{n+1}(\hat{U}_1^{n+1}) + L_{n+1} & 0 \\ 0 & -L_{n+1}(\hat{U}_1^{n+1}) + L_{n+1} \end{pmatrix}$$

$$+ q_2 k^2\begin{pmatrix} 0 & 0 \\ -\dfrac{L_{n+2}(\hat{U}_1^{n+2}) - L_n(U_1^n)}{2k} & 0 \end{pmatrix}.$$

It follows that ($\Phi \equiv \left(\begin{smallmatrix}\phi_1 \\ \phi_2\end{smallmatrix}\right)$)

$$\||(A_{n+1} - \tilde{Q}_{n+1})\Phi\||_{n+1} \le |q_1|k\|T_{n+1}^{1/2}(L_{n+1} - L_{n+1}(\hat{U}_1^{n+1}))\phi_1\|$$

$$+ |q_2|k^2(\|(L_{n+1} - L_{n+1}(\hat{U}_1^{n+1}))\phi_1\| + \|T_{n+1}^{1/2}(L_{n+1} - L_{n+1}(\hat{U}_1^{n+1}))\phi_2\|)$$

$$+ |q_2|k^2\left\|T_{n+1}^{1/2}\left(\frac{L_{n+2}(\hat{U}_1^{n+1}) - L_n(U_1^n)}{2k}\right)\phi_1\right\|. \quad (4.11)$$

From (4.8) it follows that

$$\|T_{n+1}^{1/2}(L_{n+1} - L_{n+1}(\hat{U}_1^{n+1}))\phi_1\| \le C\|u(t_{n+1}) - \hat{U}_1^{n+1}\|_{L^\infty}\|L_{n+1}^{1/2}\phi_1\|$$

and

$$\|T_{n+1}^{1/2}(L_{n+1} - L_{n+1}(\hat{U}_1^{n+1}))\phi_2\| \le C\|u(t_{n+1}) - \hat{U}_1^{n+1}\|_{L^\infty}\|L_{n+1}^{1/2}\phi_2\|.$$

Also, from (4.9),

$$\|(L_{n+1} - L_{n+1}(\hat{U}_1^{n+1}))\phi_1\| \le \frac{C}{h}\|u(t_{n+1}) - \hat{U}_1^{n+1}\|_{L^\infty}\|L_{n+1}^{1/2}\phi_1\|.$$

Since

$$\frac{L_{n+2}(\hat{U}_1^{n+2}) - L_n(U_1^n)}{2k} = \frac{L_{n+2}(\hat{U}_1^{n+2}) - L_n(U_1^n)}{2k} - \frac{L_{n+2} - L_n}{2k} + \frac{L_{n+2} - L_n}{2k},$$

it follows from (4.8) and (2.7) that

$$|q_2|k^2\left\|T_{n+1}^{1/2}\left(\frac{L_{n+2}(\hat{U}_1^{n+2}) - L_n(U_1^n)}{2k}\right)\phi_1\right\| \le Ck(\|\hat{U}_1^{n+2} - u(t_{n+2})\|_{L^\infty}$$

$$+ \|U_1^n - u(t_n)\|_{L^\infty})\|L_{n+1}^{1/2}\phi_1\| + Ck^2\|L_{n+1}^{1/2}\phi_1\|.$$

Equation (4.10) follows from the above four inequalities, (4.11), (2.3a), and (4.3) and (4.4). ∎

The next six lemmas give bounds for terms in the error equation (4.7).

LEMMA 4.3

There is a constant C which is independent of h and k such that

$$|||\tilde{P}_n E^n|||_{n+1} \le (1 + Ck)|||Q_n E^n|||_n \tag{4.12}$$

and

$$|||(Q_{n+1} - \tilde{Q}_{n+1})E^{n+1}|||_{n+1} \le Ck|||Q_{n+1}E^{n+1}|||_{n+1}. \tag{4.13}$$

Proof. These estimates follow from estimates given in Bales[10]. ∎

LEMMA 4.4

For $n = 0, 1, \ldots, M - 1$, there is a constant C which is independent of h and k such that

$$\left|\left|\left| -\tilde{Q}_{n+1}W(t_{n+1}) + \tilde{P}_n W(t_n) + p_1 k \begin{pmatrix} 0 \\ f_n \end{pmatrix} + p_2 k^2 \begin{pmatrix} f_n \\ f_n^{(1)} \end{pmatrix} \right. \right. \right.$$
$$\left. \left. \left. - q_1 k \begin{pmatrix} 0 \\ f_{n+1} \end{pmatrix} - q_2 k^2 \begin{pmatrix} f_{n+1} \\ f_{n+1}^{(1)} \end{pmatrix} \right|\right|\right|_{n+1} \le Ck(h^r + k^\nu). \tag{4.14}$$

Proof. Since

$$\tilde{Q}_{n+1}W(t_{n+1}) = \begin{pmatrix} w(t_{n+1}) \\ w^{(1)}(t_{n+1}) \end{pmatrix} + q_1 k \begin{pmatrix} w^{(1)}(t_{n+1}) \\ -PL(t_{n+1})u(t_{n+1}) \end{pmatrix} + q_2 k^2 \begin{pmatrix} -PL(t_{n+1})u(t_{n+1}) \\ -P\dfrac{d}{dt}(Lu)(t_{n+1}) \end{pmatrix}$$

and

$$\tilde{P}_n W(t_n) = \begin{pmatrix} w(t_n) \\ w^{(1)}(t_n) \end{pmatrix} + p_1 k \begin{pmatrix} w^{(1)}(t_n) \\ -PL(t_n)u(t_n) \end{pmatrix} + p_2 k^2 \begin{pmatrix} -PL(t_n)u(t_n) \\ -P\dfrac{d}{dt}(Lu)(t_n) \end{pmatrix},$$

it follows that

$$-\tilde{Q}_{n+1}W(t_{n+1}) + \tilde{P}_n W(t_n) + p_1 k \begin{pmatrix} 0 \\ f_n \end{pmatrix} + p_2 k^2 \begin{pmatrix} f_n \\ f_n^{(1)} \end{pmatrix} - q_1 k \begin{pmatrix} 0 \\ f_{n+1} \end{pmatrix} - q_2 k^2 \begin{pmatrix} f_{n-1} \\ f_{n+1}^{(1)} \end{pmatrix}$$
$$= -\begin{pmatrix} w(t_{n+1}) \\ w^{(1)}(t_{n+1}) \end{pmatrix} + \begin{pmatrix} w(t_n) \\ w^{(1)}(t_n) \end{pmatrix} - q_1 k \begin{pmatrix} w^{(1)}(t_{n+1}) \\ Pu_{tt}(t_{n+1}) \end{pmatrix} - q_2 k^2 \begin{pmatrix} Pu_{tt}(t_{n+1}) \\ Pu_{ttt}(t_{n+1}) \end{pmatrix}$$
$$+ p_1 k \begin{pmatrix} w^{(1)}(t_n) \\ Pu_{tt}(t_n) \end{pmatrix} + p_2 k^2 \begin{pmatrix} Pu_{tt}(t_n) \\ Pu_{ttt}(t_n) \end{pmatrix},$$

where P is the L^2 orthogonal projection onto S_h. It follows from estimates in Bales[1] and (3.3) that

$$\left|\left|\left| -\tilde{Q}_{n+1}W(t_{n+1}) + \tilde{P}_n W(t_n) + p_1 k \begin{pmatrix} 0 \\ f_n \end{pmatrix} + p_2 k^2 \begin{pmatrix} f_n \\ f_n^{(1)} \end{pmatrix} - q_1 k \begin{pmatrix} 0 \\ f_{n+1} \end{pmatrix} \right. \right. \right.$$
$$\left. \left. \left. - q_2 k^2 \begin{pmatrix} f_{n+1} \\ f_{n+1}^{(1)} \end{pmatrix} \right|\right|\right|_{n+1} \le Ck \sum_{j=0}^{2} \sup_{0 \le t \le \tau} \|u^{(j)}(t) - w^{(j)}(t)\| + Ck^{\nu+1}.$$

Equation (4.14) follows from this estimate and (2.11). ∎

592   L. A. BALES

LEMMA 4.5

There is a constant C which is independent of h and k such that for U_1^n and $\hat{U}_1^{n+1} \in Y$,

$$\left\|\left\| p_1 k \begin{pmatrix} 0 \\ f_n(U_1^n) - f_n \end{pmatrix} - q_1 k \begin{pmatrix} 0 \\ f_{n+1}(\hat{U}_1^{n+1}) - f_{n+1} \end{pmatrix} \right\|\right\|_{n+1}$$
$$\leq C[k(\|\|Q_n E^n\|\|_n + h^r)] + Ck\|\hat{U}^{n+1} - u(t_{n+1})\|. \quad (4.15)$$

Proof. Since

$$\left\|\left\| p_1 k \begin{pmatrix} 0 \\ f_n(U_1^n) - f_n \end{pmatrix} - q_1 k \begin{pmatrix} 0 \\ f_{n+1}(\hat{U}_1^{n+1}) - f_{n+1} \end{pmatrix} \right\|\right\|_{n+1}$$
$$\leq k(|p_1| \, \|T_{n+1}^{1/2}(f_n(U_1^n) - f_n)\| + |q_1| \, \|T_{n+1}^{1/2}(f_{n+1}(\hat{U}_1^{n+1}) - f_{n+1})\|),$$

it follows from (1.2) and (2.11) that

$$\left\|\left\| p_1 k \begin{pmatrix} 0 \\ f_n(U_1^n) - f_n \end{pmatrix} - q_1 k \begin{pmatrix} 0 \\ f_{n+1}(\hat{U}_1^{n+1}) - f_{n+1} \end{pmatrix} \right\|\right\|_{n+1}$$
$$\leq Ck(\|U_1^n - w(t_n)\| + h^r + \|\hat{U}_1^{n+1} - u(t_{n+1})\|).$$

Equation (4.15) follows from this estimate and (4.3).   ∎

LEMMA 4.6

There is a constant C which is independent of h and k such that for $n = 1, \ldots, M - 1$ and U_1^{n-1}, U_1^n, $\hat{U}_1^{n+1}$ and $\hat{U}_1^{n+2} \in Y$,

$$\left\|\left\| p_2 k^2 \begin{pmatrix} f_n(U_1^n) - f_n \\ \dfrac{f_{n+1}(\hat{U}_1^{n+1}) - f_{n-1}(U_1^{n-1})}{2k} - f_n^{(1)} \end{pmatrix} - q_2 k^2 \begin{pmatrix} f_{n+1}(\hat{U}_1^{n+1}) - f_{n+1} \\ \dfrac{f_{n+2}(\hat{U}_1^{n+1}) - f_n(U_1^n)}{2k} - f_{n+1}^{(1)} \end{pmatrix} \right\|\right\|_{n+1}$$
$$\leq C[k(\|\|Q_n E^n\|\|_n + \|\|Q_{n-1} E^{n-1}\|\|_{n-1} + (h^r + k^{\max\{\nu,3\}})$$
$$+ \|\hat{U}_1^{n+1} - u(t_{n+1})\| + \|\hat{U}_1^{n+2} - u(t_{n+2})\|)]. \quad (4.16)$$

Also,

$$\left\|\left\| -q_2 k^2 \begin{pmatrix} f_1(\hat{U}_1^1) - f_1 \\ \dfrac{f_2(\hat{U}_1^2) - f_0}{2k} - f_1^{(1)} \end{pmatrix} + p_2 k^2 \begin{pmatrix} 0 \\ \dfrac{-f_2(\hat{U}_1^2) + 4f_1(\hat{U}_1^1) - 3f_0}{2k} - f_0^{(1)} \end{pmatrix} \right\|\right\|_1$$
$$\leq C[k^4 + k\|\hat{U}_1^2 - u(t_2)\| + k\|\hat{U}_1^1 - u(t_1)\|]. \quad (4.17)$$

Proof. The estimate (4.16) follows from the estimate

$$\left\|\left\| p_2 k^2 \begin{pmatrix} f_n(U_1^n) - f_n \\ \dfrac{f_{n+1}(\hat{U}_1^{n+1}) - f_{n+1}(U_1^{n-1})}{2k} - f_n^{(1)} \end{pmatrix} - q_2 k^2 \begin{pmatrix} f_{n+1}(\hat{U}_1^{n+1}) - f_{n+1} \\ \dfrac{f_{n+2}(\hat{U}_1^{n+2}) - f_n(U_1^n)}{2k} - f_{n+1}^{(1)} \end{pmatrix} \right\|\right\|_{n+1}$$
$$\leq |p_2| k^2 \|f_n(U_1^n) - f_n\| + |q_2| k^2 \|f_{n+1}(\hat{U}_1^{n+1}) - f_{n+1}\|$$
$$+ \left\| T_{n+1}^{1/2}\left(p_2 k^2 \left(\dfrac{f_{n+1}(\hat{U}_1^{n+1}) - f_{n+1}(U_1^{n-1})}{2k} - f_n^{(1)} \right)\right.\right.$$
$$\left.\left. - q_2 k^2 \left(\dfrac{f_{n+2}(\hat{U}_1^{n+2}) - f_n(U_1^n)}{2k} - f_{n+1}^{(1)} \right) \right) \right\|$$

and Eqs. (1.2), (2.11) and (4.3), since, for $n \geq 1$,

$$p_2 k^2 \left(\dfrac{f_{n+1}(\hat{U}_1^{n+1}) - f_{n-1}(U_1^{n-1})}{2k} - f_n^{(1)} \right) - q_2 k^2 \left(\dfrac{f_{n+2}(\hat{U}_1^{n+2}) - f_n(U_1^n)}{2k} - f_{n+1}^{(1)} \right)$$

$$
\begin{aligned}
&= p_2 k^2 \left(\frac{f_{n+1}(\hat{U}_1^{n+1}) - f_{n-1}(U_1^{n-1})}{2k} - \frac{f_{n+1} - f_{n-1}}{2k} \right) \\
&\quad - q_2 k^2 \left(\frac{f_{n+1}(\hat{U}_1^{n+2}) - f_n(U_1^n)}{2k} - \frac{f_{n+2} - f_n}{2k} \right) \\
&\quad + p_2 k^2 \left(\frac{f_{n+1} - f_{n-1}}{2k} - f_n^{(1)} \right) - q_2 k^2 \left(\frac{f_{n+2} - f_n}{2k} - f_{n+1}^{(1)} \right).
\end{aligned}
$$

Note that when $\nu = 4$, $p_2 = q_2 = \frac{1}{12}$, and

$$
\left\| T_{n+1}^{1/2} \left(p_2 k^2 \left(\frac{f_{n+1} - f_{n-1}}{2k} - f_n^{(1)} \right) - q_2 k^2 \left(\frac{f_{n+2} - f_n}{2k} - f_{n+1}^{(1)} \right) \right) \right\| \leq Ck^5.
$$

Equation (4.17) follows from similar considerations.   ∎

LEMMA 4.7

If U_1^{n-1}, U_1^n, $\hat{U}_1^{n+1}$ and $\hat{U}_1^{n+2} \in Y$, then for $n \geq 0$,

$$
\begin{aligned}
\||(\tilde{Q}_{n+1} - A_{n+1})E^{n+1}\||_{n+1} \leq C \Big(\frac{k}{h}(\|\hat{U}_1^{n+1} - u(t_{n+1})\|_{L^\infty} \\
+ \|\hat{U}_1^{n+2} - u(t_{n+2})\|_{L^\infty} + \|U_1^n - u(t_n)\|_{L^\infty}) + k \Big) \||Q_{n+1}E^{n+1}\||_{n+1};
\end{aligned} \quad (4.18)
$$

for $n \geq 1$,

$$
\begin{aligned}
\||(B_n - \tilde{P}_n)E^n\||_{n+1} \leq C \Big(\frac{k}{h}(\|U_1^n - u(t_n)\|_{L^\infty} + \|\hat{U}_1^{n+1} - u(t_{n+1})\|_{L^\infty} \\
+ \|U_1^{n-1} - u(t_{n-1})\|_{L^\infty}) + k \Big) \||Q_nE^n\||_n;
\end{aligned} \quad (4.19a)
$$

and

$$
\begin{aligned}
\||(B_0 - \tilde{P}_0)E^0\||_1 \leq C \Big(\frac{k}{h}(\|U_1^0 - u(t_0)\|_{L^\infty} + \|\hat{U}_1^1 - u(t_1)\|_{L^\infty} \\
+ \|\hat{U}_1^2 - u(t_2)\|_{L^\infty}) + k \Big) \||Q_0E^0\||_0.
\end{aligned} \quad (4.19b)
$$

Proof. Since $\tilde{Q}_{n+1} - A_{n+1} = \tilde{Q}_{n+1} - Q_{n+1} + Q_{n+1} - A_{n+1}$, (4.18) follows from (4.10) and (4.13). In order to prove (4.19a) note that $B_n - \tilde{P}_n = B_n - P_n + P_n - \tilde{P}_n$. From estimates in Bales[10], it follows that $\||(P_n - \tilde{P}_n)E^n\||_{n+1} \leq Ck\||Q_nE^n\||_n$. Thus it suffices to consider

$$
\begin{aligned}
B_n - P_n &= p_1 k \begin{pmatrix} 0 & 0 \\ -L_n(U_1^n) + L_n & 0 \end{pmatrix} \\
&\quad + p_2 k^2 \begin{pmatrix} -L_n(U_1^n) + L_n & 0 \\ -\dfrac{L_{n+1}(\hat{U}_1^{n+1}) - L_{n-1}(U_1^{n-1})}{2k} & -L_n(U_1^n) + L_n \end{pmatrix}.
\end{aligned} \quad (4.20)
$$

The estimates

$$
\begin{aligned}
\||(B_n - P_n)E^n\||_{n+1} \leq &|p_1|k\|T_{n+1}^{1/2}(-L_n(U_1^n) + L_n)E_1^n\| + |p_2|k^2\|(-L_n(U_1^n) + L_n)E_1^n\| \\
&+ |p_2|k^2\|T_{n+1}^{1/2}(-L_n(U_1^n) + L_n)E_2^n\| \\
&+ |p_2|k^2 \left\| T_{n+1}^{1/2}\left(-\frac{L_{n+1}(\hat{U}_1^{n+1}) - L_{n-1}(U_1^{n-1})}{2k} \right)E_1^n \right\|,
\end{aligned}
$$

594   L. A. BALES

$$\|T_{n+1}^{1/2}(-L_n(U_1^n) + L_n)E_1^n\| \le C\|U_1^n - u(t_n)\|_{L^\infty}\|L_n^{1/2}E_1^n\|,$$

$$\|(-L_n(U_1^n) + L_n)E_1^n\| \le \frac{C}{h}\|U_1^n - u(t_n)\|_{L^\infty}\|L_n^{1/2}E_1^n\|,$$

$$\|T_{n+1}^{1/2}(-L_n(U_1^n) + L_n)E_2^n\| \le C\|U_1^n - u(t_n)\|_{L^\infty}\|L_n^{1/2}E_2^n\|,$$

and

$$k^2\left\|T_{n+1}^{1/2}\left(-\frac{L_{n+1}(\hat{U}_1^{n+1}) - L_{n-1}(U_1^{n-1})}{2k}\right)E_1^n\right\|$$
$$\le Ck(\|\hat{U}_1^{n+1} - u(t_{n+1})\|_{L^\infty} + \|U_1^{n-1} - u(t_{n-1})\|_{L^\infty})\|L_n^{1/2}E_1^n\| + Ck^2\|L_n^{1/2}E_1^n\|$$

used with (2.3a), (4.3) and (4.4) give (4.19a). The proof of (4.19b) is similar. $\blacksquare$

LEMMA 4.8

For $n = 1, \ldots, M - 1$, U_1^{n-1}, U_1^n, $\hat{U}_1^{n+1}$, and $\hat{U}_1^{n+2} \in Y$,

$$\||(\tilde{Q}_{n+1} - A_{n+1})W(t_{n+1}) + (B_n - \tilde{P}_n)W(t_n)\||_{n+1} \le Ck[(1 + k/h)\|u(t_{n+1}) - \hat{U}_1^{n+1}\|$$
$$+ \|u(t_{n+2}) - \hat{U}_1^{n+2}\| + (1 + k/h)(\||Q_nE^n\||_n + h^r)$$
$$+ \||Q_{n-1}E^{n-1}\||_{n-1} + h^r + k^{\max\{\nu,3\}}]. \tag{4.21}$$

Also,

$$\||(\tilde{Q}_1 - A_1)W(t_1) + (B_0 - \tilde{P}_0)W(t_0)\||_1 \le Ck[(1 + k/h)\|U(t_1) - \hat{U}_1^1\|$$
$$+ \|u(t_2) - \hat{U}_1^2\| + (1 + k/h)(\||Q_0E^0\||_0 + h^r)] + Ck^4. \tag{4.22}$$

Proof. Since

$$(\tilde{Q}_{n+1} - A_{n+1})W(t_{n+1}) + (B_n - \tilde{P}_n)W(t_n) = \left[q_1k\begin{pmatrix} 0 & 0 \\ -L_{n+1} + L_{n+1}(\hat{U}_1^{n+1}) & 0 \end{pmatrix}\right.$$
$$+ q_2k^2\begin{pmatrix} -L_{n+1} + L_{n+1}(\hat{U}_1^{n+1}) & 0 \\ \dfrac{-L_{n+1}^{(1)} + L_{n+1}(\hat{U}_1^{n+2}) - L_n(U_1^n)}{2k} & -L_{n+1} + L_{n+1}(\hat{U}_1^{n+1}) \end{pmatrix}\left.\right]$$
$$\times \begin{pmatrix} w(t_{n+1}) \\ w^{(1)}(t_{n+1}) \end{pmatrix}$$
$$+ \left[p_1k\begin{pmatrix} 0 & 0 \\ -L_n(U_1^n) + L_n & 0 \end{pmatrix}\right.$$
$$+ p_2k^2\begin{pmatrix} -L_n(U_1^n) + L_n & 0 \\ -\dfrac{L_{n+1}(\hat{U}_1^{n+1}) - L_{n-1}(U_1^{n-1})}{2k} + L_n^{(1)} & -L_n(U_1^n) + L_n \end{pmatrix}\left.\right]$$
$$\times \begin{pmatrix} w(t_n) \\ w^{(1)}(t_n) \end{pmatrix},$$

it follows that

$$\||(\tilde{Q}_{n+1} - A_{n+1})W(t_{n+1}) + (B_n - \tilde{P}_n)W(t_n)\||_{n+1}$$
$$\le |q_1|k\|T_{n+1}^{1/2}(-L_{n+1} + L_{n+1}(\hat{U}_1^{n+1}))w(t_{n+1})\|$$
$$+ |p_1|k\|T_{n+1}^{1/2}(-L_n(U_1^n) + L_n)w(t_n)\|$$
$$+ |q_2|k^2\|(-L_{n+1} + L_{n+1}(\hat{U}_1^{n+1}))w(t_{n+1})\|$$
$$+ |q_2|k^2\|T_{n+1}^{1/2}(-L_{n+1} + L_{n+1}(\hat{U}_1^{n+1}))w^{(1)}(t_{n+1})\|$$

$$+ |p_2|k^2\|(-L_n(U_1^n) + L_n)w(t_n)\|$$

$$+ |p_2|k^2\|T_{n+1}^{1/2}(-L_n(U_1^n) + L_n))w^{(1)}(t_n)\|$$

$$+ k^2\left\|T_{n+1}^{1/2}\left(q_2\left(-L_{n+1}^{(1)} + \frac{L_{n+2}(\hat{U}_1^{n+2}) - L_n(U_1^n)}{2k}\right)w(t_{n+1})\right.\right.$$

$$+ \left.\left. p_2\left(\frac{-L_{n+1}(\hat{U}_1^{n+1}) - L_{n-1}(U_1^{n-1})}{2k} + L_n^{(1)}\right)w(t_n)\right)\right\|.$$

From (4.8), (4.9) and (2.14) it follows that

$$\|T_{n+1}^{1/2}(-L_{n+1} + L_{n+1}(\hat{U}_1^{n+1}))w(t_{n+1})\| \le C\|u(t_{n+1}) - \hat{U}_1^{n+1}\|,$$

$$\|T_{n+1}^{1/2}(-L_n(U_1^n) + L_n)w(t_n)\| \le C(\|U_1^n - w(t_n)\| + h^r),$$

$$\|(-L_{n+1} + L_{n+1}(\hat{U}_1^{n+1}))w(t_{n+1})\| \le \frac{C}{h}\|u(t_{n+1}) - \hat{U}_1^{n+1}\|,$$

$$\|T_{n+1}^{1/2}(-L_{n+1} + L_{n+1}(\hat{U}_1^{n+1}))w^{(1)}(t_{n+1})\| \le C\|u(t_{n+1}) - \hat{U}_1^{n+1}\|,$$

$$\|(-L_n(U_1^n) + L_n)w(t_n)\| \le \frac{C}{h}(\|U_1^n - w(t_n)\| + h^r),$$

and

$$\|T_{n+1}^{1/2}(-L_n(U_1^n) + L_n)w^{(1)}(t_n)\| \le C\|U_1^n - w(t_n)\| + Ch^r.$$

Note that

$$k^2\left(q_2\left(-L_{n+1}^{(1)} + \frac{L_{n+2}(\hat{U}_1^{n+2}) - L_n(U_1^n)}{2k}\right)w(t_{n+1})\right.$$

$$+ \left. p_2\left(-\frac{L_{n+1}(\hat{U}_1^{n+1}) - L_{n-1}(U_1^{n-1})}{2k} + L_n^{(1)}\right)w(t_n)\right)$$

$$= \frac{q_2}{2}k(L_{n+2}(\hat{U}_1^{n+2}) - L_{n+2} + L_n - L_n(U_1^n))w(t_{n+1})$$

$$+ \frac{p_2}{2}k(L_{n+1} - L_{n+1}(\hat{U}_1^{n+1}) + L_{n-1}(U_1^{n-1}) - L_{n-1})w(t_n)$$

$$+ q_2k^2\left(\frac{L_{n+2} - L_n}{2k} - L_{n+1}^{(1)}\right)w(t_{n+1}) - p_2k^2\left(\frac{L_{n+1} - L_{n-1}}{2k} - L_n^{(1)}\right)w(t_n),$$

and using (4.8) and (2.14),

$$\|T_{n+1}^{1/2}(L_{n+2}(\hat{U}_1^{n+2}) - L_{n+2} + L_n - L_n(U_1^n))w(t_{n+1})\|$$
$$\le C(\|\hat{U}_1^{n+2} - u(t_{n+2})\| + \|U_1^n - w(t_n)\| + h^r)$$

and

$$\|T_{n+1}^{1/2}(L_{n+1} - L_{n+1}(\hat{U}_1^{n+1}) + L_{n-1}(U_1^{n-1}) - L_{n-1})w(t_n)\|$$
$$\le C(\|\hat{U}_1^{n+1} - u(t_{n+1})\| + \|U_1^{n-1} - w(t_{n-1})\| + h^r).$$

Also,

$$\left\|T_{n+1}^{1/2}\left(q_2k^2\left(\frac{L_{n+2} - L_n}{2k} - L_{n+1}^{(1)}\right)w(t_{n+1})\right.\right.$$
$$\left.\left. - p_2k^2\left(\frac{L_{n+1} - L_{n-1}}{2k} - L_n^{(1)}\right)w(t_n)\right)\right\| \le Ck^{\max\{v+1,4\}}$$

using the fact that $q_2 = p_2$ when $v = 4$. The above estimates and (4.3) imply (4.21), and the proof of (4.22) is similar.   ■

THEOREM 4.1

Let $\{U^n\}_{n=0}^M$ be given by Algorithm 3.3 and $k \leq C^*h$ for any constant C^*. If $v \geq 3$ and $r \geq 3$ when $N = 2, 3$, then for k and h sufficiently small there is a constant C which is independent of h and k such that for $n = 0, 1, \ldots, M$,

$$\||Q_n(U^n - W(t_n))\||_n \leq C(h^r + k^v) \tag{4.23}$$

and

$$\|U_1^n - u(t_n)\| \leq C(h^r + k^v). \tag{4.24}$$

Proof. Note that (4.24) follows from (4.23), (4.3) and (2.11).

In order to prove (4.23) when $n = 0$, note that if $U^0 = W(0)$, then $U^0 - W(t_0) = 0$ and if

$$U^0 = (\tilde{Q}_0)^{-1}\begin{pmatrix} P & 0 \\ 0 & P \end{pmatrix}\left[\begin{pmatrix} u^0 \\ u_t^0 \end{pmatrix} + q_1 k \begin{pmatrix} u_t^0 \\ -L(0)u_t^0 \end{pmatrix} + q_2 k^2 \begin{pmatrix} -L(0)u^0 \\ -\left(\dfrac{d}{dt}Lu\right)(0) \end{pmatrix}\right],$$

then $\||Q_0(U^0 - W(0))\||_0 \leq Ch^r$ using (2.11). See Bales[1] for more details. Also, $\|U_1^0 - w(t_0)\|_{L^\infty} \leq Ch^{-N/2} \||Q_0(U^0 - W(0))\||_0 \leq Ch^{-N/2}h^r$, so that $U_1^0 \in Y$ for h sufficiently small.

For step (2) of Algorithm 3.3 and, for example, $v = 4$,

$$\begin{aligned}
\hat{U}_1^1 - u(t_1) &= U_1^0 + kPu_t(0) + \frac{k^2}{2}Pu_{tt}(0) - u(t_1) \\
&= U_1^0 - w(t_0) + w(t_0) - u(t_0) \\
&\quad + u(t_0) + ku_t(t_0) + \frac{k^2}{2}u_{tt}(0) - u(t_1) \\
&\quad + k(P - I)u_t(0) + \frac{k^2}{2}(P - I)u_{tt}(0)
\end{aligned} \tag{4.25}$$

and

$$\begin{aligned}
\hat{U}_1^2 - u(t_2) &= U_1^0 + 2kPu_t(0) + 2k^2Pu_{tt}(0) - u(t_2) \\
&= U_1^0 - w(t_0) + w(t_0) - u(t_0) \\
&\quad + u(t_0) + 2ku_t(0) + 2k^2u_{tt}(0) - u(t_2) \\
&\quad + 2k(P - I)u_t(0) + 2k^2(P - I)u_{tt}(0).
\end{aligned} \tag{4.26}$$

It follows from (2.11) that

$$\|\hat{U}_1^1 - u(t)\| \leq C(h^r + k^3) \tag{4.27}$$

and

$$\|\hat{U}_1^2 - u(t_2)\| \leq C(h^r + k^3), \tag{4.28}$$

since $\|U_1^0 - w(t_0)\| \leq C\||Q_0(U^0 - W(t_0))\||_0 \leq Ch^r$. Also, from (4.25), (4.26) and (2.13) it follows that

$$\|\hat{U}_1^1 - u(t_1)\|_{L^\infty} \leq C(h + k) \tag{4.29}$$

and

$$\|\hat{U}_1^2 - u(t_2)\|_{L^\infty} \leq C(h + k), \tag{4.30}$$

since for a smooth function and $1 \leq N \leq 3$,

$$\|(P - I)v\|_{L^\infty} \leq \|(P - T_h(t)L(t))v\|_{L^\infty} + \|(T_h(t)L(t) - I)v\|_{L^\infty} \leq Ch^{-N/2}h^r + C \leq C$$

and

$$\|U_1^0 - w(t_0)\|_{L^\infty} \leq Ch^{-N/2}\|U_1^0 - w(t_0)\| \leq Ch^{-N/2}\|\|Q_0(U_1^0 - W(t_0))\|\|_0 \leq Ch^{-N/2}h^r \leq Ch$$

($r \geq 3$ when $N = 3$). Thus for h and k sufficiently small $\hat{U}_1^1$ and $\hat{U}_1^2 \in Y$. Therefore, it follows from (4.7), (4.12)–(4.15), (4.17), (4.18), (4.19b), (4.22) and the triangle inequality that, if $k \leq C^*h$ is sufficiently small, then

$$\begin{aligned}
\|\|Q_1E^1\|\|_1 \leq{}& (1 + Ck)\|\|Q_0E^0\|\| + C[k(h^r + k^\nu) + k^4 \\
&+ k(\|\hat{U}_1^1 - u(t_1)\| + \|\hat{U}_1^2 - u(t_2)\|) \\
&+ \frac{k}{h}(\|\hat{U}_1^1 - u(t_1)\|_{L^\infty} + \|\hat{U}_1^2 - u(t_2)\|_{L^\infty} + \|U_1^0 - u(t_0)\|_{L^\infty})\|\|Q_1E^1\|\|_1 \\
&+ \frac{k}{h}(\|U_1^0 - u(t_0)\|_{L^\infty} + \|\hat{U}_1^1 - u(t_1)\|_{L^\infty} + \|\hat{U}_1^2 - u(t_2)\|_{L^\infty})\|\|Q_0E^0\|\|_0].
\end{aligned} \tag{4.31}$$

Since $\|\|Q_0E^0\|\|_0 \leq Ch^r$, it follows from (4.27)–(4.31) that

$$\|\|Q_1E^1\|\| \leq C(h^r + k^4). \tag{4.32}$$

The proofs of (4.23) when $n = 1$ for $\nu = 2$ and 3 are similar. For h and k sufficiently small it is straightforward to see that $U_1^1 \in Y$.

In order to prove (4.23) for steps (3), (4), and (5) of Algorithm 3.3, note that for $\hat{U}_1^{n+1}$ and $\hat{U}_1^{n+2} \in Y$ it follows from (4.7), (4.12)–(4.16), (4.18), (4.19a), (4.21) and the triangle inequality that if $k \leq C^*h$ and k is sufficiently small, then

$$\begin{aligned}
\|\|Q_{n+1}E^{n+1}\|\|_{n+1} \leq{}& (1 + Ck)\|\|Q_nE^n\|\|_n \\
&+ C(k\|\|Q_{n+1}E^{n+1}\|\|_{n+1} + k\|\|Q_{n-1}E^{n-1}\|\|_{n-1} \\
&+ k(h^r + k^\nu) + k(\|\hat{U}_1^{n+1} - u(t_{n+1})\| + \|\hat{U}_1^{n+2} - u(t_{n+2})\|) \\
&+ \frac{k}{h}(\|\hat{U}_1^{n+1} - u(t_{n+1})\|_{L^\infty} + \|\hat{U}_1^{n+2} - u(t_{n+2})\|_{L^\infty} \\
&+ \|U_1^n - u(t_n)\|_{L^\infty})\|\|Q_{n+1}E^{n+1}\|\|_{n+1} \\
&+ \frac{k}{h}(\|U_1^n - u(t_n)\|_{L^\infty} + \|\hat{U}_1^{n+1} - u(t_{n+1})\|_{L^\infty} \\
&+ \|U_1^{n-1} - u(t_{n-1})\|_{L^\infty})\|\|Q_nE^n\|\|_n).
\end{aligned} \tag{4.33}$$

The proofs of the error estimate (4.33) for steps (3) and (4) of Algorithm 3.3 follow from (4.33) ($n = 1$ and 2) and the estimates

$$\|\hat{U}_1^{n+1} - u(t_{n+1})\| \leq C(h^r + k^{\nu-1}), \tag{4.34}$$

$$\|\hat{U}_2^{n+2} - u(t_{n+2})\| \leq C(h^r + k^{\nu-1}), \tag{4.35}$$

$$\|\hat{U}_1^{n+1} - u(t_{n+1})\|_{L^\infty} \leq Ch, \tag{4.36}$$

and

$$\|\hat{U}_1^{n+2} - u(t_{n+2})\|_{L^\infty} \leq Ch. \tag{4.37}$$

Equations (4.34)–(4.37) are proved using techniques similar to the ones used to prove (4.27)–(4.30). As before, it easily follows that U_1^2 and $U_1^3 \in Y$.

For step (5) of Algorithm 3.3, $v \leq n + 1 \leq M$, and the proof is by induction. For each $v \leq n + 1 \leq M$, it follows from (4.7), (4.12)–(4.16), (4.18), (4.19a), (4.21) and the triangle inequality that if $k \leq C^* h$ and k is sufficiently small, then

$$
\begin{aligned}
\|\|Q_{n+1}E^{n+1}\|\|_{n+1} \leq & (1 + C_1 k)\|\|Q_n E^n\|\|_n \\
& + C_1(k\|\|Q_{n+1}E^{n+1}\|\|_{n+1} + k\|\|Q_{n-1}E^{n-1}\|\|_{n-1} + k(h^r + k^v) \\
& + k(\|\hat{U}_1^{n+1} - u(t_{n+1})\| + \|\hat{U}_1^{n+2} - u(t_{n+2})\|) \\
& + \frac{k}{h}(\|\hat{U}_1^{n+1} - u(t_{n+1})\|_{L^\infty} + \|\hat{U}_1^{n+2} - u(t_{n+2})\|_{L^\infty} \\
& + \|U_1^n - u(t_n)\|_{L^\infty})\|\|Q_{n+1}E^{n+1}\|\|_{n+1} \\
& + \frac{k}{h}(\|U_1^n - u(t_n)\|_{L^\infty} + \|\hat{U}_1^{n+1} - u(t_{n+1})\|_{L^\infty} \\
& + \|U_1^{n-1} - u(t_{n-1})\|_{L^\infty})\|\|Q_n E^n\|\|_n).
\end{aligned}
\tag{4.38}
$$

Note that, for example, when $v = 4$,

$$
\begin{aligned}
\hat{U}_1^{n+1} - u(t_{n+1}) = & \; 4U_1^n - 6U_1^{n-1} + 4U_1^{n-2} - U_1^{n-3} - u(t_{n+1}) \\
= & \; 4(U_1^n - w(t_n)) - 6(U_1^{n-1} - w(t_{n-1})) \\
& + 4(U_1^{n-2} - w(t_{n-2})) - (U_1^{n-3} - w(t_{n-3})) \\
& + 4(w(t_n) - u(t_n)) - 6(w(t_{n-1}) - u(t_{n-1})) \\
& + 4(w(t_{n-2}) - u(t_{n-2})) - (w(t_{n-3}) - u(t_{n-3})) \\
& + 4u(t_n) - 6u(t_{n-1}) + 4u(t_{n-2}) - u(t_{n-3}) - u(t_{n+1})
\end{aligned}
\tag{4.39}
$$

and

$$
\begin{aligned}
\hat{U}_1^{n+2} - u(t_{n+2}) = & \; 10U_1^n - 20U_1^{n-1} + 15U_1^{n-2} - 4U_1^{n-3} - u(t_{n+2}) \\
= & \; 10(U_1^n - w(t_n)) - 20(U_1^{n-1} - w(t_{n-1})) \\
& + 15(U_1^{n-2} - w(t_{n-2})) - 4(U_1^{n-3} - w(t_{n-3})) \\
& + 10(w(t_n) - u(t_n)) - 20(w(t_{n-1}) - u(t_{n-1})) \\
& + 15(w(t_{n-2}) - u(t_{n-2})) - 4(w(t_{n-3}) - u(t_{n-3})) \\
& + 10u(t_n) - 20u(t_{n-1}) + 15u(t_{n-2}) - 4u(t_{n-3}) - u(t_{n+2}).
\end{aligned}
\tag{4.40}
$$

It follows from (4.39), (4.40), (2.11) and (4.3) that

$$
\begin{aligned}
\|\hat{U}_1^{n+1} - u(t_{n+1})\| \leq C_2(&\|\|Q_n E^n\|\|_n + \|\|Q_{n-1}E^{n-1}\|\|_{n-1} \\
& + \|\|Q_{n-2}E^{n-2}\|\|_{n-2} + \|\|Q_{n-3}E^{n-3}\|\|_{n-3} + h^r + k^4)
\end{aligned}
\tag{4.41}
$$

and

$$
\begin{aligned}
\|\hat{U}_1^{n+2} - u(t_{n+2})\| \leq C_2(&\|\|Q_n E^n\|\|_n + \|\|Q_{n-1}E^{n-1}\|\|_{n-1} \\
& + \|\|Q_{n-2}E^{n-2}\|\|_{n-2} + \|\|Q_{n-3}E^{n-3}\|\|_{n-3} + h^r + k^4).
\end{aligned}
\tag{4.42}
$$

Also, from (4.39), (4.40), (2.13), (4.3) and (2.3b) it follows that

$$
\begin{aligned}
\|\hat{U}_1^{n+i} - u(t_{n+1})\|_{L^\infty} \leq Ch^{-N/2}(&\|\|Q_n E^n\|\|_n + \|\|Q_{n-1}E^{n-1}\|\|_{n-1} \\
& + \|\|Q_{n-2}E^{n-2}\|\|_{n-2} + \|\|Q_{n-3}E^{n-3}\|\|_{n-3}) + C(h^{3/2} + k^2)
\end{aligned}
\tag{4.43}
$$

and

$$\|\hat{U}_1^{n+2} - u(t_{n+2})\|_{L^\infty} \leq Ch^{-N/2}(\||Q_nE^n|\|_n + \||Q_{n-1}E^{n-1}|\|_{n-1}$$
$$+ \||Q_{n-2}E^{n-2}|\|_{n-2} + \||Q_{n-3}E^{n-3}|\|_{n-3}) + C(h^{3/2} + k^2). \quad (4.44)$$

The induction assumptions are that $U_1^n \in Y$ and $\||Q_nE^n|\|_n \leq \alpha e^{\alpha t_n}(h^r + k^v)$ for $n = 0, 1, \ldots,$ $M - 1$, where $\alpha \geq 4vC_3 + 1$, $C_3 \geq 2C_1 + C_1C_2$, and C_1 and C_2 are the constants in (4.38), (4.41) and (4.42). From (4.43) and (4.44) it follows that if $k \leq C^*h$ and for $N = 2, 3, r \geq$ 3 and $v \geq 3$, then for h and k sufficiently small

$$\|\hat{U}_1^{n+1} - u(t_{n+1})\|_{L^\infty} \leq h/3 \quad (4.44)$$

and

$$\|\hat{U}_1^{n+2} - u(t_{n+2})\|_{L^\infty} \leq h/3. \quad (4.45)$$

Thus, for h sufficiently small $\hat{U}_1^{n+1}$ and $\hat{U}_1^{n+2} \in Y$. Also, note that $\|U_1^n - u(t_n)\|_{L^\infty} \leq Ch^{-N/2}$ $\||Q_nE^n|\|_n + Ch^{3/2} \leq h/3$ for h sufficiently small. This estimate and (4.41), (4.42), (4.44), (4.45) (or similar estimates when $v = 2$ and $v = 3$) and (4.38) imply that

$$\||Q_{n+1}E^{n+1}|\|_{n+1} \leq \||Q_nE^n|\|_n + C_3k(\||Q_{n+1}E^{n+1}|\|_{n+1} + \||Q_nE^n|\|_n$$
$$+ \cdots + \||Q_{n-v+1}E^{n-v+1}|\|_{n-v+1} + k^v + h^r). \quad (4.46)$$

It follows that for k sufficiently small,

$$\||Q_{n+1}E^{n+1}|\|_{n+1} \leq \||Q_nE^n|\|_n + 4C_3k(\||Q_nE^n|\|_n$$
$$+ \cdots + \||Q_{n-v+1}E^{n-v+1}|\|_{n-v+1} + k^v + h^r). \quad (4.47)$$

Summing (4.47) from $n = v - 1$ to $n = M - 1$ gives

$$\||Q_ME^M|\|_M - \||Q_{v-1}E^{v-1}|\|_{v-1}$$
$$\leq 4C_3k \sum_{n=v-1}^{M-1} (\||Q_nE^n|\|_n + \cdots + \||Q_{n-v+1}E^{n-v+1}|\|_{n-v+1} + k^v + h^r)$$
$$\leq 4vC_3k \sum_{n=v-1}^{M-1} \alpha e^{\alpha t_n}(h^r + k^v) + 4C_3(t_M - t_{v-1})(h^r + k^v). \quad (4.48)$$

Therefore, since $\alpha \geq 4vC_3 + 1$,

$$\||Q_ME^M|\|_M \leq (\alpha e^{\alpha t_{v-1}} + 4vC_3(e^{\alpha t_M} - e^{\alpha t_{v-1}}) + 4C_3(t_M - t_{v-1}))(h^r + k^v)$$
$$\leq (\alpha e^{\alpha t_{v-1}} + (\alpha - 1)(e^{\alpha t_M} - e^{\alpha t_{v-1}}) + (e^{\alpha(t_M - t_{v-1})} - 1))(h^r + k^v)$$
$$\leq (\alpha e^{\alpha t_{v-1}} + (\alpha - 1)(e^{\alpha t_M} - e^{\alpha t_{v-1}}) + e^{\alpha t_{v-1}}(e^{\alpha(t_M - t_{v-1})} - 1))(h^r + k^v)$$
$$\leq (\alpha e^{\alpha t_{v-1}} + \alpha(e^{\alpha t_M} - e^{\alpha t_{v-1}}))(h^r + k^v)$$
$$\leq \alpha e^{\alpha t_M}(h^r + k^v).$$

Also, note that $\|U_1^M - w(t_M)\|_{L^\infty} \leq Ch^{-N/2}\||Q_ME^M|\|_M \leq Ch^{-N/2}(h^r + k^v) \leq Ch$ so that $U_1^M \in$ Y for h sufficiently small. This completes the proof. $\blacksquare$

5. ITERATIVE METHODS

As described for the linear problem in [1], preconditioned iterative methods can be used to approximately solve the fully discrete Eq. (3.7). These iterative methods avoid the repeated work of new matrix factorizations at each time step.

Let H be a finite-dimensional Hilbert space with inner product $(\cdot, \cdot)_H$ and norm $\|\cdot\|_H = (\cdot, \cdot)_H^{1/2}$. Let A be a positive definite self-adjoint operator on H and consider approximating the solution of the linear system $Ax = b$ with b given in H. Let A_0 be another positive definite self-adjoint operator on H (the "preconditioner") with the properties that systems of the form $A_0 y = z$, $z \in H$, are easily solvable and that there exist constants $0 < \lambda_0 \leq \lambda_1$ such that

$$\lambda_0(A_0 z, z)_H \leq (Az, z)_H \leq \lambda_1(A_0 z, z)_H \tag{5.1}$$

for all $z \in H$. There are iterative methods (cf., for example, [14,7,15]) for solving the system $Ax = b$, which, given an initial guess $x^{(0)} \in H$, generate a sequence $\{x^{(j)}\}_{j \geq 1}$ of approximations to x and have the following properties.

(i) Calculating $x^{(p+1)}$, given $\{x^{(j)}\}_{j=0}^{p}$, only requires multiplying A with vectors, solving systems with A_0 and computing inner products and linear combinations of vectors.

(ii) There is a smooth decreasing function $\rho: (0, 1] \to [0, 1)$, with $\rho(1) = 0$ and a constant C such that

$$\|A_0^{1/2}(x - x^{(p)})\|_H \leq C[\rho(\lambda_0/\lambda_1)]^p \|A_0^{1/2}(x - x^{(0)})\|_H, \tag{5.2}$$

where λ_0 and λ_1 are the constants in (5.1).

For the application of these iterative methods to solving (3.7) the Hilbert space H will be $S_h \times S_h$ and for $\Phi = \binom{\Phi_1}{\Phi_2} \in S_h \times S_h$ and $\Psi \in S_h \times S_h$ $(\Phi, \Psi)_H \equiv (\phi_1, \psi_1) + (\phi_2, \psi_2)$. The equations in (3.7) can be written

$$A_1^* \begin{pmatrix} I & 0 \\ 0 & T_0 \end{pmatrix} A_1 U^1 = A_1^* \begin{pmatrix} I & 0 \\ 0 & T_0 \end{pmatrix} \left(B_0 U^0 + p_1 k \begin{pmatrix} 0 \\ f_0 \end{pmatrix} \right.$$

$$+ p_2 k^2 \begin{pmatrix} f_0 \\ \dfrac{-f_2(\hat{U}_1^2) + 4f_1(\hat{U}_1^1) - 3f_0}{2k} \end{pmatrix}$$

$$\left. - q_1 k \begin{pmatrix} 0 \\ f_1(\hat{U}_1^1) \end{pmatrix} - q_2 k^2 \begin{pmatrix} f_1(\hat{U}_1^1) \\ \dfrac{f_2(\hat{U}_1^2) - f_0}{2k} \end{pmatrix} \right),$$

and, for $n \geq 1$, $\tag{5.3}$

$$A_{n+1}^* \begin{pmatrix} I & 0 \\ 0 & T_0 \end{pmatrix} A_{n+1} U^{n+1} = A_{n+1}^* \begin{pmatrix} I & 0 \\ 0 & T_0 \end{pmatrix} \left(B_n U^n + p_1 k \begin{pmatrix} 0 \\ f_n(U_1^n) \end{pmatrix} \right.$$

$$+ p_2 k^2 \begin{pmatrix} f_n(U_1^n) \\ \dfrac{f_{n+1}(\hat{U}_1^{n+1}) - f_{n-1}(U_1^{n-1})}{2k} \end{pmatrix} - q_1 k \begin{pmatrix} 0 \\ f_{n+1}(\hat{U}_1^{n+1}) \end{pmatrix}$$

$$\left. - q_2 k^2 \begin{pmatrix} f_{n+1}(\hat{U}_1^{n+1}) \\ \dfrac{f_{n+2}(\hat{U}_1^{n+2}) - f_n(U_1^n)}{2k} \end{pmatrix} \right).$$

In (5.3) A_{n+1}^* is the adjoint of A_{n+1} in H. The operator $\left(\begin{smallmatrix} I & 0 \\ 0 & T_0 \end{smallmatrix} \right)$ is used in (5.3) so that the norm $\|A^{1/2}\cdot\|_H$ is equivalent to the norm $\||Q_{n+1}\cdot\||_{n+1}$. $A \equiv A_{n+1}^* \left(\begin{smallmatrix} I & 0 \\ 0 & T_0 \end{smallmatrix} \right) A_{n+1}$ is self-adjoint and positive definite in H.

From equivalence of the norms and Lemma 4.2 it follows that if k, $\|\hat{U}_1^{n+1} - u(t_{n+1})\|_{L^\infty}$, $\|\hat{U}_1^{n+2} - u(t_{n+2})\|_{L^\infty}$, and $\|U_1^n - u(t_n)\|_{L^\infty}$ are sufficiently small and $k \leq C^*h$ for some constant C^*, then for $0 \leq n \leq M - 1$

$$C_1 \||Q_{n+1}\Phi\||_{n+1}^2 \leq \||A_{n+1}\Phi\||_0^2 \leq C_2 \||Q_{n+1}\Phi\||_{n+1}^2, \tag{5.4}$$

for all $\Phi \in S_h \times S_h$. From estimates in Bales[1] and (5.4) it follows that the preconditioner A_0 can be

$$(I + \alpha_1 k \mathcal{L}_0)^* \begin{pmatrix} I & 0 \\ 0 & T_0 \end{pmatrix} (I + \alpha_1 k \mathcal{L}_0) \tag{5.5}$$

if $q_2 = 0$, or

$$[(I + \alpha_1 k \mathcal{L}_0)(I + \alpha_2 k \mathcal{L}_0)]^* \begin{pmatrix} I & 0 \\ 0 & T_0 \end{pmatrix} (I + \alpha_1 k \mathcal{L}_0)(I + \alpha_2 k \mathcal{L}_0)$$

if $q_2 \neq 0$, where α_1 and α_2 are any complex numbers such that Re $\alpha_1 \neq 0$ and Re $\alpha_2 \neq 0$. More details on the preconditioner are given in [1].

Accurate starting values for the iterative methods are provided by the following extrapolations of the approximate solution:

$$\begin{aligned}
Z_{n+1}^{(1)}(U) &= U^n && \text{for } 0 \leq n \leq M - 1, \\
Z_{n+1}^{(2)}(U) &= 2U^n - U^{n-1} && \text{for } 1 \leq n \leq M - 1, \\
Z_{n+1}^{(3)}(U) &= 3U^n - 3U^{n-1} + U^{n-2} && \text{for } 2 \leq n \leq M - 1,
\end{aligned} \tag{5.6}$$

and

$$Z_{n+1}^{(4)}(U) = 4U^n - 6U^{n-1} + 4U^{n-2} - U^{n-3} \qquad \text{for } 3 \leq n \leq M - 1.$$

For the application of iterative methods define U^{n+1} to be the approximation to the exact solution $\overline{U}^{n+1}$ of

$$A_1^* \begin{pmatrix} I & 0 \\ 0 & T_0 \end{pmatrix} A_1 \overline{U}^1 = A_1^* \begin{pmatrix} I & 0 \\ 0 & T_0 \end{pmatrix} \left(B_0 U^0 + p_1 k \begin{pmatrix} 0 \\ f_0 \end{pmatrix} \right.$$

$$+ p_2 k^2 \begin{pmatrix} f_0 \\ \dfrac{-f_2(\hat{U}_1^2) + 4f_1(\hat{U}_1^1) - 3f_0}{2k} \end{pmatrix}$$

$$\left. - q_1 k \begin{pmatrix} 0 \\ f_1(\hat{U}_1^1) \end{pmatrix} - q_2 k^2 \begin{pmatrix} f_1(\hat{U}_1^1) \\ \dfrac{f_2(\hat{U}_1^2) - f_0}{2k} \end{pmatrix} \right)$$

and, for $n \geq 1$, $\tag{5.7}$

$$A_{n+1}^* \begin{pmatrix} I & 0 \\ 0 & T_0 \end{pmatrix} A_{n+1} \overline{U}^{n+1} = A_{n+1}^* \begin{pmatrix} I & 0 \\ 0 & T_0 \end{pmatrix} \left(B_n U^n + p_1 k \begin{pmatrix} 0 \\ f_n(U_1^n) \end{pmatrix} \right.$$

$$+ p_2 k^2 \begin{pmatrix} f_n(U_1^n) \\ \dfrac{f_{n+1}(\hat{U}_1^{n+1}) - f_{n-1}(U_1^{n-1})}{2k} \end{pmatrix} - q_1 k \begin{pmatrix} 0 \\ f_{n+1}(\hat{U}_1^{n+1}) \end{pmatrix}$$

$$\left. - q_2 k^2 \begin{pmatrix} f_{n+1}(\hat{U}_1^{n+1}) \\ \dfrac{f_{n+2}(\hat{U}_1^{n+2}) - f_n(U_1^n)}{2k} \end{pmatrix} \right),$$

where

$$A_{n+1} = q(k \mathcal{L}_{n+1}(\hat{U}_1^{n+1})) + q_2 k^2 \frac{\mathcal{L}_{n+2}(\hat{U}_1^{n+2}) - \mathcal{L}_n(U_1^n)}{2k},$$

$$B_n = p(k \mathcal{L}_n(U_1^n)) + p_2 k^2 \frac{\mathcal{L}_{n+1}(\hat{U}_1^{n+1}) - \mathcal{L}_{n-1}(U_1^{n-1})}{2k} \qquad \text{for } n \geq 1, \tag{5.8}$$

602   L. A. BALES

and

$$B_0 = p(k\mathcal{L}_0(U_1^0)) + p_2 k^2 \frac{-\mathcal{L}_2(\hat{U}_1^2) + 4\mathcal{L}_1(\hat{U}_1^1) - 3\mathcal{L}_0(U_1^0)}{2k}.$$

As in Bales[1], it follows from (5.1) and (5.2) that for any small positive number β_{n+1} (after, say, p iterations),

$$\left\|\left\|Q_{n+1}(\overline{U}^{n+1} - U^{n+1})\right\|\right\|_{n+1} \le C\beta_{n+1}\left\|\left\|Q_{n+1}(\overline{U}^{n+1} - U_0^{n+1})\right\|\right\|_{n+1}, \tag{5.9}$$

where U_0^{n+1} is the initial guess for $\overline{U}^{n+1}$. β_{n+1} will be chosen to be k^ν for $0 \le n \le \nu - 2$ and k for $\nu - 1 \le n \le M - 1$. This choice for β_{n+1} requires on the order of $\log(1/k)$ iterations at each time step.

The following algorithm describes the procedure for approximately solving the fully discrete Eqs. (5.7) using a preconditioned iterative method.

ALGORITHM 5.1

For a given positive number γ we proceed as follows.

(1) Let

$$U^0 = \begin{pmatrix} w(0) \\ w^{(1)}(0) \end{pmatrix}$$

or let U^0 be the approximation to the solution $\overline{U}^0$ of

$$\tilde{Q}_0 \overline{U}^0 = \begin{pmatrix} P & 0 \\ 0 & P \end{pmatrix} \left[\begin{pmatrix} u^0 \\ u_t^0 \end{pmatrix} + q_1 k \begin{pmatrix} u_t^0 \\ -L(0)u_t^0 \end{pmatrix} + q_2 k^2 \begin{pmatrix} -L(0)u^0 \\ -\left(\frac{d}{dt}Lu\right)(0) \end{pmatrix} \right], \tag{5.10}$$

using zero as an initial guess and a tolerance $\beta_0 = \min\{\gamma, h^r\}$.

(2) Set

$$\hat{U}_1^1 = \begin{cases} U_1^0, & \nu = 2, \\ U_1^0 + kPu_t(0), & \nu = 3, \\ U_1^0 + kPu_t(0) + \dfrac{k^2}{2}Pu_{tt}(0), & \nu = 4, \end{cases}$$

and

$$\hat{U}_1^2 = \begin{cases} U_1^0, & \nu = 2, \\ U_1^0 + 2kPu_t(0), & \nu = 3, \\ U_0^1 + 2kPu_t(0) + 2k^2 Pu_{tt}(0), & \nu = 4, \end{cases}$$

in (5.8); let U^1 be the approximation to the solution $\overline{U}^1$ of (5.7) using U^0 as an initial guess and a tolerance $\beta_1 = \min\{\gamma, k^\nu\}$; if $\nu = 2$, go to step (5).

(3) Set

$$\hat{U}_1^2 = \begin{cases} 2U_1^1 - U_1^0, & \nu = 3, \\ 4U_1^1 - 3U_1^0 - 2kPu_t(0), & \nu = 4, \end{cases}$$

and

$$\hat{U}_1^3 = \begin{cases} 3U_1^1 - 2U_1^0, & \nu = 3, \\ 9U_1^1 - 8U_1^0 - 6kPu_t(0), & \nu = 4, \end{cases}$$

in (5.8); let U^2 be the approximation to the solution $\overline{U}^2$ of (5.7) using U^1 as an initial guess and a tolerance of $\beta_2 = \min\{\gamma, k^\nu\}$. If $\nu = 3$, go to step (5).

(4) Set $\hat{U}_1^3 = 3U_1^2 - 3U_1^1 + U_1^0$ and $\hat{U}_1^4 = 6U_1^2 - 8U_1^1 + 3U_1^0$ in (5.8); let U^3 be the approximation to the solution $\overline{U}^3$ of (5.7) using U^2 as an initial guess and a tolerance $\beta_3 = \min\{\gamma, k^\nu\}$.

(5) For $\nu \le n + 1 \le M$, set

$$\hat{U}_1^{n+1} = \begin{cases} 2U_1^n - U_1^{n-1}, & \nu = 2, \\ 3U_1^n - 3U_1^{n-1} + U_1^{n-2}, & \nu = 3, \\ 4U_1^n - 6U_1^{n-1} + 4U_1^{n-2} - U_1^{n-3}, & \nu = 4, \end{cases}$$

and

$$\hat{U}_1^{n+2} = \begin{cases} 3U_1^n - 2U_1^{n-1}, & \nu = 2, \\ 6U_1^n - 8U_1^{n-1} + 3U_1^{n-2}, & \nu = 3, \\ 10U_1^n - 20U_1^{n-1} + 15U_1^{n-2} - 4U_1^{n-3}, & \nu = 4, \end{cases}$$

in (5.8); let U^{n+1} be the approximation to the solution $\overline{U}^{n+1}$ of (5.8) using $Z_{n+1}^{(\nu)}(U)$ as an initial guess and a tolerance $\beta_{n+1} = \min\{\gamma, k\}$.

THEOREM 5.2

Let $\{U^n\}_{n=0}^M$ be given by Algorithm 5.1 and $k \le C^*h$ for any constant C^*. If $\nu \ge 3$ and $r \ge 3$ when $N = 2, 3$, then for k, h, and γ sufficiently small there is a constant C which is independent of h and k such that for $n = 0, 1, \ldots, M$,

$$\|\|Q_n(U^n - W(t_n))\|\|_n \le C(h^r + k^\nu) \tag{5.11}$$

and

$$\|U_1^n - u(t_n)\| \le C(h^r + k^\nu). \tag{5.12}$$

Proof. The proof follows from the error equations

$$Q_1 E^1 = \tilde{P}_0 E^1 + (Q_1 - \tilde{Q}_1)E^1 + A_1(U^1 - \overline{U}^1)$$
$$+ \left[-\tilde{Q}_1 W(t_1) + \tilde{P}_0 W(t_0) + p_1 k \begin{pmatrix} 0 \\ f_0 \end{pmatrix} + p_2 k^2 \begin{pmatrix} f_0 \\ f_0^{(1)} \end{pmatrix} \right.$$
$$\left. - q_1 k \begin{pmatrix} 0 \\ f_1 \end{pmatrix} - q_2 k^2 \begin{pmatrix} f_1 \\ f_1^{(1)} \end{pmatrix} \right] + p_2 k^2 \begin{pmatrix} 0 \\ \dfrac{-f_2(\hat{U}_1^2) + 4f_1(\hat{U}_1^1) - 3f_0}{2k} - f_0^{(1)} \end{pmatrix}$$
$$- q_1 k \begin{pmatrix} 0 \\ f_1(\hat{U}_1^1) - f_1 \end{pmatrix} - q_2 k^2 \begin{pmatrix} \dfrac{f_1(\hat{U}_1^1) - f_1}{} \\ \dfrac{f_2(\hat{U}_1^2) - f_0}{2k} - f_1^{(1)} \end{pmatrix}$$
$$+ (\tilde{Q}_1 - A_1)E^1 + (B^0 - \tilde{P}_0)E^0$$
$$+ (\tilde{Q}_1 - A_1)W(t_1) + (B_0 - \tilde{P}_0)W(t_0),$$

and, for $n \ge 1$, $\tag{5.13}$

$$Q_{n+1}E^{n+1} = \tilde{P}_n E^n + (Q_{n+1} - \tilde{Q}_{n+1})E^{n+1} + A_{n+1}(U^{n+1} - \overline{U}^{n+1})$$
$$+ \left[-\tilde{Q}_{n+1} W(t_{n+1}) + \tilde{P}_n W(t_n) + p_1 k \begin{pmatrix} 0 \\ f_n \end{pmatrix} + p_2 k^2 \begin{pmatrix} f_n \\ f_n^{(1)} \end{pmatrix} \right.$$

$$- q_1 k \begin{pmatrix} 0 \\ f_{n+1} \end{pmatrix} - q_2 k^2 \begin{pmatrix} f_{n+1} \\ f^{(1)}_{n+1} \end{pmatrix} \Bigg]$$

$$+ p_1 k \begin{pmatrix} 0 \\ f_n(U^n_1) - f_n \end{pmatrix} + p_2 k^2 \begin{pmatrix} f_n(U^n_1) - f_n \\ \dfrac{f_{n+1}(\hat{U}^{n+1}_1) - f_{n-1}(U^{n-1}_1)}{2k} - f^{(1)}_n \end{pmatrix}$$

$$- q_1 k \begin{pmatrix} 0 \\ f_{n+1}(\hat{U}^{n+1}_1) - f_{n+1} \end{pmatrix} - q_2 k^2 \begin{pmatrix} f_{n+1}(\hat{U}^{n+1}_1) - f_{n+1} \\ \dfrac{f_{n+2}(\hat{U}^{n+2}_1) - f_n(U^n_1)}{2k} - f^{(1)}_{n+1} \end{pmatrix}$$

$$+ (\tilde{Q}_{n+1} - A_{n+1})E^{n+1} + (B_n - \tilde{P}_n)E^n$$

$$+ (\tilde{Q}_{n+1} - A_{n+1})W(t_{n+1}) + (B_n - \tilde{P}_n)W(t_n),$$

using the techniques in Sec. 4. The term $A_{n+1}(U^{n+1} - \overline{U}^{n+1})$ in (5.13) is estimated using (5.4), (5.1), (5.2), and (5.9), and estimates similar to those in Bales[1] for linear second-order hyperbolic equations with time-dependent coefficients.   ■

REFERENCES

1. L. A. Bales, Higher order single step fully discrete approximations for second order hyperbolic equations with time dependent coefficients. To appear in *SIAM J. Numer. Anal.*
2. J. Douglas, Jr., T. Dupont, and R. E. Ewing, Incomplete iteration for time-stepping a Galerkin method for a quasilinear parabolic problem. *SIAM J. Numer. Anal.* **16**, 503–522 (1979).
3. J. H. Bramble and P. H. Sammon, Efficient higher order single step methods for parabolic problems: Part II (submitted for publication).
4. R. E. Ewing, On efficient time-stepping methods for nonlinear partial differential equations. *Computers and Mathematics with Applications* **6**, 1–13 (1980).
5. J. H. Bramble, A. H. Schatz, V. Thomée and L. B. Wahlbin, Some convergence estimates for semidiscrete Galerkin type approximations for parabolic equations. *SIAM J. Numer. Anal.* **14**, 218–241 (1977).
6. G. A. Baker, J. H. Bramble and V. Thomée, Single step galerkin approximations for parabolic problems. *Math. Comp.* **31**, 818–847 (1977).
7. J. H. Bramble and P. H. Sammon, Efficient higher order single step methods for parabolic problems: Part I. *Math. Comp.* **35**, 655–677 (1980).
8. G. A. Baker and J. H. Bramble, Semidiscrete and single step fully discrete approximations for second order hyperbolic equations. *RAIRO Anal. Numer.* **13**, 75–100 (1979).
9. G. A. Baker, V. A. Dougalis and O. A. Karakashian, On multistep Galerkin discretizations of semilinear hyperbolic and parabolic equations. *Nonlinear Analysis, Theory, Methods, and Applications* **4**, 579–597 (1980).
10. L. A. Bales, Semidiscrete and single step fully discrete approximations for second order hyperbolic equations with time dependent coefficients. *Math. Comp.* **43** (1984).
11. M. Huang and V. Thomée, Some convergence estimates for semidiscrete type schemes for time-dependent nonselfadjoint parabolic equations. *Math Comp.* **37**, 327–346 (1981).
12. P. H. Sammon, Convergence estimates for semidiscrete parabolic equation approximations. *SIAM J. Numer. Anal.* **19**, 68–92 (1982).
13. J. D. Lambert, *Computational Methods in Ordinary Differential Equations.* John Wiley & Sons (1973).
14. O. Axelsson, *On Preconditioning and Convergence Acceleration in Sparse Matrix Problems.* CERN European Organization for Nuclear Research, Geneva (1974).
15. L. A Hageman and D. M. Young, *Applied Iterative Methods.* Academic Press (1981).

Comp. & Maths. with Appls. Vol. 12A, Nos. 4/5, pp. 605–613, 1986
Printed in Great Britain.

0886–9553/86 $3.00 + .00
© 1986 Pergamon Press Ltd.

GALERKIN SINGLE-STEP METHODS FOR SECOND-ORDER HYPERBOLIC EQUATIONS

OHANNES A. KARAKASHIAN

Department of Mathematics, University of Tennessee, Knoxville, TN, U.S.A.

Abstract—Single-step methods, coupled with Galerkin discretizations in space, are applied to second-order hyperbolic equations. These methods are applied directly to the second-order equations. Optimal order convergence estimates are derived

INTRODUCTION

We shall consider Galerkin, fully discrete approximations to the solution of the following initial boundary-value problem: Let Ω be a bounded domain in R^N with smooth $\partial\Omega$; we seek a real-valued function $u(x, t)$ satisfying

$$
u_{tt} = -Lu = \sum_{i,j=1}^{N} \frac{\partial}{\partial x_i}\left(l_{ij}(x)\frac{\partial u}{\partial x_j}\right) - l_0(x)u \quad \text{in } \Omega \times (0, \tau],
$$

$$
u = 0 \quad \text{on } \partial\Omega \times (0, \tau],
$$

$$
u(x, 0) = u^0(x \quad \text{in } \Omega,
$$

$$
u_t(x, 0) = u_t^0(x) \quad \text{in } \Omega;
$$

u^0, u_t^0 given. Here, $l_{ij}(x) = l_{ji}(x) \in C^\infty(\overline{\Omega})$, and $l_0 \in C^\infty(\overline{\Omega})$ with $l_0(x) \geq 0$ in $\overline{\Omega}$. We shall assume that the operator L is uniformly elliptic, i.e. for some constant $c_0 > 0$

$$
\sum_{i,j=1}^{N} l_{ij}(x)\zeta_i\zeta_j \geq c_0 \sum_{i=1}^{N} \zeta_i^2, \quad \forall x \in \overline{\Omega}, \forall \zeta \in R^N.
$$

In [1], Baker and Bramble analyzed both semidiscrete and fully discrete Galerkin approximations to the solution of (1.1). The fully discrete methods they considered are based on rational approximations to e^{-z}, and thus require converting the semidiscrete equations into a first-order system.

Another natural approach consists in applying special single-step methods directly to the second-order equation. Among such methods are the Runge–Kutta–Nyström methods (cf. [2]). This approach is more general, and it is well known that in the absence of the first time derivative, an extra order of accuracy can be gained for the same amount of work (see also [3], [4]). For problem (1.1), Gekeler[5] has obtained quasioptimal error estimates under regularity conditions on the initial data stronger than those in [1].

In this paper, we prove optimal order convergence estimates in L^2, for a general class of Nyström methods. In addition, the regularity requirements on the initial data are identical to those of [1]. Both conditionally and unconditionally stable methods are considered.

The paper is organized as follows. In Sec. 2, notation is established. In Sec. 3, the main stability result is proved. It is then shown that the optimal bound on the error can be obtained via an adaptation of the technique used in [1]. In Sec. 4, examples of fully discrete Nyström methods are discussed.

2. NOTATION AND PRELIMINARIES

Let $0 < \lambda_1 \leq \lambda_2 \leq \ldots$ be the eigenvalues of L, and let $\{\phi_j\}_{j \geq 1}$ be the (orthonormal) set of corresponding eigenfunctions. Using these, we can define (cf. [1]) the following spaces: for $s \geq 0$,

$$
\dot{H}^s(\Omega) = \left\{ v: \|v\|_s \equiv \left\{ \sum_{j=1}^{\infty} \lambda_j^s(v, \phi_j)^2 \right\}^{1/2} < \infty \right\}.
$$

It is shown in [6] that $\dot{H}^s(\Omega) = \{v \in H^s(\Omega), L^j v = 0 \text{ on } \partial\Omega, j < s/2\}$, where $H^s(\Omega)$ are the usual Sobolev spaces.

It is well known that for $s \geq 1$, if $u^0 \in \dot{H}^s(\Omega)$ and $u_t^0 \in \dot{H}^{s-1}(\Omega)$, then a unique solution to (1.1) exists for all $t > 0$ and that the following estimate holds:

$$\|u(t)\|_s^2 + \|u_t(t)\|_{s-1}^2 = \|u^0\|_s^2 + \|u_t^0\|_{s-1}^2. \tag{2.1}$$

We shall henceforth assume that the solution u is sufficiently smooth to guarantee the convergence estimates below.

Let $T: L^2 \to L^2$ be the solution operator of the associated elliptic problem

$$Lu = f \quad \text{in } \Omega,$$
$$u = 0 \quad \text{on } \partial\Omega.$$

For integer $r \geq 2$, let $\{S_h^r\}_{h>0}$ be a one-parameter family of finite-dimensional subspaces of $L^2(\Omega)$, e.g. the space of piecewise polynomial functions defined on a triangulation of Ω. We assume the existence of a family $\{T_h\}_{h>0}$ of operators $T_h: L^2(\Omega) \to S_h^r$ possessing the following properties.

(i) T_h is symmetric, positive semidefinite on $L^2(\Omega)$ and positive definite on S_h^r.
(ii) $\|(T - T_h)f\| \leq ch^j \|f\|_{j-2}$, $2 \leq j \leq r$, where c is independent of h.
(iii) T_h has eigenvalues $0 \leq \mu_1 \leq \mu_2 \leq \cdots \leq \mu_{d(h)}$. Moreover, $\mu_{d(h)} \leq c$ for some constant c independent of h.

Several types of approximating operators T_h satisfying (i)–(iii) are well known. These include the Galerkin method, two methods of Nitsche and the Lagrange multiplier method of Babuška (see [1,7–10]).

With T_h at hand, the semidiscrete approximation $u_h(t) \in S_h^r$ of $u(t)$ is defined by

$$T_h u_{htt}(t) + u_h(t) = 0, \quad t > 0,$$
$$u_h(0) = Pu^0, \tag{2.2}$$
$$u_{ht}(0) = Pu_t^0,$$

where $P: L^2 \to S_h^r$ is the L^2 projection operator.

Denoting the inverse of T_h on S_h^r by L_h, (2.2) can be rewritten as

$$u_{htt}(t) + L_h u_h(t) = 0,$$
$$u_h(0) = Pu^0, \tag{2.3}$$
$$u_{ht}(0) = Pu_t^0.$$

It is proven in [1,11] that for some constant c independent of h

$$\|u_h(t) - u(t)\| \leq ch^r(\|u^0\|_{r+1} + \|u_t^0\|_r). \tag{2.4}$$

Now let $\langle \cdot, \cdot \rangle$ and $|\cdot|$ denote the Euclidean inner product and norm on R^2. For $\mu > 0$, we define the following inner product on R^2

$$\langle x, y \rangle_\mu = x_1 y_1 + \mu^{-2} x_2 y_2$$

and let $|\cdot|_\mu$ denote the corresponding norm.

We shall also use the following seminorm on $L^2 \times L^2$: for $u = [u_1, u_2] \in L^2 \times L^2$, $\|\|u\|\|^2 = \|u_1\|^2 + \|T_h^{1/2} u_2\|^2$; note that $\|\|\cdot\|\|$ is a norm on $S_h^r \times S_h^r$.

Remark. All norm notations will also be used to denote corresponding operator norms.

3. FULLY DISCRETE APPROXIMATIONS

We consider operators $\mathscr{R}:R^2 \to R^2$ of the form

$$\mathscr{R}(x, z) = \begin{bmatrix} r_{11}(z) & x\,r_{12}(z) \\ -x^{-1}z\,r_{21}(z) & r_{22}(z) \end{bmatrix},$$

where x is a positive parameter and r_{ij} are rational functions in z with $r_{ij}(0) = 1$, $i, j = 1, 2$. These operators are used to generate fully discrete approximations to u, the solution of (1.1) in the following way: Let $k > 0$ be the time step, w_h^n, w_{ht}^n approximations in S_h^r to $u(nk)$ and $u_t(nk)$ respectively, then

$$\begin{aligned} \begin{bmatrix} w_h^{n+1} \\ w_{ht}^{n+1} \end{bmatrix} &= \mathscr{R}(k, k^2 L_h) \begin{bmatrix} w_h^n \\ w_{ht}^n \end{bmatrix} \\ &= \begin{bmatrix} r_{11}(k^2 L_h) & k\,r_{12}(k^2 L_h) \\ -kL_h\,r_{21}(k^2 L_h) & r_{22}(k^2 L_h) \end{bmatrix} \begin{bmatrix} w_h^n \\ w_{ht}^n \end{bmatrix}, \end{aligned} \tag{3.1}$$

with $w_h^0 = Pu^0$, $w_{ht}^0 = Pu_t^0$.

Formulation (3.1) is in general appropriate only for theoretical purposes. The practical implementation of these methods must take a computationally more efficient formulation. From this point of view, Runge–Kutta–Nyström methods form a particularly interesting class,

$$w_h^{n,i} = w_h^n + k\alpha_i w_{ht}^n - k^2 \sum_{j=1}^{q} a_{ij} L_h w_h^{n,j}, \quad i = 1, \ldots, q$$

$$w_h^{n+1} = w_h^n + kw_{ht}^n - k^2 \sum_{i=1}^{q} b_i L_h w_h^{n,i} \tag{3.2}$$

$$w_{ht}^{n+1} = w_{ht}^n - k \sum_{i=1}^{q} \bar{b}_i L_h w_h^{n,i} \,;$$

here q, α_i, a_{ij}, b_i, $\bar{b}_i$ are given constants and completely determine the method.

We next characterize the accuracy and stability properties of these methods. From (2.3) and (3.1), for $n \geq 0$,

$$\begin{aligned} \begin{bmatrix} \xi^{n+1} \\ \xi_t^{n+1} \end{bmatrix} &= \left[e^{-k \begin{bmatrix} 0 & -I \\ L_h & 0 \end{bmatrix}} - \mathscr{R}(k, k^2 L_h) \right] \begin{bmatrix} \xi^n \\ \xi_t^n \end{bmatrix} \\ &= \begin{bmatrix} \cos kL_h^{1/2} - r_{11}(k^2 L_h) & L_h^{-1/2}[\sin kL_h^{1/2} - kL_h^{1/2} r_{12}(k^2 L_h)] \\ -L_h^{1/2}[\sin kL_h^{1/2} - kL_h^{1/2} r_{21}(k^2 L_h)] & \cos kL_h^{1/2} - r_{22}(k^2 L_h) \end{bmatrix} \begin{bmatrix} \xi^n \\ \xi_t^n \end{bmatrix}, \end{aligned}$$

where $[\xi^n, \xi_t^n]^T = [u_h(t_n) - w_h^n, u_{ht}(t^n) - w_{ht}^n]^T$, $n \geq 0$. This motivates us to require $\mathscr{R}$ to satisfy the following consistency condition: there exist positive constants c, σ such that

$$\sum_{i=1}^{2} \{|\cos z - r_{ii}(z^2)| + |\sin z - z\,r_{i,3-i}(z^2)|\} \leq cz^{\nu+1}, \quad 0 \leq z \leq \sigma \tag{3.3}$$

for some integer $\nu \geq 1$. We say that $\mathscr{R}$ is consistent of order ν.

We gather the stability conditions into four groups with $\sigma(\mathscr{R})$ denoting the spectral radius of $\mathscr{R}$.

Type I   (1) $\sigma(\mathscr{R}(1, z)) < 1$ for some $s < \infty$ $0 < z \leq s$
              (2) $r_{11} = r_{22}$
Type II   (1) $\sigma(\mathscr{R}(1, z)) \leq 1$ $0 \leq z \leq s$
              (2) $r_{11} = r_{22}$
              (3) $0 < \underline{b} \leq \dfrac{|r_{12}(z)|}{|r_{21}(z)|} \leq \bar{b} < \infty$ $0 < z \leq s$

Type III   (1) $\sigma(\mathscr{R}(1, z)) < 1, 0 < z \le s,$ for any finite s

           (2) $r_{11} = r_{22}$

           (3) $0 < \lim\limits_{z \to \infty} \left| \dfrac{r_{12}(z)}{r_{21}(z)} \right| < \infty$

Type IV   (1) $\sigma(\mathscr{R}(1, z)) \le 1$ $0 \le z < \infty$

           (2) $r_{11} = r_{22}$

           (3) $0 < \underline{b} \le \left| \dfrac{r_{12}(z)}{r_{21}(z)} \right| \le \overline{b} < \infty$ $0 < z < \infty.$

We next prove our main stability results.

THEOREM 3.1

Let $\mathscr{R}$ be of types I or II and suppose that $k^2\sigma(L_h) \le s$. Then there exists a constant c independent of h and k such that

$$||| \mathscr{R}^n(k, k^2 L_h) ||| \le c, \quad \forall n \ge 0.$$

Proof. Let $\{\phi_j\}_{j=1}^d \subset S_h^r$ be an orthonormal set of eigenfunctions of L_h, and $\{\lambda_j\}_{j=1}^d$ the set of corresponding eigenvalues arranged in nondecreasing order. Let v_h^i, v_{ht}^i, $i = 0, 1, \ldots, n$ in S_h^r be such that

$$\begin{bmatrix} v_h^n \\ v_{ht}^n \end{bmatrix} = (\mathscr{R}(k, k^2 L_h))^n \begin{bmatrix} v_h^0 \\ v_{ht}^0 \end{bmatrix} \equiv \mathscr{R}^n(k, k^2 L_h) \begin{bmatrix} v_h^0 \\ v_{ht}^0 \end{bmatrix}. \tag{3.4}$$

Letting $v_h^i = \sum_{j=1}^d \alpha_j^i \phi_j$, $v_{ht}^i = \sum_{j=1}^d \beta_j^i \phi_j$ by the orthonormality of $\{\phi_j\}_{j=1}^d$, we have

$$\begin{bmatrix} \alpha_j^n \\ \beta_j^n \end{bmatrix} = \mathscr{R}^n(k, k^2\lambda_j) \begin{bmatrix} \alpha_j^0 \\ \beta_j^0 \end{bmatrix}$$

$$= \begin{bmatrix} r_{11}(k^2\lambda_j) & kr_{12}(k^2\lambda_j) \\ -k\lambda_j r_{21}(k^2\lambda_j) & r_{22}(k^2\lambda_j) \end{bmatrix}^n \begin{bmatrix} \alpha_j^0 \\ \beta_j^0 \end{bmatrix}, \quad j = 1, 2, \ldots, d. \tag{3.5}$$

Hence

$$\begin{bmatrix} \alpha_j^n \\ \lambda_j^{-1/2} \beta_j^n \end{bmatrix} = \tilde{\mathscr{R}}^n(k, k^2\lambda_j) \begin{bmatrix} \alpha_j^0 \\ \lambda_j^{-1/2} \beta_j^0 \end{bmatrix}, \tag{3.6}$$

where

$$\tilde{\mathscr{R}}(k, k^2\lambda_j) = \begin{bmatrix} r_{11}(k^2\lambda_j) & k\lambda_j^{1/2} r_{12}(k^2\lambda_j) \\ -k\lambda_j^{1/2} r_{21}(k^2\lambda_j) & r_{22}(k^2\lambda_j) \end{bmatrix}, \quad j = 1, 2, \ldots, d.$$

We first consider a type II method. For each $j, j = 1, \ldots, d$, let b_j be a nonzero scalar, to be suitably chosen below. Let $B_j = \text{diag}\{1, b_j\}$. Consider $N_j = B_j \tilde{\mathscr{R}}(k, k^2\lambda_j)B_j^{-1}$. A simple computation shows that N_j is normal provided

$$|b_j|^2 = \left| \frac{r_{12}(k^2\lambda_j)}{r_{21}(k^2\lambda_j)} \right|. \tag{3.7}$$

Hence

$$\tilde{\mathscr{R}}(k, k^2\lambda_j) = B_j^{-1} U_j^* M_j U_j B_j,$$

where U_j is unitary and $M_j = \text{diag}\{\mu_{1j}, \mu_{2j}\}$; μ_{1j}, μ_{2j} being the eigenvalues of $\tilde{\mathscr{R}}(k, k^2\lambda_j)$. These are also the eigenvalues of $\mathscr{R}(1, k^2\lambda_j)$, and hence $\max\{|\mu_{1j}|, |\mu_{2j}|\} \le 1$, $j =$

$1, 2, \ldots, d$. Thus

$$
\begin{aligned}
|\tilde{\mathscr{R}}^n(k, k^2\lambda j)| &\le |B_j^{-1}|\,|B_j| \\
&\le \max\{1, |b_j|, |b_j|^{-1}\} \\
&\le \max\{\bar{b}, \underline{b}^{-1}\}.
\end{aligned}
\tag{3.8}
$$

Next, suppose the method is of type I and fix $\delta > 0$ sufficiently small, so that in view of $r_{12}(0) = r_{21}(0) = 1$, we have for $\underline{b}, \bar{b} > 0$,

$$
0 < \underline{b} \le \left| \frac{r_{12}(z)}{r_{21}(z)} \right| \le \bar{b} < \infty, \quad 0 \le z \le \delta.
$$

If $k^2\lambda_j \le \delta$, then the argument leading to (3.8) can be used. If $k^2\lambda j > \delta$, then we let

$$
\tilde{\mathscr{R}}(k, k^2\lambda_j) = U_j^* \begin{bmatrix} \mu_{1j} & c_j \\ 0 & \mu_{2j} \end{bmatrix} U_j,
$$

where U_j is unitary. Since $\delta < k^2\lambda_j \le s$, and the method is of type I, $\max\{|\mu_{1j}|, |\mu_{2j}|\} \le \mu < 1$; it also holds that $|c_j| \le c$ independently of h and k. Hence

$$
|\tilde{\mathscr{R}}^n(k, k^2\lambda_j)| = \left\| \begin{bmatrix} \mu_{1j}^n & c_j \displaystyle\sum_{l=0}^{n-1} \mu_{1j}^l \mu_{2j}^{n-1-l} \\ 0 & \mu_{2j}^n \end{bmatrix} \right\| \le c, \quad n \ge 1,
\tag{3.9}
$$

where c depends on δ and c_j but is otherwise independent of h and k.

To conclude the proof, we note that

$$
\left\| \left\| \begin{bmatrix} v_h^n \\ v_{ht}^n \end{bmatrix} \right\| \right\|^2 = \sum_{j=1}^d ((\alpha_j^n)^2 + \lambda_j^{-1}(\beta_j^n)^2).
$$

It follows then from (3.8) or (3.9) that

$$
\left\| \left\| \begin{bmatrix} v_h^n \\ v_{ht}^n \end{bmatrix} \right\| \right\|^2 \le c \sum_{j=1}^d ((\alpha_j^0)^2 + \lambda_j^{-1}(\beta_j^0)^2)
$$

$$
= c \left\| \left\| \begin{bmatrix} v_h^0 \\ v_{ht}^0 \end{bmatrix} \right\| \right\|^2.
$$

Theorem 3.2

Let $\mathscr{R}$ be of type III or IV. Then there exists a constant c independent of h and k such that

$$
\||\mathscr{R}^n(k, k^2 L_h)\|| \le c.
$$

Proof. The proof in the case of type IV methods can be done in a manner identical to the proof of a type II method, upon taking $s = \infty$.

For type III methods, the proof proceeds as follows. We write $[0, \infty) = [0, \beta) \cup [\beta, \infty)$, where β is chosen sufficiently large so that $r_{12}(z)/r_{21}(z)$ has neither zeros nor poles on $[\beta, \infty)$. In view of $0 < \lim_{z \to \infty} |r_{12}(z)/r_{21}(z)| < \infty$, we have $0 < \underline{b} \le |r_{12}(z)/r_{21}(z)| \le \bar{b} < \infty, \beta \le z < \infty$. Now on $[0, \beta)$, the proof proceeds exactly as in the case of type I methods, and on $[\beta, \infty)$ as in the case of type II methods. This completes the proof.

Remarks. (1) For type I and type II methods, a condition relating the sizes of h and k must be imposed. For subspaces S_h^r with elements satisfying inverse assumptions of the form

$$\|x\|_1 \le ch^{-1}\|x\|,$$

it is easily shown that the aforementioned condition takes the form $kh^{-1} \le c$ for some fixed c.

(2) Type II and type IV methods contain as special cases the methods considered in [1]. For these methods it can be shown that $r_{11} = r_{22}$ and $r_{12} = r_{21}$; it then follows that $\tilde{\mathscr{R}}^T(k, k^2\lambda_j)\tilde{\mathscr{R}}(k, k^2\lambda_j) = (r_{11}^2 + k^2\lambda_j r_{12}^2)I_{2\times 2}$. Now noting that $r_{11}^2 + k^2\lambda_j r_{12}^2 = (\sigma(\tilde{\mathscr{R}}(k, k^2\lambda_j)))^2 \le 1$ for $k^2\lambda_j \le s$, we can replace the constant c in Theorems 3.1 and 3.2 by 1, thus recovering the stability results of [1].

To derive the error estimates, we adopt the technique of [1]. For $t > 0$, $n \ge 1$, let

$$F_n(t, t^2y^2) = e^{-nt\begin{bmatrix} 0 & -1 \\ y^2 & 0 \end{bmatrix}} - \mathscr{R}^n(t, t^2y^2). \tag{3.10}$$

In the sequel, we shall take $s = \infty$ in the case of methods of type III or IV.

Proposition 3.3

Let $\mathscr{R}$ have order v and be of one of the types I–IV and suppose $k^2\sigma(L_h) \le s$. Then there exists a constant c^*, independent of h and k, such that

$$|F_n(t, t^2y^2)|_y \le nc^*(ty)^l, \quad n \ge 1, \tag{3.11}$$

for $ty \le s$ and $1 \le l \le v + 1$.

Proof. Let $\sigma^* = \min(\sigma, 1)$. We have after a simple calculation

$$S^{-1}F_1(t, t^2y^2)S = \begin{bmatrix} \cos ty - r_{11}(t^2y^2) & \sin ty - tyr_{12}(t^2y^2) \\ -\sin ty + tyr_{21}(t^2y^2) & \cos ty - r_{22}(t^2y^2) \end{bmatrix}, \tag{3.12}$$

where $S = \text{diag}\{1, y\}$. It follows from (3.3) that

$$\begin{aligned} |F_1(t, t^2y^2)|_y &= |S^{-1}F_1(t, t^2y^2)S| \\ &\le c(ty)^{v+1} \le c(ty)^l, \quad ty \le \sigma^*, 1 \le l \le v + 1. \end{aligned} \tag{3.13}$$

Now suppose $\sigma^* < s$. Noting that,

$$\left| e^{-jt\begin{bmatrix} 0 & -1 \\ y^2 & 0 \end{bmatrix}} \right|_y = 1, \quad j \ge 0, t, y > 0; \tag{3.14}$$

for $\sigma^* < ty \le s$, we have, using Theorem 3.1 or Theorem 3.2,

$$\begin{aligned} |F_1(t, t^2y^2)|_y &\le c = c(ty)^l \left(\frac{\sigma^*}{ty}\right)^l (\sigma^*)^{-l} \\ &\le c(ty)^l(\sigma^*)^{-(v+1)} \le c(ty)^l. \end{aligned} \tag{3.15}$$

Now for any integer $n \ge 1$, from Theorem 3.1 or Theorem 3.2, (3.13), (3.14) and (3.15), for $ty \le s$ we have

$$\begin{aligned} |F_n(t, t^2y^2)|_y &= \left| \sum_{j=0}^{n-1} \mathscr{R}^j(t, t^2y^2)F_1(t, t^2y^2)e^{-(n-1-j)t\begin{bmatrix} 0 & -1 \\ y^2 & 0 \end{bmatrix}} \right|_y \\ &\le cn|F_1(t, t^2y^2)|_y \le nc^*(ty)^l, \quad 1 \le l \le v + 1. \end{aligned}$$

PROPOSITION 3.4

Under the conditions of Proposition 3.3, there exists a constant c independent of h and k such that for all $z \in L^2 \times L^2$

$$\left\| \left| F_n(k, k^2 L_h) \begin{bmatrix} 0 & -T_h \\ I & 0 \end{bmatrix}^l z \right| \right\| \le ck^{l-1}\||z\||, \quad 1 \le l \le \nu + 1 \tag{3.16}$$

Proof. We first prove (3.16) for l even. Letting

$$z = \sum_{j=1}^{d} \begin{bmatrix} \alpha_j \\ \beta_j \end{bmatrix} \phi_j \equiv \sum_{j=1}^{d} \Gamma_j \phi_j,$$

we find

$$E_n \equiv F_n(k, k^2 L_n) \begin{bmatrix} 0 & -T_h \\ I & 0 \end{bmatrix}^l z = (-1)^{l/2} F_n(k, k^2 L_h) \begin{bmatrix} T_h^{l/2} & 0 \\ 0 & T_h^{l/2} \end{bmatrix} z$$

$$= (-1)^{l/2} \sum_{j=1}^{d} \lambda_j^{-l/2} F_n(k, k^2 L_h) \Gamma_j \phi_j.$$

It is easily verified that

$$\||E_n\||^2 = \sum_{j=1}^{d} \lambda_j^{-l} |F_n(k, k^2\lambda_j) \Gamma_j|^2_{\lambda_j^{1/2}}$$

$$\le \sum_{j=1}^{d} \lambda_j^{-l} |F_n(k, k^2\lambda_j)|^2_{\lambda_j^{1/2}} |\Gamma_j|^2_{\lambda_j^{1/2}}. \tag{3.17}$$

Since

$$\||z\||^2 = \sum_{j=1}^{d} |\Gamma_j|^2_{\lambda_j^{1/2}},$$

from (3.11) it follows that

$$\||E_n\||^2 \le \sum_{j=1}^{d} \lambda_j^{-l} n^2 (c^*)^2 (k\lambda_j^{1/2})^{2l} |\Gamma_j|^2_{\lambda_j^{1/2}}$$

$$\le (c^*)^2 k^2 n^2 k^{2l-2} \sum_{j=1}^{d} |\Gamma_j|^2_{\lambda_j^{1/2}}$$

$$\le ck^{2l-2}\||z\||^2. \tag{3.18}$$

Now for l odd, we have

$$\||E_n\||^2 = \left\| \left| F_n(k, k^2 L_h) \begin{bmatrix} 0 & -T_h \\ I & 0 \end{bmatrix}^{l+1} \begin{bmatrix} 0 & -T_h \\ I & 0 \end{bmatrix}^{-1} z \right| \right\|^2$$

$$= \left\| \left| F_n(k, k^2 L_n) \begin{bmatrix} 0 & -T_h \\ I & 0 \end{bmatrix}^{l+1} z^1 \right| \right\|^2,$$

where

$$z^1 = \sum_{j=1}^{d} \begin{bmatrix} \alpha_j^1 \\ \beta_j^1 \end{bmatrix} \phi_j \equiv \sum_{j=1}^{d} \Gamma_j^1 \phi_j, \quad \text{with } \alpha_j^1 = \beta_j, \ \beta_j^1 = -\lambda_j \alpha_j.$$

Hence, from (3.17) and (3.11),

$$|||E_n|||^2 \leq \sum_{j=1}^{d} \lambda_j^{-(l+1)} |F_n(k, k^2\lambda_j)|_{\lambda_j^{1/2}}^2 |\Gamma_j^1|_{\lambda_j^{1/2}}^2$$

$$\leq \sum_{j=1}^{d} \lambda_j^{-(l+1)} (nc^*)^2 (k\lambda_j^{1/2})^{2l} |\Gamma_j^1|_{gl_j^{1/2}}^2 .$$

Now since

$$\sum_{j=1}^{d} \lambda_j^{-1} |\Gamma_j^1|_{\lambda_j^{1/2}}^2 = |||z|||^2 ,$$

the result follows.

With Propositions 3.3 and 3.4 at hand, we can use Theorem 3.1 of [1]. So let $E_n^T = [u_h(nk) - w_h^n, u_{ht}(nk) - w_{ht}^n]^T$, where $[u_h(nk), u_{ht}(nk)]^T$ and $[w_h^n, w_{ht}^n]^T$ are given by (2.2) and (3.1), respectively. We have the following.

THEOREM 3.5

Under the conditions of Proposition 3.3, there exists a constant c independent of h and k such that

$$\sup_{0 \leq n \leq \tau/k} |||E_n||| \leq ch^r[\|u^0\|_{r+1} + \|u_t^0\|_r] + ck^v[\|u^0\|_{v+1} + \|u_t^0\|_v]. \tag{3.19}$$

Remark. Using (3.19) with (2.4) we get

$$\sup_{0 \leq n \leq \tau/k} \|u(nk) - w_h^n\| \leq ch^r[\|u^0\|_{r+1} + \|u_t^0\|_r]$$

$$+ ck^v[\|u^0\|_{v+1} + \|u_t^0\|_v]. \tag{3.20}$$

Moreover, similar convergence estimates can be obtained for any initial data $[w_h^0, w_{ht}^0]^T$ satisfying $\|u^0 - w_h^0\| + \|u_t^0 - w_{ht}^0\| \leq ch^r$.

4. EXAMPLES

We now consider two examples of Nyström methods as given by (3.2). First, letting $(A)_{ij} = a_{ij}$, $b^T = (b_1, \ldots, b_q)$, $\bar{b}^T = (\bar{b}_1, \ldots, \bar{b}_q)$, $e^T = (1, 1, \ldots, 1)$, $\alpha^T = (\alpha_1, \ldots, \alpha_q)$, after simple calculations

$$r_{11}(z) = 1 - zb^T(I + zA)^{-1}e, \quad r_{22}(z) = 1 - z\bar{b}^T(I + zA)^{-1}\alpha$$

$$r_{12}(z) = 1 - zb^T(I + zA)^{-1}\alpha, \quad r_{21}(z) = \bar{b}^T(I + zA)^{-1}e. \tag{4.1}$$

We first consider an explicit method (cf. [4,5]). Here $q = 3$, $v = 4$, $s = 6$:

$$A = \begin{bmatrix} 0 & 0 & 0 \\ 1/8 & 0 & 0 \\ 0 & 1/2 & 0 \end{bmatrix}, \quad \alpha^T = (0, 1/2, 1),$$

$$b^T = (\frac{1}{6}, \frac{1}{3}, 0), \quad \bar{b}^T = \left[\frac{1}{6}, \frac{4}{6}, \frac{1}{6}\right];$$

$$r_{11}(z) = r_{22}(z) = 1 - \frac{z}{2} + \frac{z^2}{24},$$

$$r_{12}(z) = 1 - \frac{z}{6},$$

$$r_{21}(z) = 1 - \frac{z}{6} + \frac{z^2}{96}.$$

The implementation of this method requires three matrix-vector multiplications (function evaluations) per step, and solutions of three linear systems $Gx = \rho$, where G is the Gramian matrix. The corresponding fourth-order polynomial approximation to e^{-z}, on the other hand, requires four function evaluations and solutions of four linear systems.

We next consider a two-stage fourth-order implicit method. Let $\beta = 1/6 \, (1 + \sqrt{3/2})$ (largest root of $144\beta^2 - 48\beta + 1$). For $\gamma \neq \beta$, we have the one-parameter family of methods

$$A = \begin{bmatrix} \beta & 0 \\ \gamma - \beta & \beta \end{bmatrix}, \quad b^T = (\gamma - \beta)^{-1} \left(\frac{\gamma}{2} - \frac{1}{24}, \ -\frac{\beta}{2} + \frac{1}{24} \right),$$

$$\bar{b}^T = (\gamma - \beta)^{-1} \left(\gamma - \frac{1}{6}, \ -\beta + \frac{1}{6} \right), \quad \alpha^T = \left(2\beta + \frac{1}{6}, \ 2\gamma + \frac{1}{6} \right).$$

This method requires two function evaluations and solutions of two linear systems with the same coefficient matrix $(G + \beta k^2 S)$, where S is the stiffness matrix. However, this method is only conditionally stable (type II) for $s < 12$.

REFERENCES

1. G. A. Baker and J. H. Bramble, Semidiscrete and single step fully discrete approximations for second order hyperbolic equations. *R.A.I.R.O. Analyse Numer.* **13**(2), 75–100 (1979).
2. E. Hairer and G. Wanner, A theory for Nyström methods. *Numer. Math.* **25**, 383–400 (1976).
3. J. D. Lambert, *Computational Methods in Ordinary Differential Equations.* John Wiley, New York (1973).
4. R. E. Scraton, The numerical solution of second-order differential equations not containing the first derivative explicitly. *Comput. J.* **6**, 368–370 (1954).
5. E. Gekeler, Galerkin–Runge–Kutta methods and hyperbolic initial value problems. *Computing* **18**, 79–88 (1977).
6. J. H. Bramble and V. Thomée, Discrete time Galerkin methods for a parabolic boundary value problem. *Ann. Mat. Pura. Appl. Series IV* **101** 115–152 (1974).
7. I. Babuska, Survey lectures on the mathematical foundations of the finite element method, in *The Mathematical Foundations of the Finite Element Method with Applications to Partial Differential Equations* (Edited by A. K. Aziz). Academic Press, New York (1972).
8. I. Babuska, The finite element method with Lagrangian multipliers. *Numer. Math.* **20**, 179–192 (1973).
9. J. Nitsche, Uber ein variations prinzip zur losung von Dirichletproblemem bei verwendung von teilraumen, die Keinen rand bedingungen unterworfen sind. *A.B.H. Math. Ser. Univ. Hamburg,* **37**, 9–15 (1971).
10. J. Nitsche, On Dirichlet problems using subspaces with nearly zero boundary conditions, in *The Mathematical Foundations of The Finite Element Method with Applications to Partial Differential Equations* (Edited by A. K. Aziz) pp. 603–627. Academic Press, New York (1972).
11. G. A. Baker, Error estimates for finite element methods for second order hyperbolic equations. *SIAM J. Numer. Anal.* **13**, 564–576 (1976).
12. M. Crouzeix, Sur l'approximation des equations différentielles operationelle par des méthodes de Runge–Kutta. Thèse, Univ. de Paris VI (1975).
13. T. Dupont, L^2-estimates for Galerkin methods for second order hyperbolic equations. *SIAM J. Numer. Anal.* **10**, 880–889 (1973).
14. E. Hairer, Unconditionally stable methods for second order differential equations. *Numer. Math.* **32**, 373–379 (1979).
15. R. Jeltsch and O. Nevanlinna, Stability of semidiscretizations of hyperbolic problems. *SIAM J. Numer. Anal.* **20**, 1210–1218 (1983).

Comp. & Maths. with Appls. Vol. 12A, Nos. 4/5, pp. 615–630, 1986
Printed in Great Britain.

0097–4943/86 $3.00 + .00
1986 Pergamon Press Ltd.

CONSTRUCTION OF SOLUTIONS FOR
TWO-DIMENSIONAL RIEMANN PROBLEMS

W. B. LINDQUIST
Courant Institute of Mathematical Sciences, New York University, New York, NY 10012, U.S.A.

Abstract—Solutions to the scalar quasilinear equation

$$\partial u(t, x)/\partial t + \Sigma_{i=1}^{2} \partial f_i(u(t, x))/\partial x_i = 0,$$

for $f_i \in C^2 : R \to R$ with initial data given by a two-dimensional Riemann problem, are piecewise smooth if $f_1 \equiv f_2 \equiv f$, and f has at most one inflection point. We show that the "pieces" of this solution can be classified and are expressible in terms of two-dimensional nonlinear waves in analogy with the nonlinear rarefaction and shock waves of the Riemann problem in one spatial dimension. The two-dimensional waves can be expressed in almost-closed form. Explicit solutions are constructable from these waves. An application is illustrated by calculation of the interaction of water/oil banks in two-phase incompressible flow in reservoirs.

1. INTRODUCTION

The Cauchy problem

$$\frac{\partial u(t, x)}{\partial t} + \sum_{i=1}^{2} \frac{\partial f_i(u(t, x))}{\partial x_i} = 0, \tag{1.1}$$

with initial data piecewise constant on a finite number of wedges focused on a single point in the x, y plane, is defined as a *two-dimensional Riemann problem*. Without loss of generality, this point can be taken to be the point $x = 0$, $y = 0$. In a previous paper[1] we have shown that, for $f_i : R \to R$ in C^2, if $f_1 \equiv f_2 \equiv f$, with f having at most one inflection point, the solution to the two-dimensional Riemann problem is piecewise smooth. The proof consists of two parts. First, it is shown that the two-dimensional Riemann problem for $f_1 \equiv f_2 \equiv f$ is equivalent to a generalization of the Riemann problem in one dimension. It is then proven that the generalized one-dimensional Riemann problem is piecewise smooth if f has at most one inflection point. The proof of the smoothness of the solutions to the two-dimensional Riemann problem does not provide a convenient method of constructing its solutions. In this paper we formulate a construction that is purely two dimensional and has the advantage (at least in specific cases) of generalizing to the problem $f_1 \neq f_2$. This construction consists of identifying the general two-dimensional nonlinear waves, analogous to the rarefaction and shock waves of the one-dimensional Riemann problem, and then using the method initiated by Guckenheimer[2] and Wagner[3] to piece these waves together into an entropy-obeying solution.

Existence and uniqueness of solutions to the two-dimensional Riemann problem within the class of bounded, measurable functions is due to Kružkov[4]. For piecewise-smooth solutions, Kružkov's uniqueness condition reduces to two requirements on the jump discontinuities in the solution:

$$n \cdot [u^+ - u^-, f(u^+) - f(u^-)] = 0, \tag{1.2}$$

and

$$n \cdot [k - u^-, f(k - f(u^-)] \geq 0, \tag{1.3}$$

where f stands for the vector $[f_1, f_2]$. In (1.2), the normal n to the discontinuity is oriented such that $u^- \leq u^+$ and k is any constant such that $u^- \leq k \leq u^+$. We refer to [1] for definition of the terms used here and throughout this article.

Equation (1.2) is denoted the jump condition for the discontinuity and (1.3) the entropy condition; in $R^1 \times R^+$ (1.2) is the familiar Rankine–Hugoniot condition and (1.3) is equivalent to the entropy condition of Oleĭnik[5]. The local nature of conditions (1.2) and (1.3) allow construction of piecewise-smooth global solutions by piecing together local solutions which individually obey (1.2) and (1.3).

In Sec. 2, the two-dimensional nonlinear waves are described for the class of problems $f_1 \equiv f_2$. In Sec. 3 a method for explicitly constructing the two-dimensional solutions in this class from the nonlinear waves is described. Complete solutions for two-dimensional Riemann problems for representative forms of the function f are given. Solutions to the problem of water/ oil bank interactions in two-phase, incompressible, gravity-free flow in reservoirs can be obtained from the study of these two-dimensional Riemann problems. The results for this physical problem are presented in Sec. 4. Conclusions and conjectures for the solutions to two-dimensional Riemann problems for the case f having more than one inflection points or $f_1 \neq f_2$ are presented in Sec. 5.

2. THE TWO-DIMENSIONAL RIEMANN PROBLEM

For convenience we shall use the notation $f_1 \equiv f$, $f_2 \equiv g$, $x_1 \equiv x$, $x_2 \equiv y$.

A sufficient condition for a two-dimensional Riemann problem to have a piecewise-smooth solution is given by [1].

THEOREM 2.1

The unique (in the sense of Kružkov) solution in the plane $y > x$, to (1.1) with initial data that is piecewise constant on a finite number of wedges focused on a single point in the plane, with $f_1 \equiv f_2 \equiv f$, $f \in C^2 : R \to R$, f having at most one inflection point, is

(a) piecewise smooth;
(b) composed of nonlinear waves and constant states which are the images under a continuous map M of the rarefaction and shock waves and constant states of a generalized Riemann problem in one dimension;
(c) composed of curves of irregular points corresponding to images under the map M of the irregular points in the one-dimensional Riemann problem.

For the case $g \equiv f$, under the 45° rotation $2\xi = x + y$, $2\eta = y - x$, (1.1) becomes

$$\frac{\partial u}{\partial t} + \frac{\partial f(u)}{\partial \xi} = 0. \tag{2.1}$$

From (2.1) we see that the solution can be obtained along each $\eta = $ const plane independent of other η. In particular, this implies the solutions obtained in the $\eta < 0$ half-space can be obtained independently of the solutions in the $\eta > 0$ half; by the symmetry of the problem, no new solutions will be found in the $\eta < 0$ halfspace that are not found in the upper. We therefore restrict our discussion to the half-space $\eta > 0$ $(y > x)$. The solution in the plane $\eta = 0$ is given in Lemma 2.2 below (see [1]).

LEMMA 2.2

(a) If the half-line $\eta = 0$, $\xi < 0$ is not a line of discontinuity of the initial data (i.e. is not a wedge line) the waves incident upon the corresponding half-plane $t > 0$, $\xi < 0$, $\eta = 0$ are continuous across the half-plane.
(b) If the half-line $\eta = 0$, $\xi < 0$ is a line of discontinuity (constant jump) of the initial data, it remains a half-plane $t > 0$, $\xi < 0$, $\eta = 0$ of (in general variable) jump discontinuity in the solution.

Similar statements hold for the half-line $\eta = 0$, $\xi > 0$.

Although the map M mentioned in Theorem 2.1 provides a means of constructing the solution to the two-dimensional Riemann problem, it is not a convenient method of doing so. In particular it is a method that will not generalize to cases $f \neq g$, where no appeal to a one-dimensional analysis can be made. We therefore proceed by formulating a construction that is purely two-dimensional and has the advantage of generalizing (at least in specific cases) to the problem $f \neq g$. We propose that the correct method of dealing with the general solution to the two-dimensional Riemann problem is to identify the general two-dimensional nonlinear waves. This construction method (initiated by Guckenheimer[2] and Wagner[3]) can then be used to piece these waves together into an entropy-obeying solution in a manner that places no reliance on one-dimensional analyses.

2.3. *Two-dimensional nonlinear waves for piecewise-smooth solutions*

We proceed with the analysis and definition of rarefaction and shock waves valid for piecewise-smooth, entropy-obeying solutions to the two-dimensional Riemann problem for the case $f \equiv g$. The explicit forms of these waves and the complete characterization of the irregular points are given.

The two-dimensional Riemann problem is invariant under the similarity transformation $(t, x, y) \rightarrow (ct, cx, cy)$. The solution is therefore constant along rays having the space–time origin as the vertex. Consequently the solution can be determined by its restriction to any plane $t = \text{const} (>0)$. The plane $t = 1$ is most convenient for this purpose.

In addition to this self-similarity property, at points at which u is regular, the solution is constant on characteristics $(t, x(t), y(t))$ which are the straight lines

$$\frac{x - x_0}{t - t_0} = f'(u), \qquad \frac{y - y_0}{t - t_0} = g'(u), \qquad t_0 = 0. \tag{2.2}$$

Shock waves and the entropy condition. Let (t, x, y) be a point of jump of a unique piecewise-smooth solution to a two-dimensional Riemann problem. Condition (1.2) thus states that the vector

$$[1, S_f(u^+, u^-), S_g(u^+, u^-)],$$

where

$$S_f(u^+, u^-) \equiv \frac{f(u^+) - f(u^-)}{u^+ - u^-}, \qquad S_g(u^+, u^-) \equiv \frac{g(u^+) - g(u^-)}{u^+ - u^-}$$

is a tangent vector at the point of jump t, x, y. However, by the self-similarity of the Riemann solution, the vector

$$[1, x/t, y/t]$$

must also be tangent to the point of discontinuity. Consequently, the vector

$$[0, x/t - S_f(u^+, u^-), y/t - S_g(u^+, u^-)] \tag{2.3a}$$

is tangent to the point of discontinuity. In the plane $t = 1$, (2.3a) has the simple form

$$[0, x - S_f(u^+, u^-), y - S_g(u^+, u^-)]. \tag{2.3b}$$

Let $a > b$. Consider all planes passing through the points $(1, S_f(a, b), S_g(a, b))$ and the origin $(0, 0, 0)$. Parametrize the straight lines thus defined in the $t = 1$ plane [these are just the lines along the vectors (2.3b)] by the angle θ, measured positive in the counterclockwise sense from the x axis, with the straight line oriented as shown in Fig. 1. The normal (pointing from the b side to the a side) to each plane is given by

$$n \equiv (n_t, n_x, n_y) = \pm (S_f(a, b) \tan \theta - S_g(a, b), -\tan \theta, 1), \tag{2.4}$$

where the plus sign is used for $-\pi/2 \leq \theta \leq \pi/2$ and the minus sign for $\pi/2 \leq \theta \leq 3\pi/2$.

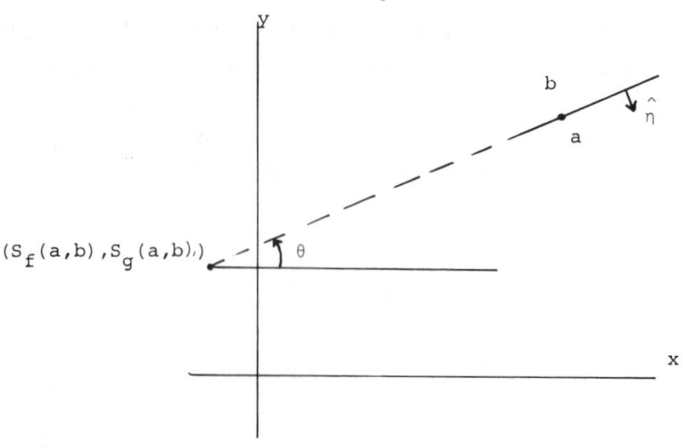

Fig. 1. The tangent line to a point of jump on a shock curve in the $t = 1$ plane. At the point the shock separates the two states a and b, $b < a$. The tangent passes through the point $(S_f(a, b), S_g(a, b))$ at angle θ with respect to the positive x axis with the orientation that the larger state a is always to the left of the shock line when viewed from the above point towards the origin.

These planes will obey the entropy condition (1.3) for the jump $a \to b$ provided

$$E(k) \equiv \pm (k - b) \{[S_g(a, b) - S_g(k, b)] - \tan \theta[S_f(a, b) - S_f(k, b)]\} \geq 0, \quad (2.5)$$

for every k such that $b \leq k \leq a$. The plus and minus signs are correlated with the angle θ as above. The following theorem summarizes this condition.

THEOREM 2.4

A point of jump $(t = 1, x, y)$, in whose neighbourhood the solution is piecewise smooth, with the one-sided limit values a and b will obey the entropy condition (1.3) if and only if its tangent vector (2.3b) is oriented so as to obey the conditions of (2.5).   □

For the case $f \equiv g$ (2.5) takes on a simpler form,

$$E(k) = \pm (k - b)[S_f(a, b) - S_f(k, b)](1 - \tan \theta) \geq 0. \quad (2.6)$$

The next theorem follows directly from (2.6).

THEOREM 2.5

Let $(t = 1, x, y)$ be a point of jump of a piecewise-smooth solution to the Riemann problem for the case $f \equiv g$. Then, we have the following.

(a) If $S_f(a, b) \geq S_f(k, b)$ for every k in $[b, a]$ the point of jump will obey the entropy condition (1.3) if and only if its tangent vector (2.3b) (oriented as in Fig. 1) lies in the angular range $- 3\pi/4 \leq \theta \leq \pi/4$. [Note the correlation for choice of plus and minus sign with angle as given by (2.4), which must be taken into account to obtain this result.]

(b) If $S_f(a, b) \leq S_f(k, b)$ for every k in $[b, a]$ the point of jump will obey the entropy condition (1.3) if and only if its tangent vector (2.3b) (oriented as in Fig. 1) lies in the angular range $\pi/4 \leq \theta \leq 5\pi/4$. [Note the correlation for choice of plus and minus sign with angle as given by (2.4) which must be taken into account to obtain this result.]

(c) If the states a and b are such that the quantity $S_f(a, b) - S_f(k, b)$ changes sign for some k_0 where $b < k_0 < a$, the point of jump cannot obey the entropy condition.   □

Thus for $f \equiv g$, the states a and b can occur as the left and right limiting values of u at a point of jump that obeys the entropy condition if and only if either $S_f(a, b) \geq S_f(k, b)$ or $S_f(a, b) \leq S_f(k, b)$ for every k in $[a, b]$.

In two spatial dimensions, shock waves are then smooth surfaces of discontinuity, each point of which is a point of jump and satisfies the entropy condition (1.3).

The rarefaction waves. Let (t, x, y) be a regular point on a smooth level surface of a piecewise solution u to a two-dimensional Riemann problem. From (1.1) we conclude that

$$[1, f'(u), g'(u)]$$

is tangent to the surface at t, x, y. However, by the self-similarity of the Riemann solution, the vector

$$[1, x/t, y/t]$$

must also be tangent to the point. Consequently, the vector

$$[0, x/t - f'(u), y/t - g'(u)] \tag{2.7}$$

is tangent to the point on the level surface.

PROPOSITION 2.6
(a) The intersection of a smooth level surface with the plane $t = 1$ is a straight-line segment defined by the vector (2.7).

   Let $u = a$ and $u = b$ denote two such level surfaces. Then
(b) if $a \neq b$, the two level surfaces cannot cross;
(c) if $a = b$, the only point that can be common to the two level surfaces and the $t = 1$ plane is $x = f'(a)$, $y = g'(a)$.

Proof. (a) This follows immediately from (2.7). The level surface is therefore a segment of a plane in t, x, y space. (b) If $a \neq b$, then the value of u on the curve along which the two level surfaces cross is not unique, contradicting the assumption that the point belongs to a level surface. (c) Follows as a consequence of the proof in part (b) and (2.7).   □

A rarefaction wave is now definable in a manner analogous to the one-spatial-dimension case (see [1]). Let n be the unit normal to a smooth level surface labelled u_1 at the point $p = (t_p, x_p, y_p)$. By the smoothness of u, there exists a $p \pm \delta_p$ n neighbourhood of level surfaces for sufficiently small $\delta_p > 0$. For some $-\delta_p < \epsilon < \delta_p$, let u_ϵ label one such level surface. In the plane defined by the vectors n and the time axis $\hat{t}$ these level surfaces appear as straight lines segments. If the lines defined by these segments intersect at some point $(\bar{t}_\epsilon, \bar{x}_\epsilon, \bar{y}_\epsilon)$ for $\bar{t}_\epsilon < t_p$ for every $0 < |\epsilon| \leq \delta_p$ then the level surface is a *rarefaction wave*. We note that a rarefaction wave is the union of characteristic lines.

Proposition 2.6 states that two rarefaction waves (of the same value $u = r$) can intersect in the $t = 1$ plane at the point $x = f'(r)$, $y = g'(r)$. This is a reflection of Lemma 2.2, which characterizes the continuity (but possible lack of differentiability) of the solution across the plane $x - y$. This situation is pictured in Fig. 2(b) for the waves of two rarefaction fans along the plane $x = y$.

As in the one-dimensional case, a *rarefaction fan* is defined as an open set, all points of which are in rarefaction waves. A *constant state* is a domain (connected open set) in space–time t, x, y over which the solution is constant.

Explicit form of the nonlinear waves. Based on the results of Sec. 3 of [1] the forms of the nonlinear waves for the case $f \equiv g$ can be explicitly characterized. The form of a typical rarefaction fan is illustrated in Fig. 2(a) as it appears in the plane $t = 1$.

Figure 2(c) displays the three general forms of shock waves. The rarefaction fans that are used to distinguish the three forms are drawn in as dotted lines. The shock form labelled Σ_{cc} separates two regions of constant u values, denoted $u = a$, and $u = b$. The form labelled Σ_{cr} separates a region of constant u value from a fan of rarefaction lines. The Γ shock line discussed by Wagner[3] corresponds to a special case of this type of shock form. The shock labelled Σ_{rr} separates two fans of rarefaction lines.

The shock form Σ_{cc} will be a straight-line segment. The line defined by this segment will pass through the point $(S_f(a, b), S_g(a, b))$ in the x, y plane.

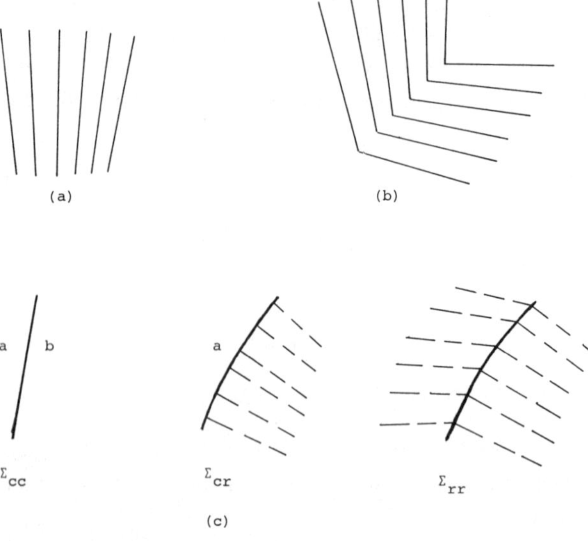

Fig. 2. (a) A typical rarefaction fan as it appears in the $t = 1$ plane. (b) Two rarefaction fans meeting continuously but not smoothly on the $x = y$ line in the $t = 1$ plane. (c) The three general forms of shock waves for the two-dimensional Riemann problem under assumptions discussed in the text. Dotted rarefaction waves are drawn to clarify the picture.

Figure 3 shows a Σ_{cr} shock separating a region of constant $u = a$ from a rarefaction fan terminating at its upper end with the value $u = b$. For clarity in the figure, we represent the rarefaction fan only by its boundary wave, $u = b$ and one interior wave. We first obtain the equation of the shock for the case $0 < \beta(r) < \pi/2$, for every r in the rarefaction fan, and for $\pi/2 < \alpha < \pi$. We parameterize the shock by the u values of the rarefaction fan. Thus the shock plane in t, x, y space has the parametric form

$$\Gamma(a, b) \equiv (t, x(r)t, y(r)t). \tag{2.8}$$

A normal to the shock surface is

$$n = (x'(r)y(r) - x(r)y'(r), \, y'(r), \, -x'(r)). \tag{2.9}$$

Letting $y(r) \equiv \gamma(r)$ be the unknown function to be solved for, we have from Fig. 3

$$x(r) = f'(r) + (g'(r) - \gamma(r)) \tan \beta(r), \tag{2.10}$$
$$y(r) = \gamma(r).$$

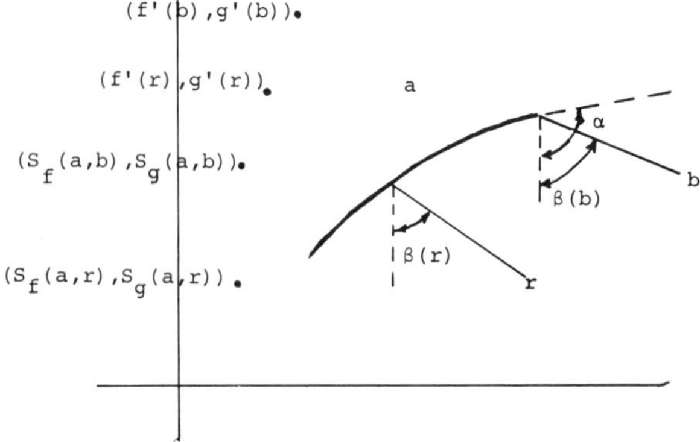

Fig. 3. The most general example of the shock type Σ_{cr}. For purposes of clarity, only two lines of the rarefaction fan that the shock separates from the constant region $u = a$ are drawn in.

The jump condition (1.2) applied to this shock gives the ordinary differential equation (for $r \neq a$) for $\gamma(r)$

$$\gamma'(r) = \frac{(\gamma(r) - S_g(r, a))[f''(r) + g''(r) \tan \beta(r) + (g'(r) - \gamma(r)) \tan' \beta(r)]}{f'(r) - S_f(r, a) + \tan \beta(r)[g'(r) - S_g(r, a)]}, \quad (2.11)$$

having boundary condition

$$\gamma(b) = \frac{S_f(a, b) + S_g(a, b) \tan \alpha - f'(b) - g'(b) \tan \beta(r)}{\tan \alpha - \tan \beta(r)}. \quad (2.12)$$

The presence of the nonlinear term due to the dependence of β on r precludes further analysis without information concerning the form of $\beta(r)$. However, if we assume all lines in the rarefaction are parallel, which in practical constructions is often the case, β becomes a constant and (2.11) reduces to

$$\gamma'(r) = \frac{(\gamma(r) - S_g(r, a))(f''(r) + g''(r) \tan \beta)}{f'(r) - S_f(r, a) + (g'(r) - S_g(r, a)) \tan \beta}. \quad (2.13)$$

Equation (2.13) has the form

$$\gamma'(r) = l(r)[\gamma(r) - S_g(r, a)],$$

thus defining $l(r)$. The solution to the system (2.12), (2.13) can be written

$$\gamma'(r) = S_g(r, a) + (\gamma(b) - S_g(b, a)) \exp\left(\int_b^r l(t)\, dt\right)$$
$$- \int_b^r \left(\frac{g'(z) - S_g(z, a)}{z - a}\right) \exp\left(-\int_r^z l(t)\, dt\right) dz. \quad (2.14)$$

We point out that our stated restriction on the angles α and β constitutes no real restriction. For any combination of angle α and β the derivation of (2.11) and solution of (2.12), (2.13) proceeds in an analogous manner. For some angles it is more advantageous to let $x(r)$ rather than $y(r)$ be the unknown function to be solved for. The cases $\alpha, \beta = n\pi/2$ represent special cases in which the equations simplify.

The derivation of the general differential equation for shocks of type Σ_{rr} proceeds analogously to that for type Σ_{cr} but leads to a much more unmanageable equation. We therefore derive the equation for this shock type for the case in which the lines in each rarefaction fan are parallel to one another (see Fig. 4).

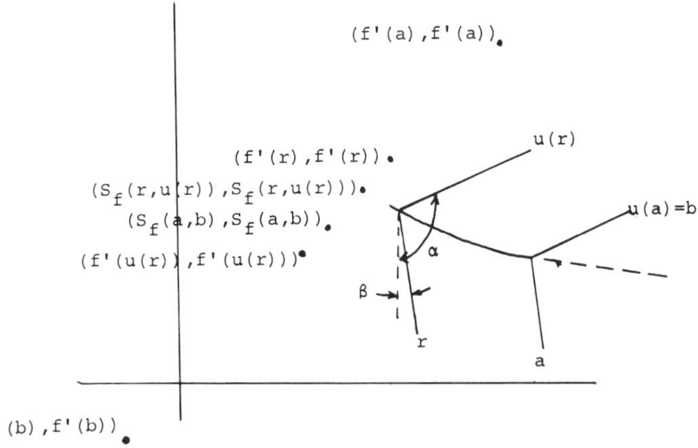

Fig. 4. A particular example of shock type Σ_{rr} where the two rarefaction fans on either side are composed of separately parallel lines. For clarity, only two of the lines in each rarefaction have been drawn in.

We again parametrize the shock plane by the u values of one of the two rarefaction fans (say for the lower fan pictured in Fig. 4). Equations (2.8) and (2.9) still hold, but now

$$x(r) = \frac{g'(u(r)) - g'(r) + f'(u(r)) \cot \alpha - f'(r) \cot \beta}{\cot \alpha - \cot \beta},$$

$$y(r) = \frac{f'(u(r)) - f'(r) + g'(u(r)) \tan \alpha - g'(r) \tan \beta}{\tan \alpha - \tan \beta}. \tag{2.15}$$

Substitution of (2.15) in (2.9) and application of the jump condition (1.2) gives an equation for $u(r)$:

$$u'(r)[g''(u(r)) \tan \alpha + f''(u(r))][f'(r) - S_f(u(r), r) + (g'(r) - S_g(u(r), r)) \tan \beta]$$

$$= [g''(r) \tan \beta + f''(r)][f'(u(r)) - S_f(u(r), r) + (g'(u(r)) - S_g(u(r), r)) \tan \alpha], \tag{2.16}$$

having the boundary condition $u(a) = b$ (see Fig. 4) which will be a known value.

For the case $f = g$, the angles drop out of (2.16), giving

$$u'(r) = \frac{f''(r)}{f''(u(r))} \frac{f'(u(r)) - S_f(u(r), r)}{f'(r) - S_f(u(r), r)}. \tag{2.17}$$

No solution is possible to (2.16) or (2.17) without knowing the form of f. If we assume $f(r) = lr^2$, the solution to (2.17) is $u(r) = -r + a + b$ where a and b are as defined in Fig. 4. Substitution of the solution to (2.16) or (2.17) in (2.15) gives the parametric equation for the shock curve.

As a final comment we note that a rarefaction line $u = a$ will terminate tangentially on a shock (Σ_{cr} or Σ_{rr}) in the plane $t = 1$ at a point p where the limiting u values on each side of the shock wave are $u = a$ and $u = b$ only if the point $(f'(a), g'(a))$ coincides with the point $(S_f(a, b), S_g(a, b))$.

In the half-spaces $x < y, x > y$, irregular points in the solution occur as intersection curves of shock surfaces, and at the discontinuities in the initial data in the plane $t = 0$. In the $t = 1$ plane these irregular points are therefore points at which N (≥ 3) shocks meet. The line $x = y$ may also contain irregular points, each corresponding to a rarefaction wave $u = a$ in the $x > y$ half-plane meeting a rarefaction wave $u = a$ in the $x < y$ half-plane continuously but not smoothly. Irregular points corresponding to the meeting of two shock waves (one from each half-plane) can also appear on this line. A third possibility is the termination of a shock wave on the line $x = y$. For this to occur the shock wave must terminate with zero strength.

3. TWO-DIMENSION METHOD OF SOLUTION CONSTRUCTION—EXAMPLES

Construction of solutions to two-dimensional Riemann problems was first approached by Guckenheimer[2], who gave the solutions to two example problems. Additionally, he listed several of the principles concerning the restriction of the solution in the plane $t = 1$. Wagner[3] has constructed solutions to (1.1) for the case that the functions f and g are both convex (analogously both concave) and that the initial data is constant in the four quadrants of the (x, y) plane. For the case $f \equiv g$, his solutions obey the entropy condition (1.3). He demonstrates that if f and g are "sufficiently close," additional restrictions on the derivatives of the functions f and g are required for the $f \equiv g$ solutions to hold for the $f \neq g$ case.

Having classified the most general nonlinear waves that can appear, the construction method these two papers have introduced can be used to construct the entropy-obeying solution to the general 2-D Riemann problem.

The method of solution construction is the following. The construction is done in the $t = 1$ plane. The curve $(f'(u), g'(u))$ and the set of points $((S_f(v, w), S_g(v, w))$, which dictate the orientation of rarefaction lines and shock curves, are identified. The nonlinear waves and their analytic forms, and the irregular points from which the solutions will be composed, are identified.

The angle restrictions which the entropy condition imposes on the tangent vectors to the nonlinear waves are obtained. In practice this imposes limits on the extent to which a particular shock-wave form may appear in the plane in a particular region and determines the "fitting together" of the various nonlinear elements. The solution in the region $x^2 + y^2 \geq a^2$ for some a is obtained as the solution of noninteracting one-dimensional Riemann problems. The solution is then extrapolated into the region $x^2 + y^2 < a^2$ by fitting together nonlinear waves. In the following subsections we display the entropy-obeying solutions for a few classes of Riemann problems for the cases $f \equiv g$.

For the case $f \equiv g$ the nonlinear wave forms and the irregular points of the solution follow from Sec. 2 and [1]. This analysis also holds for functions f and g which differ by a multiplicative constant, since this can be rotated into a problem $\tilde{f} \equiv \tilde{g}$. We note that Example 1 of [2] falls into this category.

3.1 *A convex f example*

We illustrate the case for f having no inflection points with initial data consisting of three wedges of constant states a, b and c ($a < b < c$) centered on the origin in the $x \geq y$ half of the plane. Figure 5 shows the six unique (i.e. 'topologically different') entropy-obeying solutions.

By Theorem 2.5(a) it follows that the solutions satisfy the entropy condition. By this theorem the three wedges can be changed in size relative to one another anywhere in the range $-3\pi/4 \leq \theta \leq \pi/4$ and the solutions will not change their topological form.

Completely analogous solutions and conclusions follow from Theorem 2.5(b) for the concave case of $f \equiv g$.

Shocks of type Σ_{rr} do not appear in the solutions in Fig. 5. For them to appear in the case of convex f it is necessary to go to initial data consisting of four wedges in the half-plane $x \geq y$. Twenty-four "topologically distinct" solutions exist, one of which, showing a shock of type Σ_{rr}, we sketch in Fig. 6.

3.2 *A single-inflection-point example*

We now consider the solution for $f \equiv g$ obeying the properties

(1) $f : [a, b] \to [c, d]$, $a < b$, $c < d$, $a, b, c, d \in \mathbf{R}$;
(2) f is strictly monotonic on $[a, b]$;
(3) f has a single inflection point at u_i in (a, b);
(4) $f''(u) > 0$ on $[a, u_i)$, $f''(u) < 0$ on $(u_i, b]$.

Without loss of generality we shall take $a = 0$, $b = 1$, $c = 0$, $d = 1$. f has the property that for every $u \in [0, 1]$, there exists a unique $u^* \in [0, 1]$ such that $f'(u^*) = S_f(u, u^*)$. The meaning of the iterated notation u^{**} is then to be understood as $(u^*)^*$.

Shocks of type Σ_{rr} separating a fan of rarefaction lines of values $u > u_i$ from a fan of values $u < u_i$ appear in the solution to the two-dimensional Riemann problem for this f function. If a rarefaction wave has value $u = r$ on one side of the shock, the rarefaction wave leaving on the other side will have the value $u = r^*$ (defined above) and will leave the shock line tangentially. The form derived for shocks of type Σ_{rr} in Sec. 2 does not apply since there it was assumed that the angles α and β of the two rarefaction fans were known; consequently the shape of the shock in that analysis was found by determining the state values of the rarefaction lines on one side given the state values on the other. Here knowledge of the states on either side of the shock, and the implication that one set of rarefactions leave tangentially, determines the shape of the shock curve. Therefore we determine parametrically the shape of this shock type and show that the shock obtained obeys the entropy condition (1.3) for the case $f \equiv g$.

Consider the Riemann problem shown in Fig. 7(a) with w and v as in Fig. 7(b). The solution is shown in Fig. 7(c) where the shock of type Σ_{rr} is labelled $\Psi(r, r^*)$. The analysis for the form of Ψ proceeds as in Sec. 2. Parametrize Ψ as in (2.8) with

$$x(r) \equiv \psi(r), \tag{3.1}$$
$$y(r) = f'(r) + (f'(r) - \psi(r)) \cot \alpha(r).$$

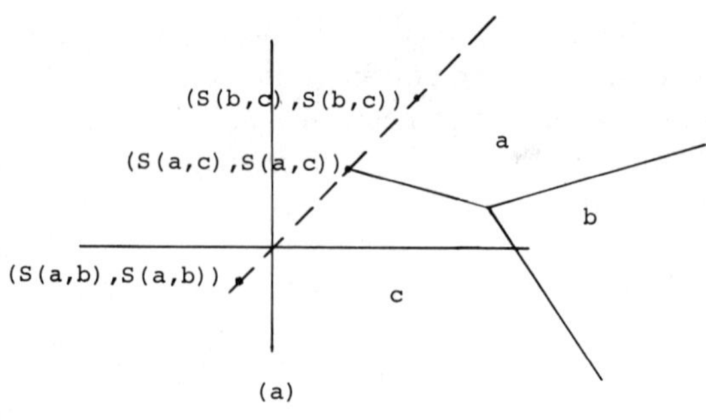

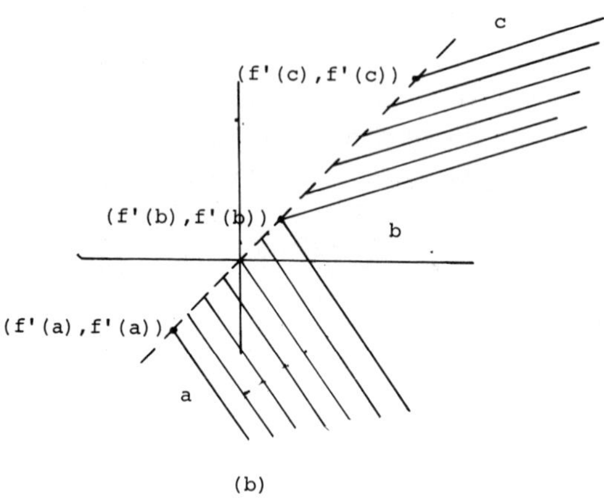

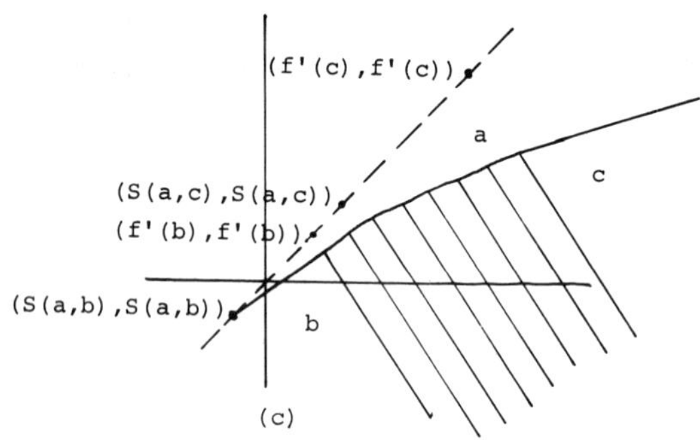

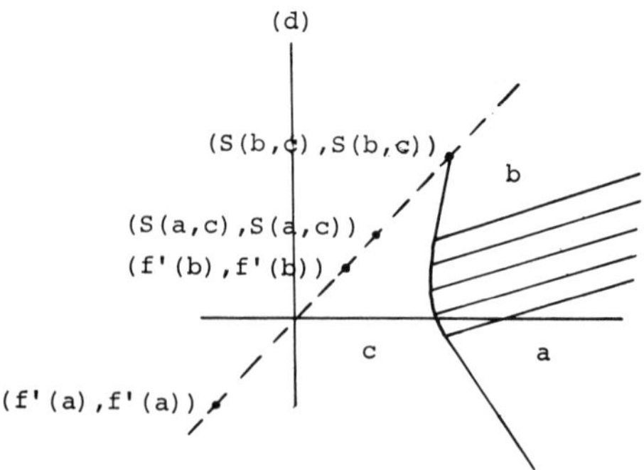

Fig. 5. The six qualitatively unique entropy-obeying solutions in the region $x \geq y$ to (1.1) for $f_1 \equiv f_1 \equiv f$, f having no inflection points, with initial data composed of three wedges of constant u value ($a < b < c$) centered on (0, 0).

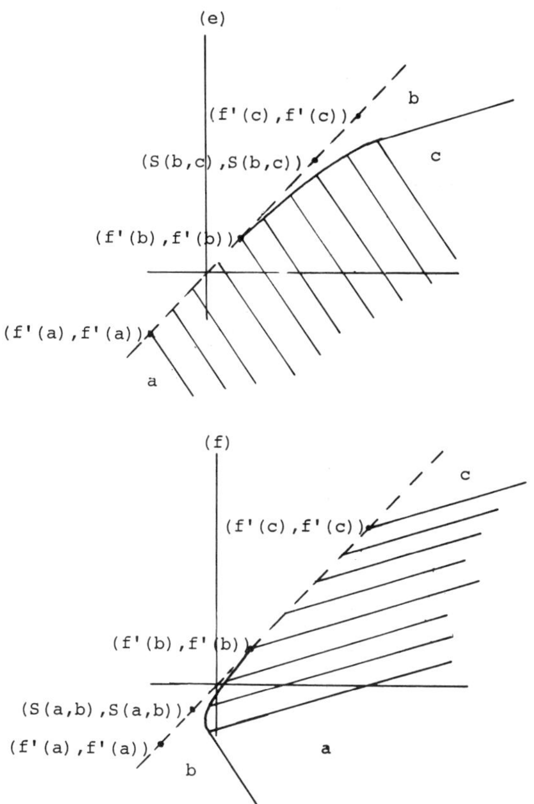

Fig. 5. (*Continued*).

Application of the jump condition (1.2) gives the equation for ψ:

$$\psi'(r) = \frac{[f''(r)(1 + \cot \alpha(r)) + (\psi(r) - f'(r)) \cot'(\alpha(r))]}{(f'(r) - S_f(r, r^*))(1 + \cot \alpha(r))}, \tag{3.2}$$

with boundary condition

$$\psi(v) = \frac{f'(v)(1 + \cot \alpha(v)) - f'(v^*)(1 + \cot \beta(v^*))}{\cot \alpha(v) - \cot \beta(v^*)}. \tag{3.3}$$

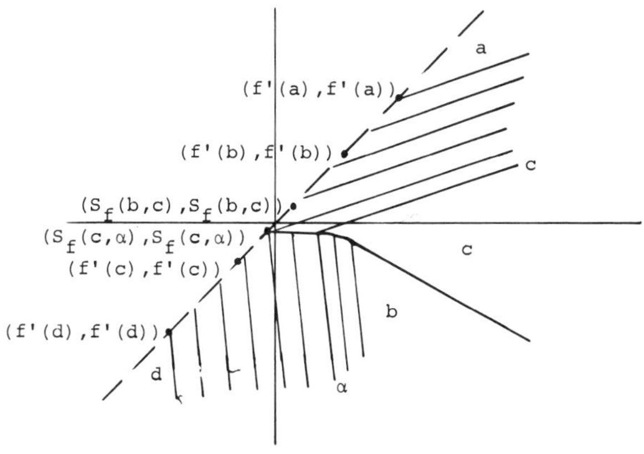

Fig. 6. A sketch of a four wedge initial data solution to (1.1) showing the appearance of a Σ_{rr} shock. The Σ_{rr} shock has been drawn as a straight line which would be the case for $f(u) = u^2/2$. Note that the Σ_{rr} shock segment joins smoothly to the Σ_{cr} segment with the same slope. The value α can be determined from the intersection of the Σ_{cr} shock and the rarefaction line $u = c$. In this example the states are ordered $d < c < b < a$.

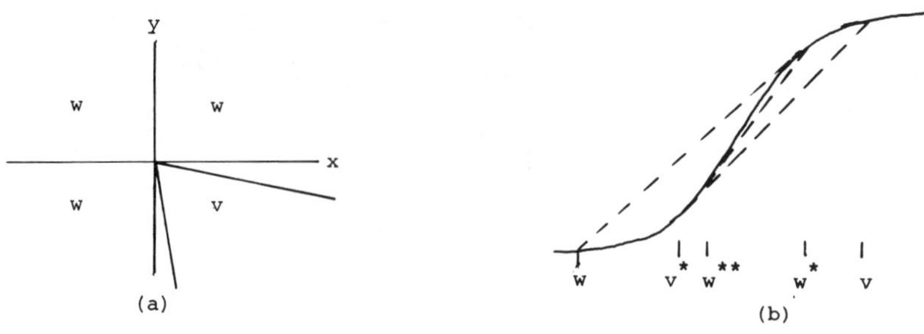

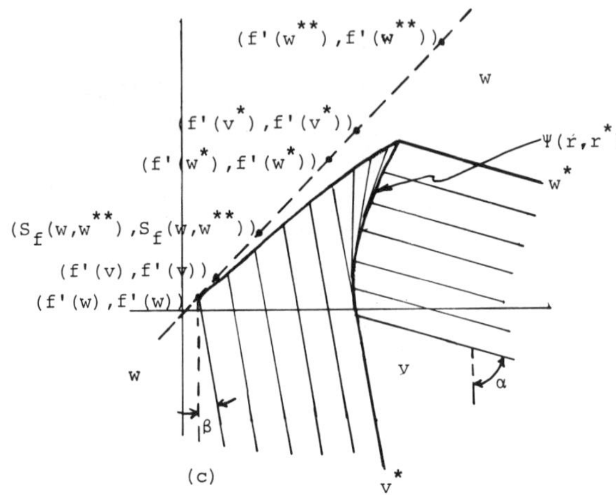

Fig. 7. A two-dimensional Reimann problem for the function f described in Sec. 3.2. (a) The initial data. (b) The function f defining various states found in the solution. (c) The solution.

As in (2.11), variation of α along the rarefaction introduces a nonlinear term into (3.2). If we assume all lines in this rarefaction are parallel [as indeed have been drawn in Fig. 7(c)], which will be the case in the solutions we display in this paper, (3.2) simplifies greatly to

$$\psi'(r) = f''(r)\,\frac{(\psi(r) - S_f(r, r^*))}{(f'(r) - S_f(r, r^*))}, \tag{3.4}$$

having the solution

$$\psi(r) = S_f(r, r^*) + [\psi(v) - S_f(v, v^*)] \exp \int_v^r l(t)\,dt$$

$$- \int_v^r \left(\exp \int_z^r l(t)\,dt\right) \frac{\partial S_f(z, z^*(z))}{\partial z}\,dz, \tag{3.5}$$

where

$$l(t) = \frac{f''(t)}{f'(t) - S_f(t, t^*)}. \tag{3.6}$$

Note that given r in $[w^*, v]$, the form of the function f guarantees the existence of r^* and for every k such that $r^* \le k \le r$ we have $S_f(r, r^*) \le S_f(k, r^*)$. From the construction, it is clear that the above shock obeys the conditions of Theorem 2.5(b) and hence obeys the entropy condition.

We return briefly to Fig. 7(c) to discuss the other curved shock, which is composed of two shocks of type Σ_{cr}, one separating the fan of parallel lines $u = w \rightarrow u = v^*$ from the constant region $u = w$, the other separating a fan of nonparallel lines $u = v^* \rightarrow u = w^{**}$ from the constant region $u = w$. The two shocks have the same tangent line where they meet. The construction and Theorem 2.5(a) guarantee that this composite shock obeys the entropy condition.

In Fig. 8 we present the entropy obeying solutions to (1.1) for $f \equiv g$ with f as described above for initial data of three wedges of constant states a, b and c ($a < b < c$) centered on the origin. We consider only the two general cases: either $a < u_i < b < c$ or $a < b < u_i < c$, otherwise the problem reduces to the purely convex or concave case discussed in Sec. 3.1. By the qualitative symmetry of the function f about its inflection point, there are only six

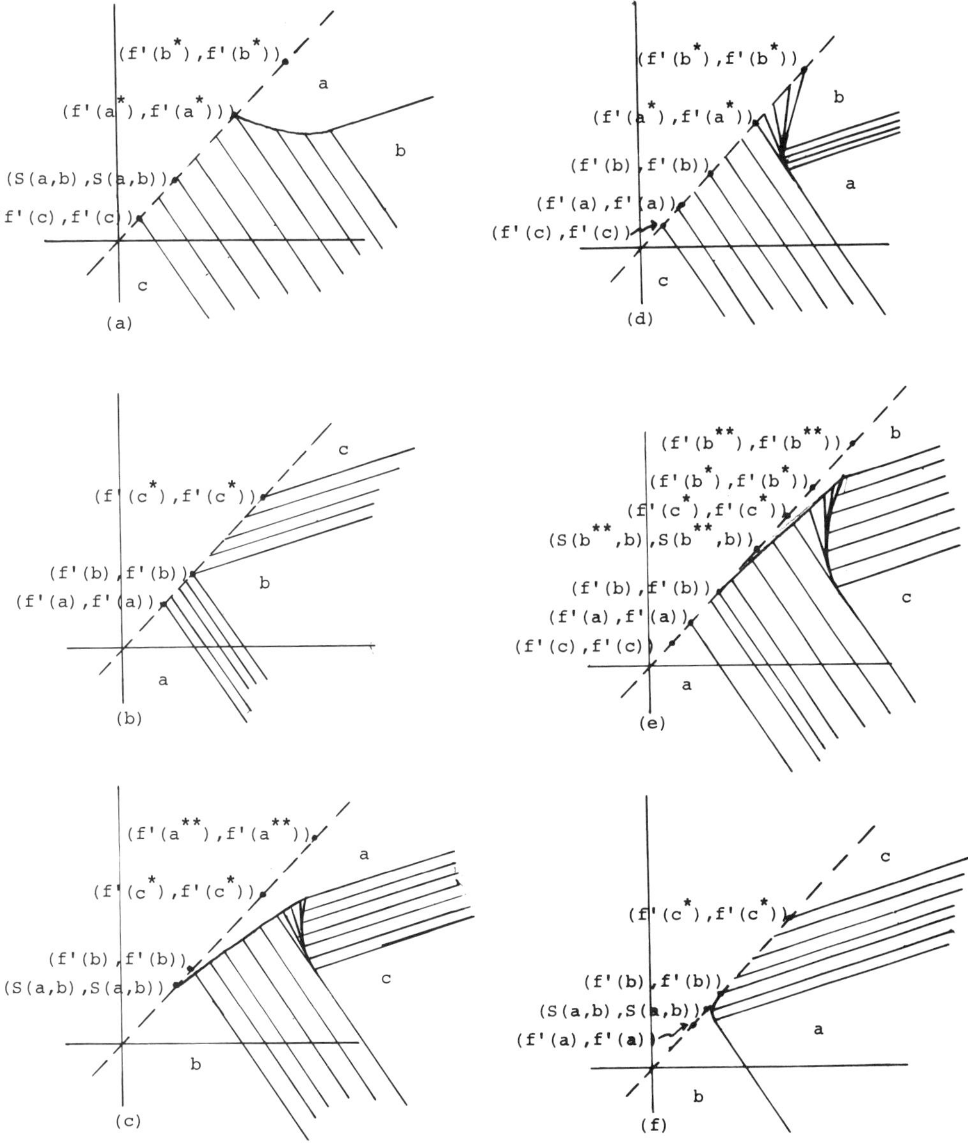

Fig. 8. The six qualitatively distinct solutions in the $x \geq y$ half-plane for the three-wedge Riemann problem discussed in Sec. 3.2. Note that one of the three initial data states a, b and c is separated from the other two by u_i, the inflection point of f.

qualitatively distinct solutions. As previously mentioned, the solutions obtained in the $x \geq y$ half of the plane can be obtained independently of the solutions in the $x \leq y$ half.

4. THE TWO-DIMENSIONAL RIEMANN PROBLEM FOR TWO-PHASE FLOW IN POROUS MEDIA

For incompressible two-phase flow in a porous medium, with capillarity effects and gravity ignored,† the system of equations to be solved in a source-free region is

$$\frac{\partial s}{\partial t} + \mathbf{v}(s, P) \cdot \nabla f(s) = 0, \tag{4.1a}$$

$$\nabla \cdot \mathbf{v}(s, P) = 0, \tag{4.1b}$$

where s is the volume fraction (saturation) of one of the two phases (the other fraction being $1 - s$) and p is the pressure field in the porous medium. The field $\mathbf{v}$ is the total fluid (consisting of both phases) velocity. The velocity $\mathbf{v}$ is generally assumed to be proportional to ∇P (Darcy's Law) and when the form of this relation is specified the system (4.1) can be solved. A form commonly used for $f(s)$ is (immiscible flow[6])

$$f(s) = \frac{s^2}{s^2 + r(1 - s)^2}, \tag{4.2}$$

where r is the ratio of the two phase viscosities (we assume constant viscosities). The fractional flow curve given above then has all the properties of the f function described in Sec. 3.2.

One method of solving (4.1) numerically consists of solving (4.1a) and (4.1b) sequentially. Thus for purposes of solving the hyperbolic equation (4.1a) the velocity field can be taken as some given vector field $\mathbf{v}(\mathbf{x})$. For two-dimensional problems, the hyperbolic problem is therefore

$$\frac{\partial s}{\partial t} + v_x(x, y)\frac{\partial f(s)}{\partial x} + v_y(x, y)\frac{\partial f(s)}{\partial y} = 0. \tag{4.3}$$

Shocks in two-phase systems are very rapid (in this model, discontinuous) changes in the phase, from a region that has largely phase 1 in its pore spaces to that having largely phase 2. To be specific we will call the phases oil and water, with $s = 0$ corresponding to pure oil and $s = 1$ corresponding to pure water. The system (4.1) describes a secondary oil-recovery procedure where injected water is used to force reservoir oil towards production wells. The oil–water shock surface is typically called a ''bank'' (water bank if the water is displacing the oil, oil bank if the oil is displacing water). We can therefore study the dynamics involved in the interaction of two such banks, for example when the water banks from two separate injection wells meet, by approximating the problem as a Riemann problem.

Figure 9(a) depicts the problem to be solved and Fig. 9(b) its approximation as a Riemann problem. Let v denote a value of v_x averaged over the interaction area; similarly let w be an averaged value of v_y. Equation (4.3) becomes

$$\frac{\partial s}{\partial t} + v\frac{\partial f(s)}{\partial x} + w\frac{\partial f(s)}{\partial y} = 0. \tag{4.4}$$

This is now of the form (1.1) with f and g differing by a multiplicative constant. We can therefore solve (4.4) in rotated coordinates $\bar{x}, \bar{y}$ where $\bar{f} = \bar{g}$.

Figure 9(c) displays the solution to the Riemann problem of Fig. 9(b) in some rotated ξ, η system (the rotation of course depends on the relative strengths of v and w; we have assumed $v > 0$, $w > 0$). Figure 9(d) displays the solution when a and c are interchanged.

---

†We have also set the rock porosity function $\phi(\mathbf{x}) = 1$.

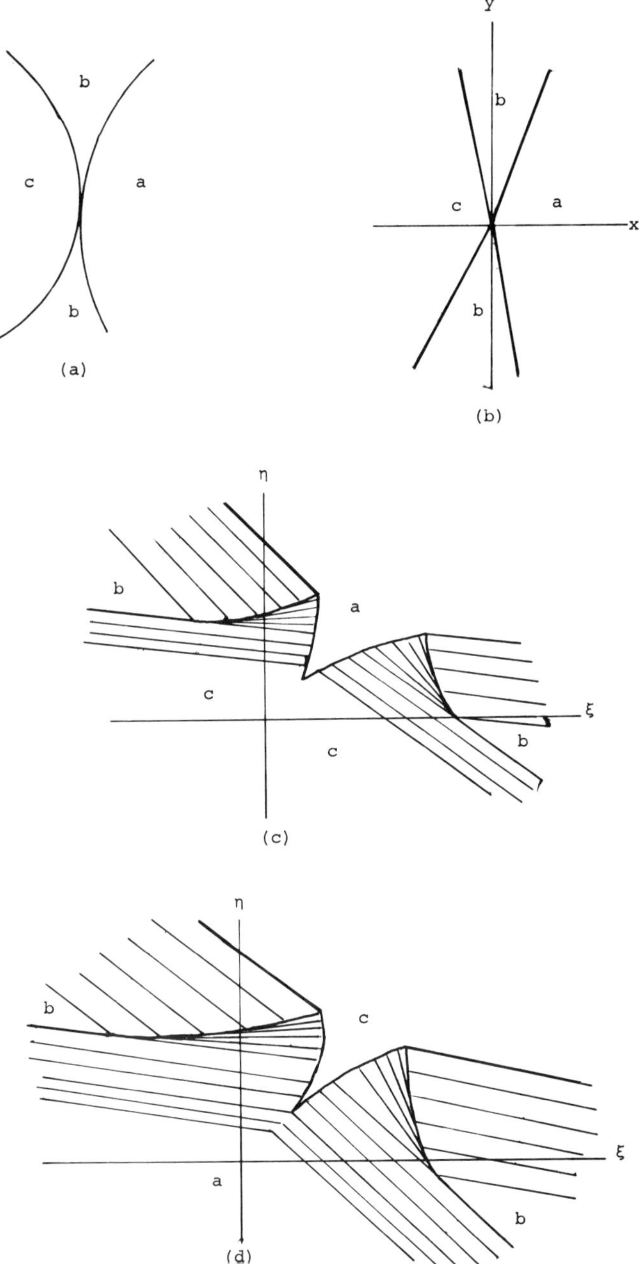

Fig. 9. (a) The interaction of two water banks in a two-dimensional oil reservoir. (b) The Riemann problem approximation to the interaction. (c,d) The solutions to the Riemann problem approximation for two example situations.

We note that for flow with gravity, the flux function associated with two-phase flow has two inflection points, as opposed to a single inflection point for the gravity-free-flow case.

5. CONCLUDING REMARKS

The general solution to the two-dimensional Riemann problem for the case $f_1 \equiv f_2 \equiv f$, $f \in C^2 : R \rightarrow R$, f having at most one inflection point, has been shown to be characterizable in terms of nonlinear (rarefaction and shock) waves in a manner analogous to the one-dimensional Riemann problem. We believe this characterization in terms of nonlinear waves is the correct formulation in which to approach the construction of solutions to two-dimensional Riemann problems for the general case $f_1 \neq f_2$. We conjecture that the set of rarefaction and shock waves described in this paper will provide the complete solutions for many cases $f_1 \neq f_2$. A regularity theorem giving conditions under which all two-dimensional Riemann problem solutions are piecewise smooth would immediately give greater generality of application to the nonlinear waveforms given here.

These solutions to two-dimensional Riemann problems also supply a set of problems for the testing of finite-difference schemes. The richness of structure of these solutions lends itself to this purpose.

Acknowledgments—The author wishes to thank Professor James Glimm for suggesting this problem.

This work is supported in part by the Applied Mathematical Sciences subprogram of the Office of Energy Research, U. S. Dept. of Energy, Contract DE-AC02-76ERO3077.

REFERENCES

1. W. B. Lindquist, The Scalar Riemann Problem in Two Spatial Dimensions: Sufficiency Condition for Piecewise Smoothness of Solutions and its Breakdown. DOE Research and Development Report DOE/ER/*3077-227* (Sept. 1984).
2. J. Guckenheimer, Shocks and rarefactions in two space dimensions. *Arch. Rational Mech. Anal.* **59**(3), 281–291 (1975).
3. D. Wagner, The Riemann problem in two space dimensions for a single conservation law. *SIAM J. Math. Anal.* **14**(3), 534–559 (1983).
4. S. N. Kružkov, First order quasilinear equations in several independent variables. *Mat. USSR-Sb* **10**(2), 217–243 (1970).
5. O. A. Oleĭnik, Uniqueness and stability of the generalized solution of the Cauchy problem for a quasi-linear equation. *Uspehi Mat. Nauk.* **14**, 165–170 (1959). English transl., *Amer. Math. Soc. Transl. Ser.* 2 **33**, 285–290 (1963).
6. S. E. Buckley and M. C. Leverett, *Trans. AIME* **146**, 107–116 (1942).

Comp. & Maths. with Appls. Vol. 12A, Nos. 4/5, pp. 631–632, 1986
Printed in Great Britain.

0097-4943/86 $3.00 + .00
© 1986 Pergamon Press Ltd.

BOOK REVIEWS

Review of Trends in Theory and Practice of Nonlinear Differential Equations, Lecture Notes in Pure and Applied Mathematics, Vol. 90, Marcel Dekker, Inc., New York and Basel, 1984, V. Lakshmikantham, Ed.

This volume contains the proceedings of an international conference held at the University of Texas at Arlington on June 14–18, 1982. The theme of the conference was recent trends in theory and applications of nonlinear differential equations. There are 73 articles in the volume devoted to various topics in this subject. Among the topics presented are spectral theory for symmetric pairs of differential operators, Lotka–Volterra systems, generalized inverses for linear manifolds, nonlinear problems at resonance, steepest descent methods, reaction-diffusion equations, Lyapunov stability, stochastic differential equations, comparison and frequency domain techniques, delay differential equations, method of upper and lower solutions, Newton-like methods, periodic solutions of nonlinear problems, population biology, effects of harvesting on population systems, models of toxicant populations, nonlinear equations of heat flow, inclusion principle for hereditary systems, vector Lyapunov functions in the analysis of dynamical properties of differential equations, recent topics on nonlinear contraction semi-groups, set valued extensions of integral inequalities, global controllability of nonlinear delay systems, cone-valued Lyapunov functions, large-scale systems, quasi-solutions, almost periodicity of solutions of parabolic equations, generalized Hopf bifurcation and exchange of stability and nonlinear elliptic problems. The volume is dedicated to Professor E. A. Coddington, who was honored by the conference participants. The volume constitutes a valuable contribution to recent developments in a great variety of research areas in nonlinear differential equations.

GLENN WEBB
Vanderbilt University
Nashville, TN

Review of Nonlinear Partial Differential Equations in Engineering and Applied Science, Lecture Notes in Pure and Applied Mathematics, Vol. 54, Marcel Dekker, Inc., New York and Basel, 1984, R. L. Sternberg, A. J. Kalinowski and J. S. Papadakis, Eds.

This volume contains 28 papers from the Conference on Nonlinear Partial Differential Equations in Engineering and Applied Science sponsored by the Office of Naval Research and held at the University of Rhode Island in June, 1979. The emphasis of the conference was on applications of nonlinear partial differential equations to problems in engineering and applied science. The topics presented included finite deformations of hyperelastic solids, rigidity and nonlinear elasticity, mathematical biology, unstable viscoelastic fluid flows, perturbed bifurcation theory, errors in mixed finite element methods, solutions of the Korteweg–deVries equations, free boundary problems in alloy solidification, spatial decay for the Navier–Stokes Equations, Riemannian metric for partial differential equations, shockless airfoils, singular perturbations of nonlinear elliptic boundary value problems, three-dimensional nonlinear evolution of water waves, numerical solution of two-dimensional advection flows, parabolic conservation laws, reaction-diffusion systems, steady-state bifurcation theory, singularities of nonlinear elliptic equations, discretizing the Sine–Gordon equation, numerical computation of capillary-gravity waves, integrable nonlinear field theories, numerical solution of stationary Navier–Stokes equations, similarity solutions of evolution equations, solutions for a finite depth stratified fluid, jump phenomena, Darcy's law for flow in porous media and variational problems in finite elasticity. The volume is a valuable summary of recent applications of nonlinear partial differential equations in a wide variety of contexts.

KEVIN P. MEADE
Illinois Institute of Technology
Chicago, IL

Review of PDE Software: Modules, Interfaces and Systems, Elsevier Science Publishing Company, Inc., New York, 1984, B. Engquist and T. Smedsaas, Eds.

This book, dedicated to the memory of N. N. Yanenko, contains the Proceedings of the IFIP TC 2 Working Conference on PDE Software held at the Söderköpings Brunn in Söderköping, Sweden from 22 to 26 August 1983. The purpose of the conference was to examine modern approaches to software for solving partial differential equations. The major emphasis was on PDE software as a modular system with due consideration of the interfaces involved. In addition to the 23 regular papers, this book includes several papers from an open session and edited versions of the discussions at the end of each paper. The topics presented included theoretical and practical aspects of the vectorization of PDE software for use on supercomputers, future forms of PDE software, multigrid methods, time-dependent PDE software and applications of PDE software to the analysis of structures, transport of pollutants and modeling of semiconductor devices. Equal treatment was given to the finite element method and the finite difference method. This book is of interest to researchers in the area of modern approaches to the numerical solution of partial differential equations.

KEVIN P. MEADE
Illinois Institute of Technology
Chicago, IL

INDEX